THE OTHER MATHEMATICS

The Other Mathematics

Language and Logic in Egyptian and in General

LEO DEPUYDT

GORGIAS PRESS
2008

First Gorgias Press Edition, 2008

Published in the United States of America by Gorgias Press LLC, New Jersey

ISBN 978-1-59333-369-0

GORGIAS PRESS
180 Centennial Ave., Piscataway, NJ 08854 USA
www.gorgiaspress.com

Library of Congress Cataloging-in-Publication Data

Depuydt, Leo.
The other mathematics : language and logic in Egyptian and in general / Leo Depuydt. -- 1st Gorgias Press ed.
p. cm.
Includes bibliographical references and index.
ISBN 978-1-59333-369-0 (alk. paper)
1. Egyptian language--Sentences. 2. Egyptian language--Conditionals. 3. Language and logic. 4. Boole, George, 1815-1864. I. Title.
PJ1201.D47 2008
493'.1--dc22
2008044028

The paper used in this publication meets the minimum requirements of the American National Standards.

Printed in the United States of America

Our bodies are given life from the midst of nothingness. Existing where there is nothing is the meaning of the phrase, "Form is emptiness." That all things are provided by nothingness is the meaning of the phrase, "Emptiness is form." One should not think that these are two separate things.

Tsunetomo Yamamoto (d. ca. 1700), *Hagakure: The Book of the Samurai* (1716), chap. 2 (trans. William Scott Wilson)

When *x* and *y* are regarded as classes we cannot but observe that not-*x* and not-*y* are themselves just as much classes as those of which they are the contradictories.

John Venn, *Symbolic Logic* (1894), 307

τὸ ... αὐτὸ ἅμα ὑπάρχειν τε καὶ μὴ ὑπάρχειν ἀδύνατον τῷ αὐτῷ καὶ κατὰ τὸ αὐτό ... αυτη δὴ πασῶν ἐστὶ βεβαιοτάτη τῶν ἀρχῶν ...

ἀδύνατον ... ὁντινοῦν ταὐτὸν ὑπολαμβάνειν εἶναι καὶ μὴ εἶναι ...

ἀξιοῦσι δὴ καὶ τοῦτο ἀποδεικνύναι τινὲς δι᾽ ἀπαιδευσίαν · ἔστι γὰρ ἀπαιδευσία τὸ μὴ γιγνώσκειν τίνων δεῖ ζητεῖν ἀπόδειξιν καὶ τίνων οὐ δεῖ · ὅλως μὲν γὰρ ἁπάντων ἀδύνατον ἀπόδειξιν εἶναι · εἰς ἄπειρον γὰρ ἂν βαδίζοι ὥστε μηδ᾽ οὕτως εἶναι ἀπόδειξιν ·

A single thing cannot at the same time possess and not possess the same attribute all else being the same. … That is the firmest of all the axioms. …

Assuming that whosoever can both be and not be the same thing is impossible.…

Some ask for proof, but only because they lack education. For not knowing of what one needs to seek proof and of what not shows lack of education. Proving everything is definitely impossible. One would just recede into infinity (in trying to prove everything by something else) and the final step would still be without proof.

Aristotle, *Metaphysics*, 4.3.9–10, 4.4.2 (L.D.'s translation)

(From the Preface, adapted)

Hiding under the human skull is the most complex structure in the universe, the human brain, the seat of thought. No concept has inspired the present investigation more than that thought is subject to absolute limitations. Yet, thought is perhaps more readily conceived as limitless. Just think of the human imagination in its various forms: literary, religious, visual, and so on. Anything seems possible when it comes to thought. Then again, the brain is a material structure that is not infinite. It therefore seems eminently reasonable to suppose that what the brain does is not infinite either. The larger aim of the present investigation is to achieve a better sense of the absolute limitations of thought and of the precise and distinct levels of thought that reach up to this final border beyond which thought is not possible. Humility is a common concept in the realms of religion and morality. But rational thought has its own kind of humility, namely the acute awareness of its own absolute limitations. What are these limitations? How smart are we really?

Max Planck recommended the study of philosophy only when conjoined to the study of more specific subjects. In this spirit, the larger concept outlined above is studied here in relation to a narrower domain. This narrower domain is the Egyptian language, whose history is the longest attested of any language. The focus is specifically on certain striking phenomena of Egyptian, along with their parallels in other languages. These phenomena lay bare some of the fundamental fiber of human thought.

Since the mid-nineteenth century, Aristotelian and scholastic logic has been fully superseded by modern scientific logic. The pioneer is George Boole (1815–1864). In the late 1930s, an M.I.T. graduate student named Claude Shannon adapted Boolean algebra for electronic circuits and the computer age began. In the present investigation, several facets of Egyptian are treated in detail in light of modern scientific logic. But no prior knowledge of logic is presupposed. A brief history of logic is provided. All that is needed from logic is defined here internally in fully explicit terms.

Topics pertaining to Egyptian treated in the present work include: sharp and simple definitions of condition and premise, of the difference between condition and premise, and of how one gets from condition to premise and back; the balanced sentence or *Wechselsatz*; the *conditio sine qua non* and the language; and the intriguing question of whether we moderns are smarter or more sophisticated than the ancient Egyptians or than ancient peoples in general. The treatments of these distinct but also interconnected topics ultimately all have these general purposes: to expose ever more clearly the basic articulation of thought into three levels and to suggest the apparent inability of thought to break out of this tripartite pattern as its absolute limitation. The three-level model is provisionally able to absorb and incorporate in complete transparency a number of abstract and much discussed terms such as "causality," "condition," "result," "consequence," "premise," "thought," "truth," "certainty," "right and wrong," and many others. Everyone senses more or less what these terms mean. But defining them precisely is another matter.

An engineering application is added at the end in support of the notion that, if a functional physical mechanism for how we draw inferences that lead us to act can exist in an electronic circuit, one of analogous structure must exist in the brain.

CONTENTS

ACKNOWLEDGMENTS

An abbreviated version of the argument of chapters 1 through 5 was presented at the Thirty-second North American Conference on Afroasiatic Linguistics, held on March 12–14, 2004, in San Diego, California. The contents of chapter 11 were read in outline at the Fifty-fifth Annual Meeting of the American Research Center in Egypt, held on April 16–18, 2004, in Tucson, Arizona. I am grateful to the editors of the *Zeitschrift für ägyptische Sprache und Altertumskunde* (Leipzig) for publishing my interpretation of condition and premise in terms of modern scientific logic (Depuydt 1999a) and to Dr. Gerd Giesler for granting permission on behalf of Berlin Akademie-Verlag GmbH, the copyright holder of *ZÄS*, to reprint this article as appendix 1. I thank Peust & Gutschmidt Verlag of Göttingen for publishing my interpretation of isolating contrast likewise in strictly logical terms (Depuydt 1999b) in the *Göttinger Beiträge zur Sprachwissenschaft* and granting permission to reprint this article as appendix 2.

There is some overlap between the two appendixes and the body of this work. However, in order to keep the line of argument of the body of this work whole and coherent, it was necessary to repeat elements that have already been discussed in those two articles. It is not possible to eliminate steps from the type of argument presented in this book without creating confusion. Some measure of repetition is therefore unavoidable. I have the impression that the main arguments defended in the two articles have not been affected in any essential way by my subsequent engagement with topics pertaining to scientific logic.

I am grateful to Professor Wolfgang Schenkel for commenting at length on an advanced draft of the present manuscript. As so often in the past, I have benefited much from his corrections and suggestions and his encouragement. I admire my colleague Schenkel's ability to keep abreast of so much of the *Fachliteratur.* To Peter Daniels I owe much editorial assistance and many perceptive comments. Duncan Burns of Forthcoming Publications made an essential contribution when the time came to give this book its final form. I also thank the late Nikita, Janet, and the late Theodore Romanoff for supporting my work on the present book. It may be called a fortunate confluence of circumstances that this book appears with Gorgias Press, whose founder, George Kiraz, is trained in computational linguistics, having receiving a doctorate in the subject from Cambridge University. The present book is hardly a work of computational linguistics, but it does use symbolic logic to a degree, albeit sparingly and with the aim of explaining certain features of the Egyptian language that would otherwise remain unexplained. I cannot see any other way of explaining the phenomena in question. One of Dr. Kiraz's passions is Syriac language and culture and Semitic languages in general. It so happens that the present author served as Senior Lector of Coptic and Syriac at Yale University (1989–91) and shares an overall interest in Syriac and Semitic languages. So it would appear that there is more than one way of combining language, logic, and Syriac.

In retrospect, I do not know to what good fortune I owe my chance encounter with a copy of the Dover edition of Boole's *Investigation* (1854) on the bookshelves of the Harvard Bookstore in Cambridge, Mass. in late 1997 or early 1998. The cost was $10.95. The sum makes one wonder about the value of things in a material world. As my understanding of logic has been developing in what seems like an open-ended quest, I was a little puzzled to read in a contemporary handbook of logic, a work not without its own merits, that "symbolic logic is largely an invention of the twentieth century" (Klenk 1983: 13). That makes Boole all but nonexistent. However, in a popularizing account of logic just published, Deborah Bennett's *Logic Made Easy*, Boole does occupy center stage. My own engagement with the history of modern scientific logic has convinced me that most of what is fundamental in the young science of scientific logic had already been formulated with precision by 1900. Beyond Boole's opus, the work that has inspired and helped me the most is John Venn's *Symbolic Logic* (2nd ed., 1894). I do not know whether Venn's book is still widely read among logicians. In all branches of the humanities, one notices on occasion a trend that what is not new may not be hip or hot. I can only encourage engagement with Venn's book. To my mind, it supports the proposition that the latest newest can sometimes be more than a century old.

Logic developed rapidly in the late nineteenth century. It will need to suffice to mention the names of many if not most of the principal contributors to an intellectual tradition in which I believe the present book stands. In addition to George Boole, John Venn, Augustus De Morgan, and Friedrich Karl Ernst Schröder, I would mention the following: Arthur Cayley, J. Delbœuf, Hermann Grassmann, Robert Grassmann, W.S. Jevons, W.E. Johnson, J.N. Keynes, Christine Ladd, F.A. Lange, Hugh McColl, A. McFarlane, O.H. Mitchell, J.J. Murphy, Charles Sanders Peirce, and Wilhelm Wundt.

I am not sure how far the application of logic can be pushed to explain phenomena that are relevant even to beginning students of Egyptian and Coptic. Recently, it appeared that one more step may be possible. In "From 'My Body' to 'Myself' to 'As For Me' to 'Me Too' in 3000 Years: A Peculiar Triple Shift in Meaning in Egyptian and Its Explanation," I first establish that a single expression undeniably changed in meaning in Egyptian from 'my body' to 'myself' to 'as for me' to 'me too'. I then propose a theoretical model to explain this evolution. This theoretical model also allows simple definitions of concepts such as "alone," "also," "as for," "even," "in turn," "only," "on (my) part," "other," "own," "same," "self," and "too." These notions are frequent in all languages. Speakers use them with confidence in their own languages. However, their meaning is anything but transparent.

Among developments that have taken place in the interval that separates the completion of the present manuscript and its appearance as a book are four papers, all planned for publication. The complete digitalization of language structure is the ultimate aim of the line of thought pursued in these papers. Boole's star seems on the rise in mathematics and may well take its rightful place in the first rank of mathematicians, to which Archimedes, Newton, Euler, Laplace, and Gauss belong. In a paper entitled "The Conjunctive in Egyptian and Coptic: Towards a Final Definition in Boolean Terms" (now in press) presented in March 2007 at the Annual Meeting of the North American Conference of Afroasiatic Linguistics held at San Antonio, Texas, an attempt is made to show that Boolean algebra fully accounts for how the Egyptian verb form called the conjunctive can be converted into a condition and *vice versa* with no change in basic purport, in the same way that "Don't drink and drive" can be converted effortlessly into "When you drink, do not drive" and *vice versa* in English. A second paper with the title "Ägyptisch in Boolscher Fassung: Die fortschreitende Digitalisierung der Sprachbetrachtung nach einfachsten und gründlichsten Begriffen" was presented in July 2007 at Berlin's Freie Universität and a third entitled "Relative Clauses in Egyptian and in General: Final Definitions of All Types, via Boole and Venn" was read in March 2008 at the Semitic Philology Workshop of Harvard University's Department of Near Eastern Languages and Civilizations. "Types of Nominal Sentences in Egyptian and Coptic: Towards Final Definitions in Boolean and Vennian Terms" is the title of a fourth paper read at the International Conference on Language Typology and Egyptian-Coptic Linguistics held at Leipzig's Max Plank Institute for Evolutionary Anthropology in October 2008.

The final draft of the present book was completed in 2004, the sesquicentenary of the appearance of Boole's *Investigation* (1854). I hope that the following statement on a border stela dating to the early second millennium B.C.E. found at Semnah (Berlin 1157,4–5; see *Aegyptische Inschriften Berlin* [1913]: 257–58 and Sethe 1928: 83, lines 22–23) might apply to the present efforts to build on some of Boole's and Venn's world of ideas.

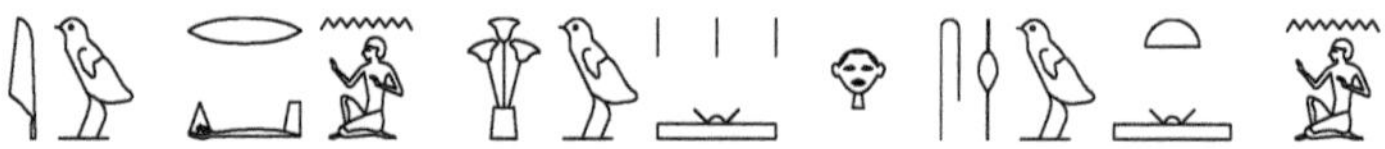

jw rd(y).n.j ḥ3w ḥr swḏt n.j

'I have added to what was handed to me'.

Leo Depuydt
Norton, Massachusetts, and Providence, Rhode Island
October 2004 (revised April 2006)

Abbreviations

Aegyptische Inschriften Berlin (1913)	
	Aegyptische Inschriften aus den Königlichen Museen zu Berlin, 1
CT	Coffin texts, as edited by de Buck (1935–61)
KRI	Kitchen, *Ramesside Inscriptions* (1975–90)
LEM	Gardiner, *Late-Egyptian Miscellanies* (1937)
LES	Gardiner, *Late-Egyptian Stories* (1932)
N	(except as noted below) Name; replaces a name in the Egyptian text
Pyramid text paragraph numbers:	
M	Version found in the pyramid of Merienre (*Mrj-n-rꜥ*)
N	Version found in the pyramid of Pepi II Neferkare (*Nfr-kꜣ-rꜥ*)
Nt	Version found in the pyramid of Queen Neith (Jéquier 1933)
Ou	Version found in the pyramid of Queen Oudjebten (Jéquier 1928)
P	Version found in the pyramid of Pepi I (*Ppjj*)
Pyr. §	Paragraph number (1–2217) in Sethe's edition (1908–22), as opposed to the spell numbers (1–714) (PT in Edel 1955/64)
T	Version found in the pyramid of Teti (*Ttj*)
W	Version found in the pyramid of Unas (*Wnjs*)
Urkunden 1	Old Kingdom documents (Sethe 1932–33)
Urkunden 4	Dynasty 18 documents (Sethe and Helck 1906–58)

Symbols

‾	Marks the supplement, supplementary class, or contradictory of the class below the stroke. For example, if *s* is the class of sheep, then $\bar{s}$ is the class encompassing anything but sheep; also denoted as 1 – *s*, or the universe (1) minus (–) sheep (*s*)
*	Marks hypothetical or non-attested forms of words or text
* in Pyr. §	Follows the conventions of Allen 1984

[ ]	Enclose text that, while completely lost, can be restored with reasonable confidence
[...]	Mark places where text is completely lost and cannot be restored with reasonable confidence
⟨ ⟩	Enclose text that, while absent from the original (for example, because the scribe mistakenly omitted it), can be restored
{ }	Enclose text wrongly included in the original that needs to be deleted, for example because the scribe presumably mistakenly wrote it out twice

INTRODUCTION: THE OTHER MATHEMATICS

> Even in ages the most devoted to material interests, some portion of the current of thought has been reflected inwards, and the desire to comprehend that by which all else is comprehended has only been baffled in order to be renewed.
>
> George Boole, *An Investigation of the Laws of Thought* (1854), 400.

0.1 HOW SMART ARE WE?

Hiding under the human skull is the most complex structure in the universe, the human brain, the seat of human thought. Comparatively little is currently known about the structure of the brain. Two fundamentally different ways of thinking are possible about what the brain produces: thought. One can think of thought as limited; or one can think of thought as boundless. Maybe thought is more readily viewed as boundless than as limited. Consider the human imagination. Imagination comes in various forms and shapes: artistic, poetic, religious, visual, and so on. More than anything, the power of imagination might easily lead one to assume that there are simply no limits to our thinking. Everything seems possible. Then again, the brain is no doubt a material structure with well-defined limitations. Why would anything the brain does not also be limited? Indeed, everyone would probably agree that we humans could be a lot smarter than we are. This limitation involves the concept of a line beyond which our thinking cannot go. In other words, our faculties are fundamentally limited. In his *Investigation of the Laws of Thought*, George Boole almost wistfully contemplates species of thought "the real nature of which it is impossible for us, with our existing faculties, adequately to conceive" (1854: 51). Once it is accepted that there are limitations to thought, the paramount question arises: Where exactly is that line beyond which thinking is not possible? The focus must be on a limit with a precise location. That limit is like a barrier or a wall. A more gloomy image presents that limit as the walls of a prison from which there is no escape. No single overarching concept has inspired the present investigation more than the awareness of the limits of human thought. Where exactly is that line? How smart are we really? Or, put differently: How dumb are we? Humility is a well-regarded notion in the realms of religion and morality. Rational thought has its own kind of humility: the acute awareness of its own absolute limitations. What are those limitations?

Thought is expressed in language. The study of thought is therefore necessarily also the study of language. However, it is hardly possible to study all languages at once. The focus of the present work is Egyptian. Of no language in the world can the history be followed for a longer period of time in written sources, for almost four thousand years. Much of the present work revolves around a most remarkable feature of earlier Egyptian: verb forms whose function is to denote events that depend on conditions. Dependency on conditions is contingency. The verb forms can therefore be called contingent tenses. Everyone knows what a condition is, but defining the term is another matter. All kinds of concepts tend to be found in the periphery of the concept of condition, including causality, consequence, logic, premise, reasoning, result, and truth. Everyone has some notion of what these concepts represent, but defining them with absolute precision is another matter.

In the present work, the examination of what is general, that is, the nature of thought, goes hand in hand with the study of what is specific, that is, the grammar of Egyptian. One is reminded of how, at the University of Berlin, Max Planck would recommend the study of philosophy only if one simultaneously also studied a specific subject of learning.

This work intends to improve the articulation of the three levels on which human thought happens: (1) the level of *things*; (2) the level of *events* (on which condition and *conditio sine qua non* are found); and (3) the level on which premise and consequence are found.

0.2 WHAT IS A CONDITION?

The principal aim of the present work is to illustrate how Boolean logic can be applied to language. No concept is more crucial to this entire investigation than that of the condition. The concept of condition has a certain abstract quality. How can one define such an abstract concept in the most simple and transparent terms? It may be useful to meet this challenge right at the outset by showing that condition can indeed be described in perfectly watertight fashion. All it takes to define condition is to realize two things. First, that two classes along with their supplement classes divide the universe—or anything we could possibly think about—into four combination classes. Second, that thinking proceeds on two levels that exhibit the same structure, namely the level of things and the level of events.

A condition is *a class on the level of events that combines with another class on the level of events (often called the apodosis) in such a way as to eliminate the existence of exactly one of exactly four combination classes.* To achieve such elimination, a conditional sentence conveys that *the class of events that is the condition or subordinate clause is fully encompassed by the class of events that is the apodosis or main clause.* The former class inhabits the latter class, like one circle inscribed inside another circle. Before illustrating this definition with an example, the level of things needs to be introduced. A thing is any entity populating reality or the imagination. The term "thing" is understood here to include human beings.

Consider, for example, on this first level, two attributes of things, namely "swallow" (*s*) and "bird" (*b*). Everything in the universe that shares the attribute "swallow" makes up the class or set of swallows. The same applies to the attribute "bird." In operating with any classes, nothing is more fundamental to modern scientific logic than the omnipresence of the supplementary classes, supplements, or contradictories. A supplement is a class whose members are united by virtue of sharing the lack of a given attribute. The members can be either things or events. In the present case, the two supplementary classes would be "non-swallows," that is, anything lacking the attribute "swallow," and "non-birds," that is, anything lacking the attribute "bird." Boole would represent the supplement "non-swallows" as $1 - s$—that is, the universe (1), or anything one could possibly think about, minus (–) swallows (*s*)—and the supplement "non-birds" as $1 - b$. He also uses the notations $\bar{s}$ and $\bar{b}$, which offer the convenience of brevity. Even before Boole, Augustus De Morgan appears to have been the pioneer in recognizing and promoting the crucial role of the contradictories.

Two attributes together with their contradictories, or absences of attributes, produce four combination classes. A combination class is a class that exhibits two or more presences or absences of attributes. In the present case, the four combination classes are as follows: (1) things that are both swallows and birds (sb); (2) things that are swallows but not birds ($s\bar{b}$); (3) things that are birds but not swallows ($\bar{s}b$); and (4) things that are neither birds nor swallows ($\bar{s}\bar{b}$). Consider, then, the statement "Swallows are birds," which means the same as "All swallows are (some) birds." This statement communicates that the class of swallows is part of the class of birds. The effect of the statement is that the existence of one of the four possible combination classes is denied, namely class (2): things that are swallows but not birds ($s\bar{b}$). By the same token, the statement clearly implies that the other three combination classes can, or do, exist. The statement "Swallows are birds" can be adequately represented in the following simple diagram. The class of swallows (*s*) is represented by circle *S*. The class of birds (*b*) is represented by circle *B*.

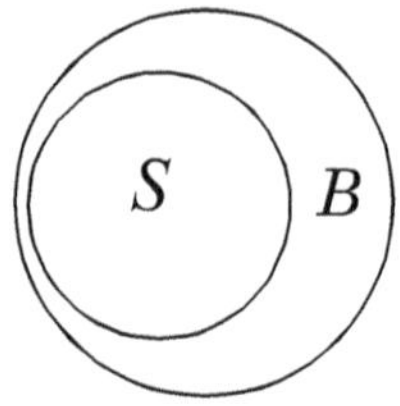

This shows that there is no such thing as swallows that are not birds ($s\bar{b}$): circle *S* is entirely encompassed by circle *B*. The other three combination classes do exist. Things that are both swallows and birds (sb) are found inside circle *S*. Things that are birds but not swallows ($\bar{s}b$) are found inside circle *B* but outside circle *S*. Things that are neither birds nor swallows ($\bar{s}\bar{b}$) are found outside circle *B*.

This much for the first of the two levels on which thought proceeds, the level of things. The concept of condition is found on the second level, the level of events. Importantly, the structure of this second level is the same as that of the first level. Consider, for example, on the level of the events, the two classes "It rains" (r) and "I stay inside" (i). As on the first level, there are again four combination classes, as follows: (1) the occasions or events when it both rains and I stay inside (ri); (2) the occasions when it rains and I do not stay inside ($r\bar{i}$); (3) the occasions when it does not rain and I stay inside ($\bar{r}i$); and (4) the occasions when it does not rain and I do not stay inside ($\bar{r}\bar{i}$). Consider, then, the statement "When(ever) it rains, I stay inside." This statement consists of the conditional clause "when(ever) it rains" and the apodosis "I stay inside." This statement on the second level, that of events, has the same structure as "Swallows are birds," a statement on the first level, that of things. The statement "When it rains, I stay inside" signifies that the class of occasions when it rains is part of the class of occasions when I stay inside, just as the statement "Swallows are birds" signifies that the class of swallows is part of the class of birds. The effect is that the existence of exactly one of the four possible combination classes is denied, namely class (2): the occasions when it rains and I do not stay inside ($r\bar{i}$). The other three combination classes do exist. This state of affairs is represented in the diagram below. The class of occasions when it rains (r) is represented by circle *R*. The class of the occasions when I stay inside (i) is represented by circle *I*.

This shows that there is no such thing as occasions when it rains and I am not inside ($r\bar{i}$): circle *R* is entirely encompassed by circle *I*. The other three combination classes do exist. Occasions when it rains and I stay inside (ri) are found inside circle *R*. Occasions when it does not rain and I stay

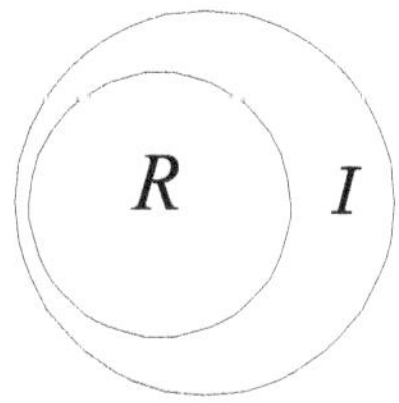

inside ($\bar{r}i$) are found inside circle *I* but outside circle R. Occasions when it does not rain and I do not stay inside ($\bar{r}\bar{i}$) are found outside circle *I*.

From "When it rains, I stay inside," it follows that "When I do not stay inside, it does not rain." This too is apparent from the above diagram. As can be seen, occasions when it does not rain fall outside circle *I* and therefore also outside circle *R*. When it does *not* rain (a class of events represented in the diagram by anything that falls outside circle *R*), there are two possibilities: I may be inside (a class of events represented in the diagram by the area falling outside circle *I* but inside circle *R*) or I may not be inside (a class of events represented in the diagram by the area falling outside circle *I* and therefore also outside circle *R*).

A condition and its apodosis both denote a class of events. Two classes along with their supplements produce exactly four combination classes. Condition-plus-apodosis serves to deny the existence of one, and only one, of the four combination classes. A condition has been defined above as *a class of events that combines with another class of events so as to deny the existence of exactly one of the four possible combinations of the two classes of events and their supplements.* Thus, "When it rains, I stay inside," a statement on the level of events, exhibits exactly the same structure as "Swallows are birds," a statement on the level of things. The two structures are never classified together in works on grammar, as far as I know. Yet, they are but two different manifestations of the very same thing.

The definition of "condition" presented above has two distinct benefits, I believe. First of all, in relying in part on this most transparent of categories of thought, number, the definition affords the benefit of clarity. Exactly one of exactly four combination classes is eliminated. Second, the existence of two levels of thought provides the benefit of mutual corroboration. The two levels exhibit the same structure. The structure of one level corroborates the structure of the other.

A sentence such as "Swallows are birds" corresponds to a nominal sentence in Egyptian. Conditions and nominal sentences appear in different chapters in grammars of Egyptian. The day may be far off when they are treated in the same chapter in Egyptian grammars. Yet that is where they belong.

0.3 TWO MAIN LINES OF RESEARCH

The present investigation lies at the intersection of two lines of research.

The dominant concept is the definition of the condition according to the laws of logic. Accordingly, the first line of research is the study of the condition in its wider context, and the second is the study of logic. In the interest of clarity and transparency, it will be useful to describe the origin and earlier stages of both lines of research and how they converged.

0.4 EARLIER RESEARCH ON CONDITIONAL CLAUSES

The origin of my first line of research, the study of the condition in its wider context, dates to the late 1980s. Five stages may be distinguished in its development.

(1) A problematic passage in Heqanakhte Papers 2.35–36 containing a condition was investigated. As a result, two distinct types of "condition" were articulated. The first type is represented in Old and Middle Egyptian by the pattern *jr* + *sḏm.f* and in Coptic by the clause conjugation **ⲉϥϣⲁⲛ-** *efšan-* 'when(ever) he'. The second type is represented in Old and Middle Egyptian by *jr* + auxiliary *wnn* + verbal tense and in Coptic by the conjunction **ⲉϣϫⲉ** *ešje* 'if (it is the case that)' (Depuydt 1988b).

(2) It was proposed that verb forms containing the morphemes *ẖr*, *k3*, and *jn* denote dependency on conditions or contingency. They can therefore be called contingent tenses. Additional articulation of the distinction between the two types of "condition" defined in (1) was permitted by the fact that contingent tenses depend on only one of the two types, that represented in Old and Middle Egyptian by the pattern *jr* + *sḏm.f*—and never on the other type (Depuydt 1989a).

(3) The distinction between the two types of "condition" defined in (1) received additional articulation as follows. First, Late Egyptian *jnn* 'if (it is the case that)' was assigned to one of the two types, the one represented in Old and Middle Egyptian by *jr* + auxiliary *wnn* + verbal tense. Second, Late Egyptian *jnn* (which is followed by a verbal tense) was identified as the direct etymological descendant of Old and Middle Egyptian *jr* + auxiliary *wnn* (which is also followed by a verbal tense) (Depuydt 1991b).

(4) It was proposed that one of the two types of "condition" is in fact not a condition at all, but rather a premise. The distinction at hand is therefore not one between two types of condition, but rather between condition and premise. Provisional definitions of condition and premise were provided (Depuydt 1995b).

(5) Final definitions were presented for both condition and premise, that is, definitions in accordance with the laws of thought (Depuydt 1999a).

Additional details about these five stages are as follows.

(1) *Heqanakhte Papers 2.35–36.* The present line of research began with the form *ḫr.k d(y).k* (Depuydt 1988b). I proposed that that sequence is not an instance of the compound verb form *ḫr.f sḏm.f*, contrary to what is normally assumed. Rather, the sequence consists, I believe, of the verboid *ḫr.k* 'you say' plus a future *sḏm.f*. Meanwhile, the following relevant passage has come to my attention, which leaves no doubt that the verboid *ḫr.f* is at least possible at the end of an *jr*-clause. The passage documents the use of *ḫr.f* in line 4 of Papyrus Berlin P10063 (edited by Luft 1992): *jr h3b.k ḥr.s ḫrw.fy sw* (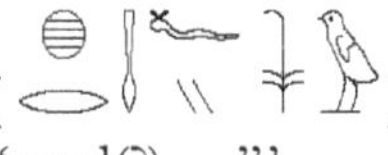) *h3b*[...] '"When you send a letter about it," he said, "send(?) ..."'.

The Heqanakhte Papers have now been newly edited in admirable fashion by Allen (2002). Allen defends the traditional interpretation of *ḫr.k d(y).k* as an instance of the form *ḫr.f sḏm.f.* He adduces a new argument in its favor: that my interpretation of *ḫr* as "parenthetic *ḫr* 'says'" "assumes an *unattested* use of the parenthetic after something other than a direct quotation" (26 [emphasis added]; similarly at p. 45). However, if the nature of the form *h3b.k* is in any way influenced by *jr* 'if, when' in line 4 of Papyrus Berlin P10063 cited above, which is quite possible, then *h3b.k ḥr.s* is no longer a fully direct quotation. *Jr* is followed as a rule, I believe, by the future substantival *sḏm.f* (*sḏmw.f*). *H3b.k* would only be part of a direct quotation if the literal translation of *jr h3b.k ḥr.s* were something like 'As for (his statement), "May you send (a letter) about it," he says, ...'. I suppose this interpretation is not completely impossible. The letter is fragmentary and the context of the passage at hand is lost. There are two more examples of the sequence *ḫr.f sḏm.f* in the Heqanakhte Papers, namely *ḫr.ṯn šm.ṯn* at 1.9 and *ḫr.k d(y).k* at 2.40. In the first case, *ḫr* is also preceded by a clause introduced by *jr* 'if, when', namely *jr tm.ṯn gm* 'when you do not find' at 1.8. Allen translates the three verb forms as follows: *ḫr.k d(y).k* at 2.35 and 2.40 as 'You'll have to put (him in charge …)' and 'You'll have to have (N brought to me)'; and *ḫr.ṯn šm.ṯn* at 1.9 as 'You will have to go'. These translations describe specific, one-time events. The translations are also marked for what may be called mood ('you'll have to'). Yet, verb forms containing the affix *ḫr* otherwise always denote the aorist, which is a kind of zero-tense or absence of tense (see already Junge 1972). The aorist is sometimes also described as the general present, as in English "He (generally) hears." The aorist therefore as a rule denotes events that generally happen. But the expression "you'll have to" in Allen's translations does not denote such general events.

Furthermore, at Heqanakhte Papers 2.35–36, *ḫr.k d(y).k* follows a subordinate clause of the type *jr* + auxiliary *wnn* + verbal tense. Verb forms containing the morpheme *ḫr* elsewhere never follow clauses of the type *jr* + auxiliary *wnn* + verbal tense. Clauses headed by *jr* preceding such verb forms are elsewhere *always* of the type *jr* + *sḏm.f.* This pattern of compatibility versus incompatibility found in Old and Middle Egyptian is confirmed strikingly by Coptic. The Coptic aorist **ϣⲁϥⲥⲱⲧⲙ̄** *ša=f-sôtm* 'he (generally) hears' behaves exactly like its etymological forebears, namely Demotic *ḫr jr.f sḏm*, Late Egyptian *ḫr sḏm.f*, and Old and Middle Egyptian *ḫr.f sḏm.f* and *sḏm.ḫr.f.* Coptic **ϣⲁϥⲥⲱⲧⲙ̄** *ša=f-sôtm* can follow **ⲉϥϣⲁⲛⲥⲱⲧⲙ̄** *e=f-šan-sôtm* 'when(ever) he hears', the functional equivalent of *jr* + *sḏm.f.* But it does not follow **ⲉϣϫⲉ** *ešje* 'if (it is the case that)', the functional equivalent of *jr* + auxiliary *wnn* + verbal tense. Instances in which that does happen are so rare that they hardly contradict the rule (for details, see Depuydt 2002a: 814–15). As a result of this limited investigation of Heqanakhte Papers 2.35–36, a distinction between two types of "condition" common to all stages of Egyptian, if not to all languages, came into focus. One type is expressed in Old and Middle Egyptian by the pattern *jr* + *sḏm.f*, the other type by the pattern *jr* + auxiliary *wnn* + verbal tense.

(2) *Contingent Tenses.* In a second stage of this same line of research, it was proposed that certain Old and Middle Egyptian verb forms exist exclusively in order to denote dependency on conditions (Depuydt 1989a, reprinted with minor alterations in Depuydt 1993a). But there is a limitation. Contingent verb forms are preceded only by clauses of the type *jr* + *sḏm.f.* As a rule, they are not preceded by clauses of the type *jr* + auxiliary *wnn* + verbal tense. Both types of clause have traditionally been called conditional clauses. But if contingent tenses are incompatible with the latter type, can the latter type properly be called a conditional clause? In this first main line of research, the need for an unequivocal definition of the concept of condition became ever more pressing. Postulating contingent verb forms turns the spotlight automatically on the concept of condition itself. At the very least, a condition can be re-defined as an event on which events denoted by verb

forms such as *sḏm.ḫr.f* and *sḏm.kꜣ.f* depend. At the same time, the question became increasingly more urgent as to why the Old and Middle Egyptian pattern *jr* + the auxiliary *wnn* + verbal tense does not seem to behave like a condition. If this pattern is not a condition, then what is it?

(3) *Late Egyptian jnn 'if (it is the case that)'*. In this third stage of the present line of research, attention turned to the Late Egyptian conjunction *jnn* (Depuydt 1991b). It was proposed that Late Egyptian *jnn* is the etymological descendant of Old and Middle Egyptian *jr wn*(*n*) but not the etymological antecedent of Demotic *jn-nꜣ* 'when(ever)'. That clearly makes *jnn* the functional equivalent of Coptic **ⲉϣϫⲉ** *ešje* 'if (it is the case that)' but not of Coptic **ⲉϥϣⲁⲛⲥⲱⲧⲙ̄** *e=f-šan-sôtm* 'when(ever) he hears'. Late Egyptian *jnn* is therefore not a condition like the Old and Middle Egyptian pattern *jr* + *sḏm.f*, on which contingent tenses can depend. If *jnn* is not a condition, then what is it?

(4) *Condition and Premise*. In this fourth stage, it was proposed that Old and Middle Egyptian *jr* + the auxiliary *wnn* + verbal tense, Late Egyptian *jnn*, and Coptic **ⲉϣϫⲉ** *ešje* all denote what may be called a *premise*, in sharp distinction with a condition (Depuydt 1995b). Provisional definitions of condition and premise were presented.

(5) *Final Definitions of Condition and Premise*. In a contribution reprinted in full below as appendix 1 (Depuydt 1999a), final definitions of the concepts of condition and premise were presented. Concepts are here defined as final if they can formulated on a level beyond which there is no level.

0.5 EARLIER RESEARCH ON LOGIC

Boole's *Investigation* (1854) is the foundation of modern scientific logic. My first chance encounter with this work dates to 1997/98. Closer study soon made it apparent to me that Boole's work affords an explanation of two phenomena that had held my attention for quite some years. Both phenomena occur in Egyptian as well as in most if not all languages. The first goes by the name of contrastive emphasis, among others; the second is the distinction between condition and premise. The two phenomena were treated in two articles, Depuydt 1999a, 1999b, reprinted in two appendixes to the present work. A few years then passed before I returned to the study of logic. The only application of logic to language that I have suggested in the interval concerns the Coptic particle **ⲣⲱ** *rô*, which was interpreted as an expression of isolating contrast (or contrastive emphasis) on the level of events, as opposed to the level of things. This interpretation was presented in an appendix to an article written around 1992, but published a decade later (Depuydt 2001a).

A brief illustration of isolating contrast on the level of things may be useful. It is a true observation that "sheep" and "*sheep!*" refer to the same reality, that is, a certain kind of animal. However, "*sheep!*" denotes isolating contrast whereas "sheep" does not. Logically speaking, the difference between "sheep" and "*sheep!*" is that "*sheep!*" denotes the supplementary class of the supplementary class of "sheep" whereas "sheep" does not. The supplement of the supplement is obviously the original class. Let "sheep" be represented by the symbol s. "Anything but sheep" then corresponds to $1 - s$. In short: "not-sheep." "*Sheep!*," then, means 'sheep and not anything else'. "And not anything else" obviously is the same as "not-sheep" in this case, that is, $1 - s$. Substituting "anything else" for "not sheep" in the expression "sheep and not anything else," one obtains "sheep and not not-sheep." This may be simplified to "not-not-sheep." If "not-sheep" is $1 - s$, then "not-not-sheep" must be $1 - (1 - s)$. Obviously, the reference is again to sheep. Or also: $1 - (1 - s) = 1 - 1 + s = 0 + s = s$.

What about isolating contrast on the level of the events? "*Sheep!*" denotes isolating contrast on the level of things. So does the Coptic sentence **ⲉϥⲙⲟⲟⲩⲧ̄** *e=f-mowt* 'he is *dead*', in the sense of 'it is dead that he is (and not not-dead)'. 'Dead' is contrasted with its opposite. By contrast, the particle **ⲣⲱ** *rô* denotes isolating contrast on the level of events, as in **ϥⲙⲟⲟⲩⲧ̄ ⲣⲱ** *fmowt rô* 'he *is* dead', that is, 'it *is* the case that he is dead', or also 'it is not not the case'. The event is contrasted with its supplement, that is, any other possible event. I have not found isolating contrast interpreted anywhere else in this manner, whether on the level of things or on the level of events. Helpfully, the eminent British logician W. E. Johnson (1892: 10) observes that "the denial of the denial of a proposition is equivalent to its affirmation." He calls this the "Law of Reciprocity." Logically speaking, this law is valid. But linguistically speaking, affirmation, as in "he is dead," and denial of denial, as in "he *is* dead," are not fully identical. The two are in that sense not "equivalent."

Boolean algebra is also used in computing. To provide context to my efforts to apply Boolean algebra to Egyptian grammar and grammar in general, I assembled some materials on the history of logic beginning with Boole and up to the late 1930s, when Claude E. Shannon first applied Boolean algebra to electronic circuits. Section 0.9 below is devoted to the difference between Boole's algebra and the Boolean algebra used in electronic circuits.

In the course of my investigations into the history of logic, the importance of John Venn became increasingly apparent. I fully agree with Alfred North Whitehead's (1897: 11) early assessment of Venn's role in the history of logic.

> The task of giving thorough consistency to Boole's ideas and notation, with the slightest possible change, was performed by Venn in his [*Sym bolic Logic* (Venn 1894)].

Styazhkin (1969: 214) voices much the same sentiment:

> An instructor in logic and morals at Cambridge, John Venn (1834–1923), first used the term "symbolic logic." He revealed the essence of the secret of success of Boole's procedures, thus marking an important contribution to the study of the history of symbolic logic (see his book *Symbolic Logic*, London, 1881).

The present work is much inspired by Venn's *Symbolic Logic.* Venn's influence most distinguishes the present attempt at applying logic to language from my earlier efforts. Inspiration from Venn is still absent in my recent attempt to define the difference between questions for specification and questions for corroboration (Depuydt 2007). Isolating contrast is intimately connected with questions for specification, that is, those that are marked by question-words such as "who?," "what?," "which?," and "where?" Contrastive emphasis or isolating contrast as a rule affects the question-word. Consequently, in Egyptian, emphatic verb forms or cleft sentences are the norm in questions for specification. If isolating contrast can be defined in terms of logic (see Depuydt 1999b), and if isolating contrast and questions for specification are related, then it seems natural to define questions for specification as well in terms of logic. Such is the aim of Depuydt 2007.

0.6 OUTLINE OF THE PRESENT INVESTIGATION

The present work extends the two converging lines of research described above: the condition and all that comes with it; and logic. The convergence concerns the fact that logic provides the ultimate source for defining the condition. Much of what follows revolves around the phenomenon of the balanced sentence or *Wechselsatz*. The balanced sentence is an Old and Middle Egyptian sentence pattern often translated as a conditional sentence. The question arises: What is the difference between the regular conditional sentence containing a conditional clause introduced by *jr* and the balanced sentence? Defining the difference as sharply as possible is a principal aim of this work. There are twelve chapters. Chapters 1–5 form a unit. Their aim is to define the difference between two sentence types that are normally rendered as conditional sentences. Earlier Egyptian exhibits two clearly distinct constructions that are both normally translated as conditional sentences in English, that is, as 'if ..., then ...' or 'when(ever) ..., then ...' sentences. The two constructions differ strikingly in form. The first construction consists of two juxtaposed substantival verb forms. This structure is often called the *Wechselsatz* ("exchange" sentence or "rotation" sentence). This is a term adopted for Egyptian grammar by H. J. Polotsky (1964: 281 [1971: 66] with n. 2) from §360 of Heinrich Ewald's (1803–1875) *Ausführliches Lehrbuch der hebräischen Sprache.* It is obvious, however, that the Hebrew *Wechselsatz* is not the same as the Egyptian *Wechelsatz*, because Hebrew does not exhibit finite substantival verb forms. Another term is "balanced" sentence. An explanation of the metaphor implied in the term "balanced" is suggested in §4.1 on p. 51. The following facts leave no doubt about the substantival character of both verb forms in balanced sentences. Both can exhibit (1) gemination with third-weak verbs or (2) the past tense affix *n* with an intransitive verb of motion. The second construction looks quite different. The subordinate clause is introduced by the particle *jr* 'if, when'. The main clause often contains verb forms whose function is to denote contingency or dependency on conditions. These contingent verb forms contain the characteristically contingent morphemes *k3* or *ḫr*.

The purpose of chapters 1–5 is to define as sharply as possible the difference between the two constructions. The key concept is condition. Condition is a concept belonging to the domain of logic. Logic is therefore where the answer to the problem at hand needs to be sought. Since about the mid nineteenth century, symbolic logic has superseded Aristotelian and scholastic logic. If symbolic logic has any claim to validity, it cannot be the exclusive domain of logicians. Scholars in other fields therefore need to inform themselves in sufficient measure about what logic has to offer, first to obtain a general sense of who deserves credit for what in the history of logic, and second to acquire the ability to extract from logic that which is needed for the successful treatment of narrowly defined problems pertaining to language. Logic concerns the fundamental laws of reasonable thought. It describes the ways in which we always inevitably reason. The laws of thought also define the limitations of reason. There is no reasoning beyond them. Any answers in terms of these laws has therefore hit a final frontier beyond which comprehension cannot possibly go. In that sense, such answers may be called final. The term "final" does not imply that criticism is unwelcome or that the proposed solutions might not turn out to be erroneous, even if they are proffered in challenge to the reader as correct.

Chapter 1 reviews essential facts regarding the two conditional constructions, the conditional sentence with *jr* 'if, when' and the balanced sentence. Chapter 2 introduces logical concepts needed to define the difference between the two constructions. Chapter 3 presents a definition of the difference between the two conditional constructions in light of the logical concepts introduced in chapter 2. Chapter 4 lays out what needs to be proven. Chapter 5 adduces empirical proof in the form of undeniable phenomena appearing in one construction but absent from the other.

Chapters 6–11 concern related matters. I have tried to show elsewhere that premises can be derived unfailingly from conditions by a fixed and stepwise procedure (see Depuydt 1999a). Each derivation is definite. That is, each derivation from conditions has one—and only one—solution in terms of premises. But is the inverse also definite? Does each derivation from premises also have one—and only one—solution in terms of conditions? It will be shown in Chapter 6 that this is the case. It may thus be considered proven that both processes, deriving premises from conditions and deriving conditions from premises, are definite in the sense to be defined below.

Chapter 7 defines the so-called *conditio sine qua non* and its relation to the normal conditional sentence in precise terms. Chapter 8 is a compilation of examples of balanced sentences, for documentary purposes. Chapter 9 surveys the role of the balanced sentence in the history of Egyptian grammar. Chapter 10 is a brief discussion of the nature of the *sḏm.f*-form in the main clause that follows a subordinate clause introduced by *jr* 'when, if'.

The topic of chapter 11 is broader. Civilization has evolved over the centuries. But are later civilizations superior intellectually to earlier civilizations? A suitable term for the process of evolution is "growth." All stages of a growth process are indispensable—and therefore of equal importance. Civilization in some ways resembles an organism. Organisms need to pass through certain successive stages. Growth is observed in all facets of civilization. The problem is to document and positively verify it. Chapter 11 focuses on one facet, logical thought as expressed in language. No language is attested for a longer period than Egyptian, almost four millennia. Egyptian *changed* from Old Egyptian to Coptic. But did it also *grow* in logical sophistication? When language and thought first originated, tens of thousand of years ago, the first manifestations must have been less sophisticated. However, it is generally assumed, perhaps by default, that the earliest attestations of language are as sophisticated as anything that comes later. The aim of chapter 11 is to show that Egyptian did grow in logical sophistication in certain verifiable ways in the course of the historical period.

Finally, Chapter 12 describes ghost phenomena, phenomena widely believed to exist while they may well not.

0.7 THE OTHER MATHEMATICS: LOGIC AND MATHEMATICS PROPER AS TWO MEMBERS OF DEDUCTIVE THOUGHT

0.7.1 The separate histories of logic and mathematics

The word "logic" appears in the subtitle of this work. Words like "logic" and "math" come easily to the lips. Most everyone, amateur or professional, probably assumes by default that logic is not math and math is not logic, without giving the matter much further thought. The main purpose of the present section, however, is to emphasize that logic and math are essentially the same. The reason for devoting some space to emphasizing this identity is as follows. The language concept of condition is central to the present work. Condition is defined here according to the simple principles of logic. Much of this book's argument rests on this definition of condition and everything that comes with it. It may therefore be useful to locate logic itself more precisely in the larger realm of thought. A better sense of where exactly logic belongs should buttress this work's foundations. In fact, in the nineteenth century, the young discipline of rigorous mathematical logic was not always well received and even subject to hostility. Whitehead (1897: vi) described this state of affairs as follows:

> [Mathematical logic] has been peculiarly unfortunate, for it has been disowned by many logicians on the plea that its interest is mathematical, and by many mathematicians on the plea that its interest is logical.

Other factors have contributed to perpetuating the unspoken assumption that logic and math are somehow separate. One factor is historical. The development of solid mathematics began already in antiquity. By contrast, the development of a truly reliable science of logic dates to the mid nineteenth century C.E. What is called logic and what is called mathematics are therefore separated from one another in the history of science. This separation has tended to obscure the unity of the two fields. On the other hand, the general sense that logic and math are somehow related has spawned a proliferation of terms that attempt to bring the two more closely together. Among those terms are "algebra of logic," "mathematical logic," "pure mathematics," and "symbolic logic." These terms presumably all denote more or less the same thing. But while they bring logic and mathematics closer together, they do not go as far as declaring logic to be a form of mathematics. Such is the more radical view proposed here.

0.7.2 Deduction

If logic and math are the same, there is a need for a one-word name that denotes both logic and math together as one thing. No such one-word name exists. A two-word term used by Whitehead (1897) is "universal algebra." It is not my intention to propose a new name here. The area of thought to which both logic and mathematics belong therefore remains nameless. However, it still needs to be clarified how logic and mathematics are the same. And how does the concept of condition relate to this identity?

Mathematics and logic are the same in both being deductive. Deduction is the process of thought by which one derives one proposition with absolute necessity and inevitability from another in a chain of two or more propositions. Most of us are familiar with this process from mathematical proofs. In a mathematical proof, one step always inevitably follows from another. That inevitability is what deduction is all about; it involves some kind of internal consistency and has nothing to do with external influence from outside reality. It is not necessary to observe reality to draw deductive conclusions. Whitehead (1897: vi), who worked prominently in both logic and mathematics, describes deductive reasoning as follows:

> Mathematical reasoning is deductive in the sense that it is based upon definitions which … need only the test of self-consistency. Thus no external verification of definitions is required in mathematics, as long as it is considered merely as mathematics.

0.7.3 Axiomatic observations of reality

Obviously, a chain of deductive inferences must have a beginning—something has got to be first. It is also obvious that the first step cannot follow from a previous step, precisely because it is the first. Instead, the first step is rooted in reality or observable fact: it is a general observation about reality that seems so overwhelmingly obvious that we cannot possibly deny it. Such first steps are called axioms or postulates. They cannot be proven. One just accepts them because they seem so clear. The first step of a deductive chain of reasoning is therefore itself not deductive.

If logic and mathematics are the same in both being deductive, how do they differ? It is said above that deductive chains of thought are rooted in general observations about reality. These observations are unprovable axioms. Logic and mathematics therefore both begin with a clear, hard look at reality around us. The aim is to make observations that are so obvious that we cannot possibly deny them. Indeed, we cannot count on proof to accept them. They are impossible to prove. Because logic and mathematics are identical in being deductive, the difference between the two must lie in the first step. Logic and mathematics begin with different types of observations about reality, but both types of first step set into motion deductive chains of thought in which reality is left behind. The chains are developed by a rigorous, internal consistency called deduction. So what are the two types of root observations that set the deductive chains of logic and mathematics into motion?

0.7.4 Quantity

Consider any class or set of things: for example, people, cars, houses, trees, and so on. Two types of observation of the most general kind are possible. The first type is that the classes are capable of increase or diminution. There can be more of them or less of them. This property of things is called quantity. From the fundamental observation of quantity, an infinite number of propositions can be developed deductively. The resulting body of deductive thinking is generally known as arithmetic and algebra. A simple example of a development follows. It is possible to divide a group of six cars into equal groups larger than one, namely into three groups of two cars or two groups of three cars. The same is not possible with seven cars. Seven is therefore called a prime number. Seven is divisible only by itself and by one. By contrast, six is not a prime. It is divisible by three and two as well as by itself and one.

0.7.5 Attribute

A second type of root observation can be made about things, as obvious as the first concerning quantity: things have attributes and all things sharing an attribute together form a class or set. This lies at the root of the only kind of mathematics that is not concerned with quantity. Whitehead called this kind of mathematics the "algebra of symbolic logic" (1897: 35).

> The Algebra of Symbolic Logic is the only known member of the non-numerical genus of Universal Algebra. "Non-numerical" basically means the same as "bearing no relation to quantity."

The second type of root observation concerns the *attributes* of things. Two examples are "blue" and "car." Importantly, both presence and absence of an attribute can be observed. As noted on page 3 above, the great champion of the pivotal role in all human thought of the *absence* of attributes was Augustus de Morgan. But it was George Boole who brought the concept fully to fruition. In support of the concept, I have myself ventured to suggest (see appendix 2 below) that the omnipresent phenomenon of contrastive emphasis, or isolating contrast, or focus, or whatever one may wish to call it, lays bare the absence of attributes as part of the fundamental fiber of thought.

It is now clear that one *attribute* produces *two* classes or sets of things. Thus, the attribute "blue" produces the class of things that are blue and the class of things that are not blue. The attribute "car" produces the class of things that are cars and the class of things that are not cars. Quantity, that is, how much blue or non-blue there is or how many cars or non-cars there are, is irrelevant to attributes. A class or set can also be defined by two or more attributes, as in the class of blue cars.

0.7.6 Rules valid with attributes but not with quantities

Thinking deductively about attributes develops in an entirely different direction than thinking deductively about quantity. Two fundamental differences are readily apparent. In Boolean algebra attributes obey the following rules:

$x + x = x$; and

$x = xx$.

Clearly, these same rules do not apply when one reasons deductively about quantity. Indeed,

$5 + 5 \neq 5$.

$5 \neq 5 \times 5$.

Yet these rules do apply to attributes. Consider the attribute c "car." It is obvious that

$c + c = c$.

If one adds to the class of things exhibiting the attribute "car" the class of all the things exhibiting the attribute "car," what one obtains in total is still the same class of things exhibiting the attribute "car." This is obvious. It can therefore be accepted without explicit proof. In fact, there is no possible proof. Acceptance without proof makes the rule into an axiom.

Then there is also the rule expressed by the equation

$c = cc$ (also represented as $c \times c$).

This equation signifies that the class of things that exhibits two attributes, namely c "car" and again c "car," is exactly the same as the class exhibiting the single attribute c "car." This too is abundantly obvious. No proof is required—or in fact possible. As one can see, quantity plays no role in this other kind of mathematics, which is called symbolic logic by some. The class c is exactly as large as the class $c + c$ and the class cc ($c \times c$).

Two fundamental postulates of this other, quantity-less, mathematics are as follows:

(i) $c\bar{c} = 0$ or ($c \times \bar{c} = 0$);

(ii) $c + \bar{c} = 1$.

According to (i), there exists no such thing (= 0) that is at the same time a car (c) *and* (×) not-a-car ($\bar{c}$). The symbol $\bar{c}$ is an abbreviation of $1 - c$—that is, the universe (1), or anything one could possibly think about, minus (–) cars (c). Equation (i) has been known since Aristotle as the fundamental axiom of philosophy. Presumably we are unable to think about anything if there is an assumption that something can be itself and something else at the same time. According to equation (ii), the universe (1) consists entirely of c "cars" and (+) $\bar{c}$ "anything but cars." Equation (ii) is rendered more obvious if one replaces $\bar{c}$ by $1 - c$. Equation (ii) then becomes

$$c + 1 - c = 1,$$

which is the same as

$$c - c + 1 = 1,$$

and therefore also as

$$0 + 1 = 1. \text{ Or: } 1 = 1.$$

Starting from such simple equations, deductive thought about attributes can be developed to great complexity, just like deductive thought about quantity.

The difference between deductive thought about attributes, commonly called "logic," and deductive thought about quantities, commonly called "mathematics," may also be illustrated by the equation

$$x + bx = x.$$

This equation is valid for attribute mathematics. Indeed, consider the two classes c "cars" and b "blue (things)." It is abundantly obvious that

$$c + bc = c.$$

The class of cars (c) plus the class of blue cars (bc) is still the class of cars (c). This is known in logic as the "rule of absorption." The class "blue cars" (bc)

does not make the class of "cars" (*c*) larger by being added to it but is rather "absorbed" into it. But obviously, the same equation is false in thought about quantity, or mathematics proper. For example,

$$5 + 3 \times 5 \neq 5.$$

3×5 is not "absorbed" into 5.

0.7.7 Combination classes

A principal development in deductive thought about attributes is the distribution into combination classes. For example, *two* attributes, such as "blue" and "car," along with their supplements or contradictories, "non-cars" and "non-blue," produce *four* combination classes. Thus, the *two* attributes "blue" and "car" divide the universe into *four* classes: (1) blue cars; (2) non-blue cars; (3) blue things that are not cars; and (4) things that are neither blue nor cars. This writer belongs to the fourth combination class.

In addition, thought happens on *two levels*. The two levels allow thought to be highly complex. Another two-level hierarchy allows great complexity in language. In this other two-level hierarchy, the two levels are the level of sounds and the level of words. Languages consist of many words, but there is no need to express each word by its own sound. Instead, words consist of different *combinations* of sounds selected from a limited set of sounds. An economy of sounds is therefore possible. Sounds are the basic building-blocks of language. The two-level hierarchy, which consists of the level of sounds and the level of words as combinations of sounds, allows great complexity by means of a limited set of basic elements.

Thought about attributes likewise proceeds on two levels. The first level concerns relations between things. On this first level, one might state that "Swallows are birds." This statement eliminates one of the four combination classes, namely swallows that are not birds. By the same statement, the other three combination classes do exist, namely (1) swallows that are birds, (2) birds that are not swallows, and (3) things that are neither birds nor swallows. On the second level, the relations between things of the first level are themselves related to one another. In other words, just as words are not different sounds but rather different *combinations* of sounds, so statements on the second level of thought are different *combinations* of statements on the first level of thought. As in the case of sounds and words, the two-level hierarchy allows greater complexity by means of fewer elements.

Furthermore, the structure of the first level is replicated on the second level. For example, the same type of elimination occurs. Conditions exist in order to eliminate one of four possible combination classes. For example,

"When(ever) the sun shines, I go to the beach" eliminates the combination of the sun shining and my *not* going to the beach. The other three combinations *do* exist: (1) occasions when the sun shines and I go to the beach, (2) occasions when the sun does not shine and I go the beach, and (3) occasions when the sun does not shine and I do not go to the beach.

0.7.8 Other axiomatic observations about reality

There are other types of root observations in addition to quantity and attribute. These other root observations also serve as the axioms of further types of deductive thought. For example, geometry starts from certain characteristic root observations about space such as the point, the line, and the plane. At some level, these concepts about space can also be associated with quantity because space too is capable of increase or diminution. Another example of a root observation concerns trigonometry. The root observation at the origin of trigonometry is the relation between *angle* and *distance*. A mere glance at reality makes it apparent that there are such things as angle and distance. Trigonometry is about the relation between the two. Consider a person raising a pole that lies flat on the ground upward by an *angle* of 90° to its maximum height or vertical *distance*. It is a fact of nature that the pole reaches *half* of its maximum vertical distance or height after one *third* of the angle has been traversed. Half the upward *distance* corresponds to a third of the upward *angle*. Mathematicians state that the sine of 30° is 0.5. But this formulation may obscure the root observation that trigonometry is about how angle relates to distance.

0.7.9 Condition and deduction

The fundamental language concept of condition has been defined above in the simplest of terms by means of attribute mathematics. Moreover, the concept of condition has been embedded in a larger conceptual framework in which logic and mathematics are two members of the same species of thought. This overarching species of thought may be called deductive thought. Deductive thought progresses independent of reality. But it is rooted in certain general and obvious observations about reality.

0.8 ON LOGIC AND ITS PROPER DOMAIN, AND ON TRUTH, CERTAINTY, RIGHT AND WRONG, AND KNOWLEDGE ITSELF

0.8.1 Logic in relation to truth and thought

It is commonly said that logic is about truth itself. The following statement by Schröder (1890: 1) at the outset of his monumental work on logic is typical.

> *Logic*, in the broad sense of the term, is concerned with all the rules whose application promotes knowledge of the truth.*

It is also commonly said that logic is about thought or reasoning itself. In this view, logic is the science that examines how we think and reason. Schröder's statement continues as follows:

> [Logic] is concerned with the methods of research in themselves. It seeks to answer the question: *how* do we gain knowledge, by what way do we attain truth? Consequently, because grasping the truth is an act of thought, we may denote as object of logic in itself: thought, to the extent that its final goal is knowing.†

Similarly, Boole begins his *Laws of Thought* (1854) as follows:

> The design of the following treatise is to investigate the fundamental laws of those operations of the mind by which reasoning is performed; to give expression to them in the symbolical language of a Calculus, and upon this foundation to establish the science of Logic and construct its method.

It would probably be easy to collect statements from various sources that convey a similar message. There seems to be a tacit assumption that logic occupies a privileged position in relation to truth and thought. It is a kind of super-thought, a type of thought that transcends all other kinds of thought. We think about all kinds of subjects: history, chemistry, literature, physics, and so on. By the view expressed above, logic would be seem to be devoted to studying how we do all that. Logic almost does not seem to have a

* Die *Logik*, im weiteren Sinne des Wortes, beschäftigt sich mit all' den Regeln, durch deren Befolgung die Erkenntnis der Wahrheit gefördert wird.

† [Die Logik] hat es demnach mit den Methoden der Forschung überhaupt zu thun. Sie sucht die Frage zu beantworten: *wie* gewinnen wir Erkenntnisse, auf welchem Wege gelangen wir zur Wahrheit? Mithin, da Erfassen der Wahrheit ein Akt des Denkens ist, dürfen wir als Gegenstand der Logik überhaupt bezeichnen: das *Denken*, sofern es das *Erkennen* zum Endzweck hat.

domain of its own. Its domain is rather to examine what all other types of thought have in common.

I would like to object to this seemingly privileged status of logic in relation to truth and thought. It is crucial to assign logic to its rightful place in the realm of thought. Boole and Schröder pay lip-service to the age-old privileged status of logic. But closer inspection reveals that the innovations found in their own work in fact contradict that very status. The following suggestions about the rightful place of logic and its relation to such concepts as truth, certainty, right and wrong, and knowledge itself are provisional. These suggestions convey initial impressions.

0.8.2 The domain of logic

From the preceding section, it should be clear that logic has a well-defined domain of its own. In classifying fields of learning, logic belongs together with mathematics: both are deductive; both begin with observations that are eminently obvious yet difficult to prove. Such observations are called axioms. Starting from axioms, thought progresses by means of rigorous, tightly controlled, incremental steps taken independent of outside reality. By contrast, inductive learning proceeds differently. In inductive thought, repeated observation allows one to establish certain regular types of behavior. These regularities are formulated as laws. Increased observation leads to an ever higher probability that a certain law is true. But absolute certainty is never attained. Schröder (1890: 9, note) phrases it well when he states that what has happened 1000 times does not *ipso facto* have to happen the 1001st time. The special domain of intellectual activity that logic can call its own has been outlined above. Mathematics has much to do with quantity, with what is capable of increase or diminution. Quantity is expressed by number. By contrast, logic is about combinations of attributes, their absences and presences, and their combinations. From the levels of things and events, it is possible to rise up to a third level, that of premise and result. Consider someone operating by the tenet "When it rains, I stay inside." Exactly two premises lead to two results or consequences: "If it is in fact raining, then I am staying inside" and "If I am in fact not staying inside, then it is not raining." Exact details about these derivations follow below (see especially chap. 6).

From contemplating the second and third levels of deductive thought about attributes, one can begin to see how it was possible for the name "logic" to become associated with what is often called thought itself, reason itself, inference itself, causality itself, or all kinds of related names. Yet all the intellectual processes involved are nothing but a special kind of mathematics, much of it pertaining to what have been called attributes above. That also means that terms such as "thought," "inference," and "causality" are utterly without definition. Any degree of definition can only be gained if one specifies precisely what kind of root observations as axioms one is thinking about and if one specifies precisely which deductive steps one derives from these axioms. Modern scientific logic is such a precise system. But it is only a system among systems. It is in a sense no more deserving of the name "laws of thought" than mathematics itself.

In this connection, it may be useful to reflect briefly on three other concepts: truth, certainty, and right or wrong. What is their nature, and how do they differ? Truth and certainty seem related. But the difference between the two seems apparent when one considers their opposites. The opposite of truth is falsehood. The opposite of certainty is doubt.

0.8.3 What are truth and certainty?

Truth is about the relation between what is inside the mind and what is outside the mind. If something inside the mind *matches* something outside the mind, then what is inside the mind can be described as true. If something inside the mind does *not* match what is outside the mind, then what is inside the mind is said to be false. What is inside the mind is externalized by means of language. Accordingly, one can make true statements and false statements.

By contrast with truth, *certainty* is an internal state of the mind. This state of the mind pertains to the relation between presence and absence of the above-mentioned match between what is inside the mind and what is outside the mind. There are two basic possibilities: certainty and doubt (or uncertainty). On the one hand, one can sense or feel that one can confidently choose between presence and absence of the match. That is the sense or feeling of certainty. Something is either true (presence of match) or false (absence of match). On the other hand, one can sense or feel that one cannot confidently make that choice. That is the feeling of doubt or uncertainty.

The difference between truth and certainty is underscored by the fact that one can be certain about something that is not true. The fact that I can state "I am certain" but not "I am true" also indicates that certainty is a sense or feeling whereas truth is not. Furthermore, one can ask, "Are you certain that it is true?" This question suggests that certainty and truth are two different things.

0.8.4 Right and wrong

That the distinction between right (correct) and wrong (incorrect) is separate from the two others described in sections §0.8.2 and §0.8.3 already appears from a simple example like the following. If one is *certain* that something is *true*, there are still two possibilities: one may be *right* or one may be *wrong*. Certainty and doubt are terms that describe a state of mind regarding what is true or false. Right and wrong are terms that in turn evaluate that state of mind. The certainty and the doubt themselves may be judged to be either right or wrong. Someone may be certain that something is true and yet be wrong.

0.8.5 What is knowledge?

It is often claimed that logic is concerned with thought and knowledge itself. Statements in support of this view have been quoted above. By this claim, logic somehow occupies a privileged status. That claim has been contested above. It is true that logic is concerned with thought. But logic is no more or less concerned with thought than are other intellectual domains, such as mathematics. What can one say, then, about thought and knowledge in general? Quite simply but provisionally, thought seems to be in large part an effort to assimilate what is inside the mind to what is outside the mind, with knowledge as the result. Consider the simple example of the street-plan of a city. The street-plan exhibits a certain structure. By assimilating the structure of the mind somehow to the structure of the street-plan, it becomes possible to find one's way around in the city. At the outset, the structure of the city's streets was absent from the mind and the mind's biochemical configuration. Through a certain kind of effort, the structure of the city's streets acquired a certain kind of imitation inside the mind. The proof of the acquisition is that one can find one's way around town. The effort involved is ultimately entirely one of assimilation. In this effort, certain facets of reality outside the mind are replicated inside the mind. Logic and mathematics are each concerned with facets of reality outside the mind that are reproduced inside the mind. Logic is concerned primarily with attributes and their combinations. Mathematics is concerned in large part with quantity.

0.9 BOOLE'S ALGEBRA AND THE BOOLEAN ALGEBRA OF COMPUTING

The introduction to the acts of a recent conference devoted to Boolean thought (Gasser 2000) rightly suggests that there is a "buzz" about Boole. "Boolean algebra" is a term that now has a familiar ring to many a layman. It is not clear to what extent this familiarity coincides with deeper acquaintance with the role of George Boole (1815–1864) in the history of thought. It seems fair to assume, however, that the first thing that comes to the minds of many in contemplating the term "Boolean" is pairs such as "On" and "Off," "One" and "Zero," and "True" and "False." There is also a sense that Boolean algebra is the language of the computer. All this is basically true. But the pair "On" and "Off" does not appear in Boole's foundational *Investigation* (1854). The question arises, how does the algebra of Boole's *Investigation* compare to the Boolean algebra of electronic circuits? It appears that there is a clear difference between the two (on this same topic, but from a different angle, see Hailperin 1981). Then again, it is also true that the latter and later is firmly rooted in the former and earlier. The need is for a calibrated comparison. I claim no competence in either electrical engineering or computer science. However, providing a general sense of the basic difference between Boole's algebra and the Boolean algebra of electronic circuits should require no such competence.

But first, a historical note. In part owing to Charles Sanders Peirce (1839–1914), Boole's logic was taught at American universities in the late nineteenth and early twentieth centuries. In 1936, equipped with a knowledge of symbolic logic, Claude E. Shannon (1916–2001) arrived in Cambridge, Massachusetts, from the University of Michigan to begin graduate study at the Massachusetts Institute of Technology. He tended to Vannevar Bush's (1890–1974) "differential analyzer." Shannon saw the potential of Boole's algebra to replace the movements of shafts, gears, and disks of this mechanical behemoth by electronic circuitry. The result was an essay entitled *A Symbolic Analysis of Relay and Switching Circuits* (Shannon 1938). Goldstine (1972: 119–20) writes the following about this essay:

> Shannon, the founder of what is called Information Theory, in his master's thesis showed in a masterful way how the analysis of complicated circuits for switching could be effected by the use of Boolean algebra. This surely must be one of the most important master's theses ever written .... The paper was a landmark in that it helped to change digital circuit design from an art into a science.

The crucial work was accomplished in the summer of 1937. Shannon describes the events in straightforward fashion as follows (Shannon 1993: xxv–xxvi [reprinted from Shannon 1987: 65]):

> I knew about symbolic logic at the time from a course at Michigan, and I realized that Boolean algebra was just the thing to take care of relay circuits and switching circuits. I went to the library and got all the books I could on symbolic logic and Boolean algebra, started interplaying the two, and wrote my Master's thesis on it.

Shannon's search for works on logic must have been greatly facilitated by Alonzo Church's detailed bibliography of logic that had just appeared (Church 1936). Shannon cites Church's bibliography as a reference at the end of his M.A. thesis. It lists Boole with four items in no. 19 of the bibliography's 547 numbers. On the connection between a relay circuit and Boolean algebra, Shannon (ibid.) writes as follows:

> Trivial, actually, once you make it. The connection was not the main thing. The more important, harder part was working out the details, how to interleave the topology of the switching circuits, the way the contacts are connected up and so on, with the Boolean algebra expressions. Working that out was a lot of fun. I think I had more fun doing that than anything else in my life, creatively speaking. It worked out so well.

Shannon also collaborated with Alan Turing in the 1930s. The computer age had begun. The disciplines of information theory and communication theory were born. The classic manifesto of communication theory is Shannon's *The Mathematical Theory of Communication* (Shannon 1948; also Shannon and Weaver 1963). The new universal unit was the binary digit, "bit" for short, a term first suggested by John W. Tukey.

How does Boole's algebra differ from Boolean algebra as Shannon understood it? In all brevity, the Boolean algebra of electronic circuits is Boole's algebra or symbolic logic *stripped of all its specifics.* What are those specifics? And what is the generality that remains after the specifics have been stripped away?

Boole first clearly showed how logical thought proceeds on two levels, the level of things and the level of events. In his algebra, a symbol can stand for a thing. Thus, c might stand for cars and bc ($b \times c$) for blue (b) cars (c). But a symbol can also stand for an event. This, r might stand for occasions when it rains and ri ($r \times i$) for occasions when it rains (r) and I also stay inside (i).

Boole's operations for the first time made it clear that it is impossible to think about any thing or any event without also thinking implicitly about that thing's or that event's supplementary class or supplement or contradictory. Hardly anything is more typical of human thought than the contradictories. The universe (1), or anything one could possibly think about, consists of $bc + \bar{b}c + b\bar{c} + \bar{b}\bar{c}$, that is, "blue cars ($bc$), plus non-blue cars ($\bar{b}c$), plus blue things that are not cars ($b\bar{c}$), plus things that are neither blue nor cars ($\bar{b}\bar{c}$)." Therefore, $1 - bc$ equals $\bar{b}c + b\bar{c} + \bar{b}\bar{c}$. This combination of three classes can be written in short as $\bar{b} + \bar{c}$. Indeed, $\bar{b}$ naturally divides into $\bar{b}\bar{c} + b\bar{c}$ (i.e., $\bar{b}$ that is c and $\bar{b}$ that is not c), and $\bar{c}$ likewise naturally divides into $\bar{b}\bar{c} + b\bar{c}$. Therefore, $\bar{b} + \bar{c}$ corresponds to $b\bar{c} + \bar{b}\bar{c} + \bar{b}\bar{c} + b\bar{c}$. In this sum, $\bar{b}\bar{c} + \bar{b}\bar{c}$ can be reduced to just $\bar{b}\bar{c}$. The four-part sum $\bar{b}c + \bar{b}\bar{c} + \bar{b}\bar{c} + b\bar{c}$ can therefore be written as the three-part sum $\bar{b}c + \bar{b}\bar{c} + b\bar{c}$. This much for the supplements of things on the level of things. The same applies to the supplements of events on the level of events.

Boole did not explicitly stress the capital role of supplementary classes in the progression of thought. But it is clear that the supplementary classes are all-pervasive of his analysis of logic. It would seem that the supplementary classes constitute the fundamental fiber of thought. But the supplements are not visible at the surface of logic as it is expressed in language. The supplements are mostly hidden from view, as it were. I have suggested elsewhere (Depuydt 1999b/appendix 2 below) that the phenomenon of isolating contrast lays bare this fundamental fiber of human thought. Indeed, the most convenient way to understand an instance of isolating contrast, as in "*Sheep!*," is simply as "not not-sheep" or "not something else than sheep." Symbolically speaking, not not-sheep corresponds to $1 - (1 - s)$, which is the same as $1 - 1 + s$, or $0 + s$, and hence simply s.

It has been suggested above that Boole's algebra needs to be stripped of specifics in order to obtain the Boolean algebra of electronic circuits, that is, the On/Off-type of popular imagination. What are the specifics that need to be stripped away, and how does one obtain mere skeletal On/Off-ness after stripping away all the specifics?

A number of steps may be distinguished in illustrating how one moves from the specific to the general. Consider just one class and its supplement, for example the class "cars," and the two basic operations pertaining to classes, addition and combination. Addition is represented by Boole with the symbol "+," and combination with the symbol "×." The representation $b + c$ denotes the class of things that encompasses all the things that are blue and all the things that are cars. The representation $b \times c$ (bc) denotes the class of things that are at the same time blue and cars, in other words, blue cars.

Third, one may now apply these two operations of addition ("union") and combination ("intersection") to a single class. Let us consider all the possible combinations of the class of cars (c) with itself or with its supplement according to the two operations of addition (+) and combination (×); there are eight.

THE EIGHT COMBINATIONS OF A CLASS AND ITS SUPPLEMENT

$c + c = c$	"Cars plus cars is still cars."
$c + \bar{c} = 1$	"The universe (1) consists of cars plus everything that is not cars."
$\bar{c} + c = 1$	*ditto*

$\bar{c} + \bar{c} = \bar{c}$	"All non-cars plus all non-cars are still all non-cars."
$cc = c$	"Things that belong to the class of cars and again to the class of cars are the same as the things that belong to the class of cars."
$c\bar{c} = 0$	"Nothing (0) can be both a car and not a car at the same time."
$\bar{c}c = 0$	*ditto*
$\bar{c}\bar{c} = \bar{c}$	"Things that do not belong to the class of cars and also do not belong to the class of cars are the same as the things that do not belong to the class of cars."

It is possible to move from the more specific to the more general by stripping away a specific element. What can be stripped away is the identity of the class itself. What is then left is presence of the class and absence of the class, whatever the class may be. Presence of the class may be represented by "On" and absence of the class may be represented by "Off." The following table of correspondences applies.

c	("cars")	= On	(belongs to class "cars")
$\bar{c}$	("non-cars")	= Off	(does not belong to class "cars")
1	(everything)	= On	(belongs to class "cars")
0	(nothing)	= Off	(does not belong to class "cars")

This table of correspondences can be applied to the table of eight combinations listed above.

THE EIGHT COMBINATIONS OF A CLASS AND ITS SUPPLEMENT, ONCE THE SPECIFIC CLASS HAS BEEN STRIPPED AWAY

On	+	On	=	On
On	+	Off	=	On
Off	+	On	=	On
Off	+	Off	=	Off
On	×	On	=	On
On	×	Off	=	Off

THE EIGHT COMBINATIONS OF A CLASS AND ITS SUPPLEMENT,
ONCE THE SPECIFIC CLASS HAS BEEN STRIPPED AWAY

Off	×	On	=	Off
Off	×	Off	=	Off

Instead of "On" and "Off," one can also use "1" and "0." In the first table above, "1" denoted "the universe of thought" and "0" denoted "nothing." If one makes "1" and "0" synonyms of "On" and "Off," then one changes their signification. The substitution is performed in the following table.

THE EIGHT COMBINATIONS OF A CLASS AND ITS SUPPLEMENT,
ONCE THE SPECIFIC CLASS HAS BEEN STRIPPED AWAY

1	+	1	=	1
1	+	0	=	1
0	+	1	=	1
0	+	0	=	0
1	×	1	=	1
1	×	0	=	0
0	×	1	=	0
0	×	0	=	0

The things and events that characterize logic as it is expressed in language have now been stripped of all specificity. All that is left is Yes and No, or On and Off, or 1 and 0, along with their eight combinations and their eight results.

On and Off can be embodied physically by presence or absence of an electric current. An electric current is either On (1) or Off (0). If one installs two switches in series, then both switches need to be On (1) for the current to pass. Installation of switches in series is represented by the multiplication sign (×) in the table above. But if one installs two switches in parallel, then only one switch needs to be On (1) for the current to pass. Installation of switches in parallel is represented by the addition sign (+) in the table above. The operation of each switch is strictly binary. There are no positions other than On (1) and Off (0). The eight combinations represented in the tables above can therefore be seen as the workings of two electronic switches.

The steps outlined above reveal part of how the binary operations of computers are rooted in simple empirical observations about things or events in what may be called attribute mathematics.

In moving from logic as it is expressed in language to the skeletal binary On and Off of computers, certain specifics of logic as it is expressed in language are lost. Three such specifics may be noted. First is the distinction between the two levels, the level of things and the level of events. The distinction simply vanishes. Second is the specific identity of the class, whether it is a class of things or a class of events. Third is the contrast between a class and its supplement. This contrast has been stripped down to a contrast between presence and absence as pure abstractions.

To conclude, it will be useful to quote, for additional illustration, the postulates listed by Shannon at the outset of his *Symbolic Analysis* (Shannon 1938 [1993]: 472). The raised dot is equivalent to ×.

POSTULATES

1. *a.* $0 \cdot 0 = 0$	A closed circuit in parallel with a closed circuit is a closed circuit.
b. $1 + 1 = 1$	An open circuit in series with an open circuit is an open circuit.
2. *a.* $1 + 0 = 0 + 1 = 1$	An open circuit in series with a closed circuit in either order (i.e., whether the open circuit is [first in line or second in line]) is an open circuit.
3. *a.* $0 + 0 = 0$	A closed circuit in series with a closed circuit is a closed circuit.
b. $1 \cdot 1 = 1$	An open circuit in parallel with an open circuit is an open circuit.

Shannon goes on to state the following (474):

> We are now in a position to demonstrate the equivalence of this calculus with certain elementary parts of the calculus of propositions. The algebra of logic, originated by George Boole, is a symbolic method of investigating logical relationships. … Due to this analogy any theorem of the calculus of propositions is also a true theorem if interpreted in terms of relay circuits.

0.10 BOOLE AND THE THEORY OF PROBABILITIES

It is not prominently remembered that Boole designed his algebra with an ulterior motive: to establish foundations for the theory of probabilities. No one has ever contested the validity of the considerable technical developments of the numerical kind in the theory of probabilities. But problems concerning the very foundations of the theory have never gone away. The theory of probabilities was about two centuries old in Boole's time. Its pioneer was Blaise Pascal (1623–1662), when he was "not yet withdrawn from the interests of science by the more distracting contemplation of the 'greatness and misery of man'," as Boole (1854: 243) describes it. The theory of probabilities is not directly relevant to the purpose of the present work; still, it may be useful to briefly mention this dimension of Boole's work here. Boole's views on the foundations of probabilities were much criticized and were met with incomprehension by such luminaries as the mathematician Arthur Cayley (1821–1895) and the economist John Maynard Keynes (1883–1946). As a result, Boole's views on probabilities have been largely ignored—"simply by-passed by the history of the subject," states Hailperin (1976). Nevertheless, there is recognition in some quarters of the fundamental correctness of Boole's approach. Boole was concerned with problems of a kind not addressed in the classic theory of probabilities. Consider two independent events, rain and thunder (see Boole's letter to John William Lubbock of 22 February 1849, in Grattan-Guiness and Bornet 1997: 198–99). The two events can occur separately or together. Suppose that one knows two things: the probability that it will rain and the probability that it will at the same time rain and thunder. The question is: What is the probability that it will thunder? A slightly more complex but related problem is derived by Boole from the following observation (1854: 321, note):

> Opposite the window of the room in which I write is a field, liable to be overflowed from two causes, distinct, but capable of being combined, viz. the floods from the upper sources of the River Lee, and tides from the ocean.

Boole assumes that the field cannot be inundated if *both* causes are absent. Suppose one knows four things: the probability that the tide will rise, the probability that the field will be inundated when the tide rises, the probability that the river will swell, and the probability that the field will be inundated when the river swells. The question is: What is the total probability that the field will be inundated? It is this kind of question that led to Boole's investigations into logic.

0.11 "IF" AND "WHEN" IN ENGLISH

Perhaps the first English word that comes to mind when one thinks of conditions is "if." However, the association between "if" and the condition is much weaker than may appear. In fact, in many if not most cases it is comparatively easy to view "if" as a synonym of "if it is the case that." Such equivalence makes "if" into a premise rather than a condition (see §6.1.1, as well as appendix 1, as well as in appendix 1).

If "if" should not be associated typically with the condition, then "when" definitely should, but its association is obscured somewhat, because tense matters. When "when" is followed by a past tense, as in "when he came," it denotes a temporal clause, not a conditional clause. "When" is therefore equally typical of temporal clauses. "When" denotes a condition when it is followed by the simple present, as in "When it rains, I stay at home." Thus "if" typically does not denote a condition and "when" can also denote a temporal clause. The question arises: Is there any English word that is exclusively associated with the condition? It appears that "whenever," as a variant of "when," is such a word. "When" can be replaced by "whenever" whenever "when" marks a condition, but never when "when" introduces a temporal clause.

0.12 ON STYLE AND ON COPTIC TRANSLITERATIONS

A general remark on expository style is in order. In view of the notion that a line of argument is probably strongest when it moves on a single track, footnotes have been consciously and assiduously avoided in the present work.

For the benefit of readers who may be less familiar with the Coptic alphabet, Coptic words have been transliterated following the conventions of the *Journal of Coptic Studies* except that *j* is used instead of *c* with hachek. Note that the transcription symbol *š* is pronounced like *sh* in *shoe*. The symbol *c* transcribes a character pronounced in origin probably roughly like *ky*, later probably more like *ch* in *chose*. The symbol *j* also corresponds to *ch* in *chose*; the modern classroom pronunciation is *j* as in *jaw*.

The transcription used in the present work departs in two other ways from the conventions of the *Journal of Coptic Studies*, as an experiment. The first exception concerns the written characters **ⲟⲩ**. **ⲟⲩ** has been transcribed as *u* when it presumably functions as a vowel, as in **ⲟⲩϣⲏ** *ušê* 'night', and as *w* when it presumably functions as a consonant, as in **ⲟⲩⲟⲛ** *won* 'someone'. **ⲟⲩ** is alternatively written as just **ⲩ** after certain vowels. It is then always a consonant, as in **ⲛⲏⲩ** *nêw* 'coming'. It cannot be excluded that **ⲟⲩ** could at some times be either a vowel or a consonant. An example is **ϩⲓⲱⲟⲩ** 'on them'. When pronounced more slowly, **ϩⲓⲱⲟⲩ** may have sounded more like *hi/ô/u*, with / marking syllables. When pronounced faster, it may have sounded more like *hyôw*.

The second exception concerns the written character **ⲓ** (though not when **ⲓ** is part of the digraph **ⲉⲓ**). It has been transcribed as *i* (pronounced as in *machine*) when it presumably functions as a vowel, as in **ⲡⲓⲥⲧⲉⲩⲉ** *pistewe* 'believe', and as *y* (pronounced like *y* in *yes*) when it presumably functions as a consonant, as in **ⲡⲉⲓ-** *pey-* 'this'.

Like **ⲓ**, the digraph **ⲉⲓ** can be pronounced both as a vowel (*i* as in *machine*) and as a consonant (*y* as in *yes*). The digraph **ⲉⲓ** has been transcribed as *y* when it is pronounced as a consonant, as in **ⲉⲓⲱⲧ** *yôt* 'father'. When it is pronounced as a vowel, **ⲉⲓ** has been transcribed as *ei*. The vowel *i* is thus written as either **ⲉⲓ**, transcribed *ei*, or as **ⲓ**, transcribed *i*. On the other hand, **ⲉⲓ** is a writing of the sounds *ey*, as is **ⲉⲉⲓ**.

1 Two Conditional Sentence Types

Grammatical distinctions of earlier Egyptian are often obscured by the defective nature of hieroglyphic writing. Fortunately, the distinction presently under investigation—between two types of sentences that are as a rule translated as conditional sentences—is reasonably accessible to observation. There is no doubt as to the existence of the two types (even though not every aspect of the two constructions is fully clear and not every problem pertaining to them is resolved). Some of the remaining problems are discussed below. Two phenomena distinguish the two types without doubt. The first is the element *jr* 'if, when(ever)', which is present in one type, as in *jr jḳr.k grg.k pr.k* 'If you excel, you will establish your house(hold)' (Wisdom of Ptahhotep [Jéquier 1911] 325), but absent from the other. *Jr* makes every instance of the first type instantly and unambiguously recognizable but leaves the second type defined only by the absence of a feature. That is unsatisfactory. A second phenomenon is needed that defines the other type in a positive way: To the extent that the two verb forms of the second type are positively identifiable, both are clearly substantival. The identity of the verb forms of the second type is not always certain. That is the perennial problem plaguing the study of Egyptian grammar. A majority of Old and Middle Egyptian verb forms are not positively recognizable in hieroglyphic writing and never will be. Three empirical features unambiguously characterize verb forms in balanced sentences as substantival verb forms: (1) gemination, (2) the past tense affix *n* with verbs of motion, and (3) the affix *w*.

(1) AORIST TENSE: Gemination in third-weak verbs, as in both verb forms in

prr.ṯn r pt m nrwt prr.j ḥr tpt ḏnḥwy.ṯn (CT 3.61f–g, Coffins B1C, B2L [B2L has *ḏnḥw.ṯn*]; cited by Niccacci 1980: 199 and many others)

'When(ever) you (pl.) depart for heaven as vultures,
I depart on top of your wings',

lit. '(The fact) that you depart for heaven as vultures equals (the fact) that I depart on top of your wings'.

An alternative translation recommended here (see §5.1.4 below) is as follows: 'You depart …, I depart …'.

In this and similar expressions, the dual *ḏnḥwy.sn* occurs at CT 3.61g and 3.115h, Coffin B1C. Elsewhere, the plural *ḏnḥw.sn* is mostly found, at CT 3.25b, 3.100i, 3.115h (Coffin B2L), 6.283g, and 6.302p.

(2) PAST TENSE: The affix *n* with verbs of motion, as in both verb forms in

pr.n.sn r pt m bjkw pr.n.j ḥr ḏnḥwy.sn (CT 3.115g–h, Papyrus Gardiner II)

'When(ever) they have departed for heaven as falcons,
I have departed on their wings',
lit. '(The fact) that they have departed for heaven as falcons equals (the fact) that I have departed on their wings'.

The identity comes out well in the following idiomatic translation in French by Vernus (1981: 76): 'Si les voilà partis au ciel en tant que faucons, me voilà parti sur leurs ailes'. The translation recommended here (see §5.1.4 below) is 'You have departed …, I have departed …'.

(3) FUTURE TENSE: The affix *w*, as in the first verb form of both balanced sentences in

mrw.f mt.ṯn mt.ṯn mrw.f ʿnḫ.ṯn ʿnḫ.ṯn (Pyr. §153c; cited by Edel 1955/64: 246 §528)

'When(ever) he wants you (pl.) to die, you will die.
When(ever) he wants you to live, you will live',
lit. 'That he will want you to die means that you will die. That he will want you to live means that you will live."

An alternative translation recommended here (see §5.1.4 below) is 'He will want you to live, you will live. He will want you to die, you will die'.

and as in the second verb form of

ḥms.j r wnm{.j} t ḥmsw Rʿ r wnm t (CT 3.185c–d, Coffin S1C; cited by Niccacci 1980: 210–11)

'When(ever) I sit down to eat bread,
Re will sit down to eat bread',
lit. 'That I will sit down to eat bread means that
Re will sit down to eat bread'.

No certain example in which both verb forms exhibit *w* is known to me. A possible example is Pyr. §1459c M (cited by Allen 1984: 131 §223), *jṯw N jnw.f*, if it means 'Whenever king N takes, he also brings'. Moreover, an example cited in §5.1 below (Pyr. Nt §193) exhibits two balanced sentences that are parallel. In the first sentence, the second verb form exhibits *w*. In the second sentence, the first verb form does. If one takes the two balanced sentences together as one, both verb forms exhibit *w*.

Not a single grammarian of Egyptian nowadays seems to doubt the existence of the balanced sentence featuring two substantival verb forms. Yet H. J. Polotsky placed this type of sentence on the map of Egyptian grammar only as recently as the early 1960s, more than a century and a quarter after the decipherment of hieroglyphic writing in 1822. The place of the balanced sentence in the history of Egyptian grammar is further detailed in chapter 8 below. Its tardy appearance on the scene of Egyptian grammar is just one more illustration of how the study of Old and Middle Egyptian, as opposed to the study of the later stages of Egyptian, is more an intellectual problem of a very peculiar kind than it is the study of a language.

2 Basic Concepts Of Logic

> At the time when the first edition of this work was composed [1881] it would scarcely be too much to say that the conception of a Symbolic Logic was either novel or repugnant to every professional logician. An amount of explanation and justification was therefore called for which does not seem to be quite so necessary now.
>
> Venn, *Symbolic Logic* (1894), vii.

> There were cliques, mostly based on fields. *The* clique at the top of the hierarchy was the topology clique .... Then came analysis .... Then came algebra .... Logic, for some reason, was not highly regarded.
>
> Nasar, *A Beautiful Mind* (1998), 64
> (on mathematics at Princeton University in the 1940s).

2.1 On the History of Logic

For clarity's sake, an effort is made here to render the present application of logic self-sufficient. No prior knowledge of symbolic logic is assumed on the part of the reader.

Full-fledged logic begins with Aristotle. By the Age of Enlightenment, logic had attained a significant level of sophistication, so much so that the philosopher Immanuel Kant (1724–1804) in the 1780s believed that the development of logic had reached an endpoint. Kant writes as follows in his *Critique of Pure Reason* (1787: 20):

> What is also remarkable about [logic] is that it has also now not been able to move a step forward and therefore appears to be in all respects closed and completed.*

According to Venn (1894: xxxv), Kant's conviction that the development of logic had reached an end-point was "disastrous" to the development of logic. In spite of their complexity and high degree of validity, Aristotelian

* Merkwürdig ist noch an [der Logik], daß sie auch jetzt keinen Schritt vorwärts hat tun können, und also allem Ansehen nach geschlossen und vollendet zu sein scheint.

logic and scholastic logic were never watertight and comprehensive in the way that a mathematical system is. Revolution came to logic in the mid nineteenth century with the advent of what Venn called symbolic logic. It was noted above that symbolic logic is nothing other than a kind of mathematics. The advent of symbolic logic was therefore nothing more or less than the advent of a new kind of mathematics.

Since the second half of the nineteenth century, Aristotelian logic and scholastic logic have been fully supplanted by symbolic logic. Symbolic logic is therefore indispensable to an exact definition of concepts such as condition both in general and in Egyptian specifically. A definition of condition and contingency (that is, dependency on conditions) in Egyptian, as expressed by the morphemes *ḫr*, *k3*, and *jn*, has been presented in the article (Depuydt 1999a) reprinted in appendix 1 (see also 1999d: lxxi–lxxiii, 2003b, and 2007).

The foundational figure is George Boole (1815–1864). Boole spent his entire university career at Queen's College (now University College) in Cork, on the southeast coast of Ireland. A biography of Boole appeared in 1985 (MacHale 1985). A sesquicentenary conference was held at Geneva in 1997 (Gasser 2000), 150 years after the appearance of Boole's *The Mathematical Analysis of Logic* (1847). These two recent works contain copious bibliographical references and many useful leads. Two other books testifying to a Boole revival are Grattan-Guiness and Bornet 1997 and Smith 1982. Before this revival, Boole's work had been neglected for decades. This is not the place to attempt a full account of the reasons for this neglect, but some of the reasons that contributed to it may be identified provisionally.

First, Boole's *Laws of Thought* (1854) was the subject of much criticism—I believe wrongly, and so does van Evra (1977). This criticism must have discouraged engagement with Boole's work to some degree. Second is what happened to Boole's Nachlass, that is, the unpublished written materials that he left behind. The events in question did not do much either to contribute to the reputation of Boole's work. Grattan-Guiness and Bornet (1997: xxiv) have recently communicated some details regarding these events. Third, Boole's logic has successfully been applied by electrical and computer engineers, but students of linguistics, philosophy, and logic do not move much in those circles. Boole's success in the world of the computer did therefore not translate into an equivalent success in the humanities. Fourth, Gottlob Frege (1848–1925) took logic in a different direction after Boole. Bertrand Russell did much to make Frege's difficult writings better known. It even became fashionable to think that it was just too bad that Boole could not have read Frege (Gasser 2000: vii). On the other hand, Frege could have read Boole; but Schröder (1880: 83) rightly criticized Frege for disregarding Boole. There seems to be two Freges: the Antiboole, whom better acquaintance with Boole's work might have benefited, and the Unfrege, a philosopher of all of language and mathematics who never existed but many think did (cf. Grattan-Guinness 2000: 177–99). There is a reason why "Boolean" caught on while "Fregean" did not.

Boole's insights rendered Aristotelian and scholastic logic obsolete, even if not unworthy of continued study. In Boole's words (1854: 241–42),

> [Scholastic logic] is not a science, but a collection of scientific truths, too incomplete to form a system of themselves, and not sufficiently fundamental to serve as the foundation upon which a perfect system may rest. It does not, however, follow, that because the logic of the schools has been invested with attributes to which it has no just claim, it is therefore undeserving of regard. A system which has been associated with the very growth of language, which has left its stamp upon the greatest questions and the most famous demonstrations of philosophy, cannot be altogether unworthy of attention.

Two books are especially informative about the history of logic. The first is Nikolai Styazhkin's *History of Mathematical Logic* (1964, 1969). This work is materialistic in outlook, owing to its Soviet and Marxist setting; this rather boosts confidence in the work's reliability in regard to a subject such as symbolic logic. The second is Venn's *Symbolic Logic* (1894). Venn consolidates Boole's system, advances it in some respects, and explains it with great clarity and patience. Moreover, Venn had read almost everything written on how the language of symbols can serve the cause of logic. The result is copious historical notes in the introduction, the footnotes, and an appendix. Venn invented the diagrams named after him in 1880. The first edition of his *Symbolic Logic* appeared in 1881. Much of the foundations of symbolic logic was established by 1900, and the concept of the Venn diagram is an integral part of these foundations. A. W. F. Edwards (2004) has published an informative appreciation of the place of Venn diagrams in the history of logic (Edwards 2004: ix [Foreword], for "1 + 1 = 0" read "1 + 1 = 1" as a "strange" rule of Boolean algebra).

The following observations about the history of logic can be derived from Venn's and Styazhkin's works. First, before Boole (1815–1864), the most advanced efforts in symbolic logic appear to have been those of Johann Lambert (1728–1777). Second, Augustus De Morgan (1806–1871) shines as an immediate precursor of Boole. I have already singled out De Morgan's role as true champion of the all-important supplementary classes or contradictories in logic. But the breakthrough to an all-comprehensive system was clearly Boole's, as De Morgan himself acknowledged. Third, John Venn (1834–1923) and Ernst Schröder (1841–1902) are among the most significant of Boole's successors in the nineteenth century. Styazhkin rightly stresses the importance of the lesser known Platon Poretskiy (1846–1907) of the University of Kazan. W. Stanley Jevons (1835–1882) also deserves mention. Jevons built a "logical machine" that is preserved at Oxford University. Venn (1894: xxviii) reports that Jevons excised from Boole's procedure everything which he found an "obscure form," "anomalous," "mysterious," or "dark and symbolic." As a result, Venn suggests, "nearly everything which is most characteristic and attractive in the system is thrown away." On occasion, one also encounters conflicting statements about the merits of the work done in logic by Charles Sanders Peirce (1839–1914) (on this, see the preface to Peirce 1883/1983 by Achim Eschbach). Whitehead (1897: 3, note) refers to "a more obscure analysis … by C. S. Peirce, *Proc. of the American Academy of Arts and Sciences*, 1867, vol. vii, p. 291." It seems significant that an eminent authority like Whitehead, who was thoroughly versed in both logic and mathematics, would find certain aspects of Peirce's logic obscure.

A fourth observation on the history of logic that I would derive from Styazhkin's and Venn's works is that Boole's system was never surpassed in any essential way. It was only refined and expanded. With regard to notation systems, I find Boole's still perfectly usable. Many other notation systems have been proposed, a couple of them complex to the point of pedantry.

Finally, a fifth observation is that related efforts by Gottlob Frege (1848–1925), Giuseppe Peano (1858–1932), Bertrand Russell (1872–1970), Alfred North Whitehead (1861–1947), and more recently by Rudolf Carnap (1891–1970) and Kurt Gödel (1906–1978) and many others, differ from logic strictly speaking. These efforts seem more about deriving insight from logic to explain the foundations of mathematics. This involves combining logic with categories such as number or quantity. Strictly speaking, logic in principle concerns relations on two levels, namely (1) the level of things and (2) the level of events.

In conclusion, I do not have the slightest reason to disagree with the following assessment by De Morgan of Boole's role in the history of logic (De Morgan 1872: 301):

> The unity of the forms of thought in all the applications of reason, however remotely separated, will one day be matter of notoriety and common wonder; and Boole's name will be remembered in connexion with one of the most important steps toward the attainment of this knowledge.

One might recall that the computer age in a sense began when Claude E. Shannon ran with Boolean algebra by applying it to electronic circuits. In subscribing to De Morgan's assessment, I am most influenced by the importance of one characteristic of Boole's system: the mind simply cannot conceive of or reason about any thing or event without implicitly also considering that thing's or event's supplementary class. Nothing is more characteristic of the way we think. Focalization, as in "*sheep!* (of all things, as opposed to other animals, etc.)," exposes the fundamental fiber of human thought. The mind focalizes when it conceives of the supplement of the supplement, obviously again the original class. "*Sheep!*" ("not-not-sheep") and "sheep" therefore refer to the same reality. But in "*Sheep!*," focus is achieved by presenting the class as the supplement's supplement. This definition holds, it seems, the complete meaning of focalization. It is in Boole's work that supplements for the first time are omnipresent in symbolic logic and in fact totally dominate it.

Some of the earliest speculations on supplementary relations date back to Anaximander of Miletus in the sixth century B.C.E. In Egypt, the Saite dynasty or Dynasty 26 ruled at the time. Examples of complementary relations are "hot" versus "cold" ("not hot") and "water" versus "fire." Supplementary relations are also fundamental to logic. They are the single characteristic that most differentiates modern scientific logic from any prior efforts at logic. Is it a coincidence, therefore, that Anaximander also has a reputation of being the first prose writer? The manner of writing prose as we know it in modern times has not always existed. It is something that human beings began doing at some point in the past and as a result are still doing basically in the same way today. Is it a coincidence, furthermore, that the earliest manifestations of the fundamental axiom of logic, namely the Law of Contradiction, also date roughly to the time of Anaximander? The Law of Contradiction states that something cannot be itself and something else at the same time, that is, providing that we human beings want to think at all profitably about anything. In other words, if we think such a contradiction possible, thought itself becomes impossible.

2.2 TWO LEVELS OF MAKING STATEMENTS AND THE PARALLELISM BETWEEN THE TWO

Thought happens mainly on two levels, even if premise and result can be assigned to a third level (Depuydt 1999a). Statements can be made on both levels. The primary level pertains to the relation between *things.* The secondary level pertains to the relation between *events.* The secondary level combines statements of the primary level. The two levels are separate. But they exhibit the same structure. Possible differences of detail between the two levels do not affect the present line of argument.

The similarity in structure of the two levels may be illustrated by repeating an obvious example. The parallelism between the two levels may be symbolized as follows.

s = swallows	$\bar{s}$ = non-swallows
b = birds	$\bar{b}$ = non-birds
r = it rains	$\bar{r}$ = it does not rain
i = I stay inside	$\bar{i}$ = I do not stay inside

PRIMARY LEVEL "Swallows (*s*) are birds (*b*)."

therefore:

$s\bar{b} = 0$ "Swallows (*s*) that are not birds ($\bar{b}$) do not exist (= 0)."

but also(!):

$\bar{s}b \neq 0$ "Birds (*b*) that are not swallows ($\bar{s}$) do exist (≠ 0)."

SECONDARY LEVEL "When it rains (*r*), I stay inside (*i*)."

therefore:

$r\bar{i} = 0$ "It does not happen (= 0) that it rains (*r*) and I do not stay inside ($\bar{i}$)."

but also(!):

$\bar{r}i \neq 0$ "It does happen (≠ 0) that I stay inside (*i*) and it does not rain ($\bar{r}$)."

The parallelism between level (1) and level (2) is obvious. Both *things* on level (1) and *events* on level (2) form *classes* or *sets*. Thus, *s* is the class of all the swallows and *r* is the class of all the occurrences that it rains.

Boole (1854: 162–64) associates level (2) with the concept of time. In his view, "When it rains, I stay inside" means that the time in which it rains falls entirely inside the time in which I stay inside, just as the class of swallows falls entirely inside the class of birds. Boole's view of the role of time in logic has not been widely accepted. However, there is an argument from language that can be adduced in favor of associating level (2) at least to some degree with the concept of time: that conditional and temporal clauses exhibit affinities in many languages. For example, English "when" can be either temporal or conditional. It is *temporal* when followed by the *past* tense, as in "when he arrived." It is *conditional* when followed by the *present* tense, as in "when he arrives." By contrast, "when*ever*" is always conditional, so "when" is conditional whenever it can be replaced by "whenever." "If" is usually portrayed as the quintessential conjunction of condition; but in fact, "if" as a rule introduces premises. It is then followed by a verbal tense—that is, past ("if he has arrived"), present ("if he is arriving"), or future ("if he will arrive"). Also when followed by the simple present ("if he arrives"), "if" denotes a premise. The simple present denotes the unmarked tense, that is, absence of a tense. "He arrives" means that he generally does so, without regard to a specific tense. There are other illustrations of the affinity between temporal clauses and conditional clauses. For example, German "als" is temporal but the related "als" in Dutch is conditional.

Nothing contributes more to the ability of human thought to process complexity than the hierarchy of the two levels, the level of things and the level of events. This hierarchy allows complexity by means of a limited number of elements, in a striking parallel to the structure of language. Language also displays two levels that allow complexity through limited means: sounds and words. Words are not distinguished from one another by each having a sound of its own—if so, there would need to be as many distinctive sounds as there are words. Such a large number of sounds would be impossible for the speech organs to produce and the ear to distinguish. Rather, on the first level, all languages exhibit a limited set of distinctive sounds. On the second, each word consists of a different combination of sounds taken from the limited set. A small number of sounds can thus efficiently represent a huge number of words. Hierarchies make it possible to produce greater complexity with fewer elements. Human thought owes its complexity in large part to the two-level hierarchy of things and events.

2.3 TYPES OF STATEMENT: BOOLE'S THREE TYPES AND FIVE TYPES DISCUSSED BY VENN

At issue in the present section is the difference in meaning between two types of statements on level (2), the relation between events. The two types are the statement containing *jr* 'if, when' and the balanced sentence. Both types have been associated with the concept of condition. But what is the meaning of the concept of condition? Obviously, the meaning of condition differs from 'house' as the meaning of the Middle Egyptian word *pr* or 'past tense' as the meaning of the Middle Egyptian verbal affix *n*. As an item of meaning, condition has a certain abstract quality. What we need, therefore, is a clear definition. Such a definition has already been anticipated in §0.2 above. But there is more. *Jr* and the balanced sentence are both associated with the concept of condition. Both are normally translated as conditional sentences. What we need, then, is a definition of condition that also allows a clear distinction between at least two types of what might be called a condition. Logic is the domain where such a definition must be sought.

All students of logic have always agreed that, in a logical perspective, there are different types of statement or proposition. But how many? The answers have differed. One well-known division dates to before the advent of modern scientific logic in the mid nineteenth century and is still widely used, the division into four types: the universal affirmative statement ("all X are Y"), the universal negative ("no X are Y"), the particular affirmative ("some X are Y"), and the particular negative ("some X are not Y"). This division holds much that is true. But it ultimately lacks the rigor and complete internal consistency that characterizes modern symbolic logic and has therefore been superseded.

By the modern age, a division into five types of statement had become widespread. They can be represented by five diagrams that resemble Venn diagrams but are strictly not (on the characteristic difference, see §A3.1). If anything, they are rather Euler diagrams. Still, Venn diagrams offer no added benefit as an aid to the eye when relations between only two classes are considered.

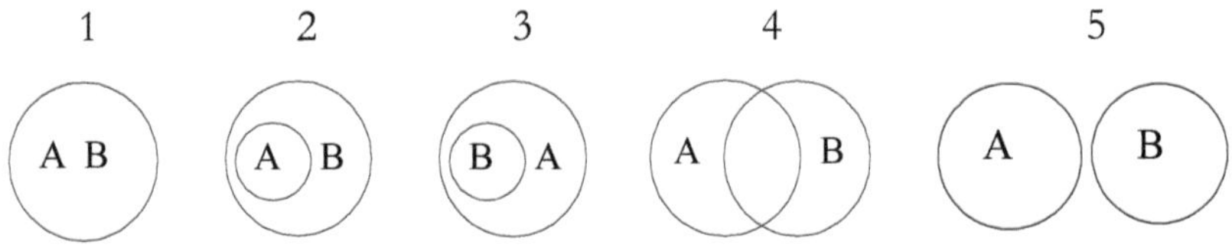

(Drawing based on Venn 1894: 7)

These are the only five ways in which two classes A and B can relate to one another in terms of inclusion and exclusion. Venn (1894: 31) paraphrases them as follows:

Type 1: All *A* is all *B*.

Type 2: All *A* is some *B*.

Type 3: Some *A* is all *B*.

Type 4: Some *A* is some *B*.

Type 5: No *A* is any *B*.

By contrast, Boole distinguishes only three types. He represents them as follows.

Type 1:	$x = y$	"All *x* is all *y*."
Type 2:	$x = vy$	"All *x* is some *y*."
Type 3:	$vx = vy$	"Some *x* are some *y*."

Boole's use of the indefinite symbol *v* poses problems that cannot be addressed here, but they should not affect the coherence of the argument that follows. One problem is that Boole himself defines *v* in two different ways:

(1) "*all*, *some*, or *none*" (we don't know which of the three) (1854: 90, 92)

(2) "*some* ... [or] *all*" (we don't know which of the two) (1854: 124)

Boole explicitly states that definition (2) "does not include *none*." Here, *v* will be interpreted according to definition (2), "some or all." Examples of Boole's three types on the primary level are:

Type 1:	"Paris is the capital of France."
Type 2:	"The Alps are mountains in Europe."
Type 3:	"Some Frenchmen are doctors."

At first sight, the fact that Boole's and the above classifications of types of statements differ does not inspire confidence in the trustworthiness of either classification. However, it is relatively easy to relate the two classifications to each other. Indeed, it is easy to interpret them as amounting to the same thing. Boole's three types can be related to the five types above as follows.

Boole	Discussed by Venn
Type 1	Type 1, Type 5
Type 2	Type 2, Type 3
Type 3	Type 4

First, Types 2 and 3 to the right can readily be combined into Boole's Type 3 if one looks at the types from a different angle. Venn (1894: 6 n. 1) himself allowed for the possibility of combining the two types. He quotes without objection the logician A. D. C. Twesten, who suggested combining them. Moreover, the representations of Types 2 and 3 to the right are the same (see above). In both types, a larger circle entirely encompasses a smaller circle. The *two* types are obtained by taking class *A* alternately as the smaller circle and as the larger circle. Still, the relationship between the two circles is the same in both cases. From that point of view, there is really just one type, Boole's Type 2.

Second, Types 1 and 5 to the right can also be associated with one another. In Type 1, the classes *A* and *B* completely overlap. In Type 5, none of *A* overlaps with *B*. The purport of Types 1 and 5 is the same in that *all* of *A* relates *in the same way* to *all* of B. In one case, all of *A* is *B*. In the other case, all of *A* is non-*B*. An essential, albeit ultimately partial, property of Boole's Type 1 is that *all* of one class relates *in the same way* to *all* of another class. In that sense, Boole's Type 1 and Venn's Types 1 and 5 are set apart from all the other types, in which one part of *A* behaves in one way to *B* and the other part of *A* behaves in another way to *B*.

In what follows, the focus is on the distinction between Boole's Types 1 and 2.

3 Conditional Sentence with *jr* and the Balanced Sentence

It is reasonable to assume that there is a difference between sentences with *jr* and the balanced sentence. It is clear that the definition of the difference must have something to do with condition and with the need to make a distinction within the concept of condition. Such a definition might be expected to be very subtle. However, the following hypothesis associates the difference with one of the fundamental distinctions of logic, between certain clearly definable fundamental types of logical statements: The balanced sentence is associated with Boole's logical statement of Type 1. The sentence with *jr* is associated with Boole's Type 2.

Association	Boole	Discussed by Venn
Balanced Sentence	Type 1	Types 1 and 5
Sentence with *jr*	Type 2	Types 2 and 3

The term "association" requires clarification. I would hesitate to declare full identity between the two Egyptian sentence types and the two types of logical statements. Language is not the same as logic. Two examples: languages change over time in the way that logic does not; language is not as precise as a logical theory. However, I believe that these two associations can be proven with the help of positive evidence (see chap. 5).

4 LOGICAL PROPERTIES OF TYPES 1 AND 2

4.0 BRIEF COMPARISON OF TYPES 1 AND 2

In chapters 2 and 3, two clearly distinct types of logical statements were described, as follows:

Type 1: $x = y$

Type 2: $x = vy$, in Venn's preferred representation, $x\bar{y} = 0$, or "There is no (= 0) x that is not-y ($\bar{y}$)."

The two types can also be represented simply and transparently by the following diagrams.

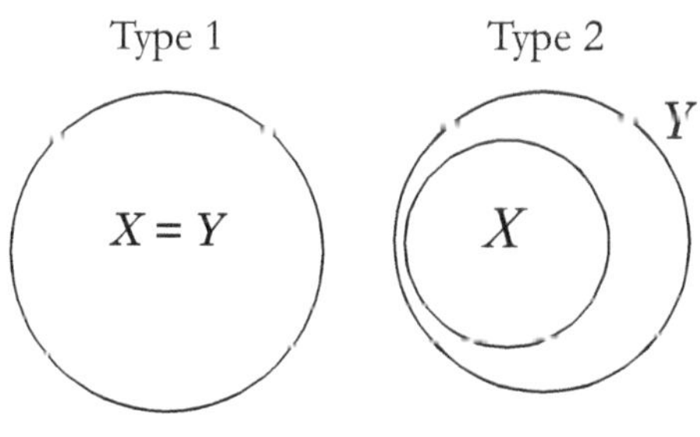

In what follows, three logical properties will be derived from Type 1 in pure deductive fashion. It will also be shown that, in contrast to Type 1, Type 2 does not and cannot exhibit these three logical properties.

4.1 REVERSIBILITY OF THE EQUATION (TYPE 1 ONLY)

Type 1: $y = x$ (reversibility) follows from $x = y$.

Type 2: $y = vx$ does not follow from $x = vy$.

Type 1 exhibits reversibility. Type 2 does not. From Type 1, $x = y$, it follows in Boolean algebra by inversion that $y = x$.

An example of $x = y$ on the level of primary propositions is "Paris is the capital of France." It indeed follows naturally by inversion that "The capital of France is Paris." An example of $x = y$ on the level of secondary propositions is "I stay inside only and always when it rains." It indeed follows naturally by inversion that "It rains only and always when I stay inside." The metaphor implied in the term "balanced sentence" can be interpreted as one of weights that are balanced on a scale. If two equal weights are switched between the two sides of a balance, the position of the balance does not change.

By contrast, Type 2 is not reversible. From Type 2, $x = vy$, that is, "x is some y," it does not follow that $y = vx$. Boole's use of v ("some") is problematic (see §2.2 above). But v is effective as far as it goes and poses no problems for the present argument.

An example of Type 2 is "The Alps are mountains in Europe." That is, "The Alps are some (not all) of the mountains of Europe." This is a statement on the first level of thought, the level of things. Boole calls this a primary proposition. An example of Type 2 on the second level of thought, the level of events, is "When it rains, I stay inside." That is, "The occasions when it rains are some (not all) of the occasions when I stay inside." Boole calls a statement on the second level of thought a secondary proposition. Furthermore, Boole interprets secondary propositions in terms of time: the *time* that it rains falls entirely inside the *time* that I stay inside. Both the primary proposition and the secondary proposition cited above exhibit the structure

$$x\bar{y} = 0$$

This means

> "Alps (x) that are not mountains in Europe ($\bar{y}$) do not exist (= 0)."

and

> "The co-occurrence of it raining (x) and my not staying inside ($\bar{y}$) does not exist (= 0)."

Most importantly, it does *not* follow by inversion that

$$y = vx.$$

Thus, it does not follow that

> "(All) mountains in Europe are Alps."

There are other mountains in Europe. Nor does it follow that

"When I stay inside, it rains."

When I stay inside, it may or may not rain.

4.2 BALANCED NEGATION (TYPE 1 ONLY)

Type 1: $\bar{x} = \bar{y}$ follows from $x = y$.

Type 2: $\bar{x} = v\bar{y}$ does not follow from $x = vy$.

Type 1 exhibits balanced negation. That means that both sides of the equation can be negated at the same time, with an equivalent logical statement as the result. Indeed, from

$$x = y$$

it follows by a simple operation that applies equally to algebra and to logic that

$$1 - x = 1 - y$$

$1 - x$ can also be written as $\bar{x}$. By this convention, the previous equation can be written as

$$\bar{x} = \bar{y}$$

On the level of primary propositions, if "Paris is the capital of France," then "Anything else but Paris is the same as anything else but the capital of France." An example of $x = y$ on the level of secondary propositions is "I stay inside only and always when it rains." It follows naturally from this statement that "Any occasion when it does *not* rain is also an occasion when I am *not* inside."

By contrast, Type 2 does not exhibit balanced negation. From

$x = vy$, e.g. "When it rains, I stay inside,"

it does not follow that

$\bar{x} = v\bar{y}$, e.g."When it does not rain, I do not stay inside."

Indeed, when it does not rain, I may or may not be inside.

4.3 RESTRICTION ON THE COMBINATION OF CLASSES (TYPE 1 ONLY)

Type 1: If $x = y$, then $\bar{x}$ is accompanied by $\bar{y}$ and only $\bar{y}$.

Type 2: If $x = vy$, then $\bar{x}$ can be accompanied by either y or $\bar{y}$.

Consider the class "blue (things)." Its supplement is "non-blue (things)," which is also a class. Therefore, from the perspective of the class "blue" and its supplement, the universe is divided into two classes, blue things and non-blue things.

When there is more than one class, the same division involves combination classes. Consider, in addition to "blue," the class "car." Its supplement is "things other than cars," which is also a class. From the perspective of the two classes "blue" and "car" and their supplements, the universe of thought is divided into four combination classes, which themselves are also classes.

FOUR COMBINATION CLASSES

(a) bc blue cars;

(b) $\bar{b}c$ non-blue cars;

(c) $b\bar{c}$ blue objects that are not cars;

(d) $\bar{b}\bar{c}$ things that are neither blue nor cars.

This fundamental property of thought affects logical statements of Types 1 and 2 differently. Both Types relate two classes x and y to each other. However, it follows from Type 1, $x = y$, that $\bar{x}$ *must be accompanied* by $\bar{y}$. That is a *crucial restriction*. By contrast, it follows from Type 2, $x = vy$, that $\bar{x}$ can be accompanied by either y or $\bar{y}$. In other words, the same restriction does not affect Type 2.

Examples may clarify this restriction. An example of Type 1 is "I stay inside only and always when it rains" ($r = i$). It follows that every occasion when it does *not* rain ($\bar{r}$) must be accompanied by the circumstance that I am *not* inside ($\bar{i}$). An example of Type 2 is "When it rains, I stay inside" ($r = vi$). It follows from this statement that, when it does *not* rain ($\bar{r}$), I may either stay inside (i) or not stay inside ($\bar{i}$). There is no way of knowing. The restriction to Type 1 does not apply in the case of Type 2.

Common sense alone makes the distinction between Type 1 and Type 2 sufficiently transparent. But the distinction can be additionally formalized. One way is in light of the totality of possible classes. There are four combination classes:

$ri =$ the occasions when it rains and I stay inside;

$\bar{r}i =$ the occasions when it does not rain and I stay inside;

$r\bar{i} =$ the occasions when it rains and I do not stay inside;

$\bar{r}\bar{i} =$ the occasions when it does not rain and I do not stay inside.

From $r = i$ "I stay inside only and always when it rains" (Type 1), it follows that $\bar{r} = \bar{i}$ "I am outside only and always when it does not rain." It is easy to see that, of the four possible classes, two classes exist (that is, add up to 1 or the universe) and two classes do not exist (that is, add up to 0 or nothing). The algebraic presentation is as follows.

$ri + \bar{r}\bar{i} = 1$ "Either it rains and I stay inside (ri)
or it does not rain and I do not stay inside ($\bar{r}\bar{i}$)."

$\bar{r}i + r\bar{i} = 0$ or $\bar{r}i = 0$ and $r\bar{i} = 0$

A different pair of equations corresponds to Type 2, $r = vi$ "When it rains, I stay inside":

$r\bar{i} = 0$ "It does not happen that it rains and I do not stay inside."

$ri + \bar{r}i + \bar{r}\bar{i} = 1$
"Either it rains and I stay inside, or
it does not rain and I stay inside, or
it does not rain and I do not stay inside."

The difference between Type 1 and Type 2 concerns the combination $\bar{r}i$. It equals zero with Type 1: "It never happens that it does not rain and I stay inside." But the same combination $\bar{r}i$ does exist in the case of Type 2. A full discussion exceeds the scope of this book.

The above equations can all be obtained in rigorous and absolute fashion by means of Boole's formula of development, $f(x) = f(1)x + f(0)(1 - x)$. A partial and abbreviated operation is laid out in the excursus to this section.

The next step is to match the above equations to the empirical evidence in an effort to prove that the balanced sentence is associated with Type 1. As a concomitant consequence, statements containing *jr* can then be associated with Type 2.

Excursus. *The equivalence of* x = y *to the combination of* $x\bar{y} = 0$ *and* $\bar{x}y = 0$ *in Boole's algebra: An abbreviated account.*—The purpose of this excursus is to illustrate the exact nature of the operations of symbolic logic. Details are found in Boole's *Laws of Thought* (1854).

First of all, by subtracting y from both sides of the equation

$$x = y,$$

one obtains

$$x - y = y - y.$$

Or also

$$x - y = 0.$$

$x - y$ can be developed with Boole's formula of development, which is

$$f(x) = f(1)x + f(0)(1 - x).$$

Because there are *two* classes, x and y, the formula needs to be expanded:

$$f(x,y) = f(1,1)xy + f(1,0)x\bar{y} + f(0,1)\bar{x}y + f(0,0)\bar{x}\bar{y}.$$

The four "coefficients" in this formula are $f(1,1)$, $f(1,0)$, $f(0,1)$, and $f(0,0)$. The formula is applied as follows. The task is to replace x and y in $x - y$ by either 1 or 0. x and y are replaced by 1 when the coefficient is immediately followed by x or y. They are replaced by 0 when the coefficient is immediately followed by $\bar{x}$ or $\bar{y}$. For example, in the case of the coefficient $f(1,1)$, both x and y are equated with 1. The full operation is as follows:

Coefficient	Substitution of 1 and 0 in $x - y$		
(1,1)	(goes with xy)	$= 1 - 1$	$= 0$
(1,0)	(goes with $x\bar{y}$)	$= 1 - 0$	$= 1$
(0,1)	(goes with $\bar{x}y$)	$= 0 - 1$	$= -1$
(0,0)	(goes with $\bar{x}\bar{y}$)	$= 0 - 0$	$= 0$

It appears that the two classes xy and $\bar{x}\bar{y}$ are eliminated from consideration because their coefficients, $f(1,1)$ and $f(0,0)$, after substitution into $x - y$, both result in zero. That leaves the two classes $x\bar{y}$ and $\bar{x}y$ with coefficients $f(1,0)$ and $f(0,1)$. The latter coefficient, $f(0,1)$, after substitution into $x - y$, comes out as –1. In Boolean algebra, –1 means that the corresponding class, $\bar{x}y$, does not exist. Only $x\bar{y}$ is left, therefore. Its coefficient after substitution is 1. 1 means that the class $x\bar{y}$ exists. $x - y$ has thus been developed into $x\bar{y}$. Since

$x - y = 0$, and $x\bar{y}$ is the same as $x - y$ after development, it is also true that $x\bar{y} = 0$.

4.4 EXPRESSING THE RELATION BETWEEN TYPE 1 AND TYPE 2

It has long been recognized that every logical statement of Type 1 is in fact a combination of two logical statements of Type 2 (see, e.g., Venn 1894: 31 and elsewhere). Thus the statement

$$x = y \quad \text{(Type 1)}$$

is a logical combination of

$$x = vy \quad \text{(Type 2)}$$

and

$$y = vx \quad \text{(Type 2).}$$

For example, "I stay inside only and always when it rains" (Type 1) is a combination of "When(ever) it rains, I stay inside" (Type 2) and "When(ever) I stay inside, it rains" (Type 2). In other words, a balanced statement is logically a combination of two non-balanced statements.

The same property obviously also applies to the negation of Type 1, which is also a balanced statement, namely

$$\bar{x} = \bar{y} \quad \text{(Type 1),}$$

which is a combination of two non-balanced statements, namely

$$\bar{x} = v\bar{y} \quad \text{(Type 2)}$$

and

$$\bar{y} = v\bar{x} \quad \text{(Type 2).}$$

For example, "I am not inside only and always when it does not rain" (Type 1) is a combination of "When(ever) it does not rain, I am not inside" (Type 2) and "When(ever) I am not inside, it does not rain" (Type 2). Again, a balanced statement logically combines two non-balanced statements.

5 Balanced Sentences and Logical Types 1 and 2

5.0 Four Properties in Relation to Types 1 and 2

In §§4.1–4.3, three logical *properties* were derived in purely deductive and theoretical fashion from Type 1. Logical statements of Type 1 exhibit these three properties, but logical statements of Type 2 do not. In this chapter, three *phenomena* are adduced. Balanced statements can be accompanied by each of these phenomena, but sentences with *jr* cannot. Each of the three *phenomena* associated with balanced sentences can be clearly associated with one of the three *properties* associated with logical statements of Type 1. The conclusion seems inevitable: *The balanced sentence and the logical statement of Type 1 can be associated with one another.* Moreover, the three phenomena are absent from sentences with *jr*, just as the three properties are absent from the logical statement of Type 2. A companion conclusion therefore follows: *The sentence with jr and the logical statement of Type 2 can be associated with one another.*

In addition, a fourth theoretical property of the logical kind was described in §4.4, expressing the relation between Types 1 and 2. Evidence is adduced to show that this property is reflected in the facts.

5.1 Reversibility as a Fact of Balanced Sentences

Compare the first logical property of Type 1, §4.1.

5.1.1 Examples of reversibility

5.1.1.1 An instance of explicit reversibility

Only one fully explicit example of reversibility in balanced sentences is known to me.

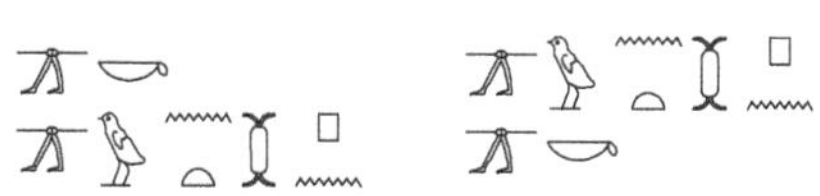

s.k	*sw Nt pn* (so, instead of *tn*)
sw Nt pn	*s.k*

'You will go (*or:* perish), this Neith will go.
This Neith will go, you will go'.

(Pyr. Nt §193c, cited by Allen 1979a: 23 bottom, 1984: 21 §34, 131 §225, 178–79 §286; see also Edel 1955/64: 241 §517; Jéquier 1933: pl. XXI, line 564)

Versions W and P exhibit *s* instead of *sw*. The affix *w* apparently marks *sw*, and hence by parallelism also *s.k*, as an instance of the future substantival form belonging to the *sḏm.f* formation, the *sḏmw.f*. The substantival character of *s.k* seems confirmed by the variant *sjj.k* in Middle Kingdom copies. Allen (1984: 132 §225) writes: "Some Middle Kingdom copies show the spellings *zw* N vs. *zjj.k*, the latter perhaps from original **zw.k* (T9C, TT 319, B6C)." The ending *jj*, as in *sjj.k*, does indeed appear instead of *w* in the *sḏmw.f*, before suffix personal pronouns consisting of a single consonant (Schenkel 2000: 27).

With regard to this spelling, we quote Schenkel's summary statement (ibid.), which results in part from his own observations and in part from others'. The examples are Schenkel's [bracketed passages added].

The "prospective" [which is, in my opinion, nothing but the substantival future *sḏm.f*] and the "subjunctive" [that is, in my opinion, the adverbial future *sḏm.f*] exhibit forms with the endings –*w* and –*jj* [double reed leaf] and forms without endings. The following rules [mostly] apply:

[*The sḏmw.f or substantival future sḏm.f.*] The original ending of the prospective is -*w*. The ending -*jj* [double reed leaf] [can] replace -*w* in the pronominal state before monoconsonantal suffix personal pronouns [as, e.g., in *hꜣjj.j* at CT 3.48b (Coffin B1C), in *hꜣjj.k* at CT 1.189b (Coffin B10Cꜥ), and in *hꜣjj.f* at CT 6.288n (Coffin B2Bo)]. The ending -*w* is retained in (1) the nominal state [as, e.g., in *hꜣw* at CT 3.48b (Coffin B3C)] and (2) the pronominal state before biconsonantal suffix personal pronouns [as, e.g., in *sšmw.sn* at CT 5.106a (Coffin Sq2Sq)]. The following verb classes can exhibit an ending: (1) causative verbs; (2) in part verbs with a weak consonant [*y* or *w*] as last root consonant; (3) in part strong verbs with more than three root consonants. Some last-weak verbs never exhibit an ending and strong verbs with two or three consonants hardly ever do. Certain other last-weak verbs do not clearly belong to either group of last-weak verbs.

[*The adverbial future sḏm.f.*] The subjunctive exhibits forms with the ending -*jj* and forms without an ending. Only verbs, including causatives, with a weak last consonant can exhibit the ending.*

5.1.1.2 *Instances of ṯs pẖr 'and vice versa'*

In all the other instances known to me, reversibility is marked by the expression *ṯs pẖr* 'and vice versa (lit. inverted statement)'.

ẖw.j sw ẖw.f wj ṯs pẖr (CT 3.347h, Coffin S1C[a] and similarly in three other versions)

'I protect him, he protects me. And vice versa'.

It seems clear that this writing is an abbreviated version of the following statement.

**ẖw.j sw ẖw.f wj ẖw.f wj ẖw.j sw*

'Whenever I protect him, he protects me.
Whenever he protects me, I protect him'.

Three further examples:

wḏ3.j *wḏ3.f* *ṯs pẖr* (CT 4.246a)

'I'm fine, he's fine, and vice versa'.

ʿnẖ.j *ʿnẖ Rʿ* *ṯs pẖr* (CT 7.207k, Papyrus Gardiner II)

'I live, Re lives, and vice versa'.

rwjj.f *rwjj.j* *ṯs pẖr* (Book of the Dead, chap. 99, Papyrus of Iouiya)

'He will leave, I will leave, and vice versa'.

The third example is cited by Niccacci (1980: 210). The original edition of this text is by Naville (1908: plate XXIII, line 5 from the right). *Ṯs pẖr* is absent from five versions of the same text at CT 5.198a. Two of these five

* Prospektiv und Subjunktiv zeigen Formen mit der Endung -*w* und -*y* sowie Formen ohne Endung. Es gelten die folgenden Regeln: Die Endung des Prospektivs ist ursprünglich -*w*. -*y* tritt an die Stelle von -*w* im Status pronominalis mit einkonsonantigem Suffixpronomen. Im Status nominalis und vor zweikonsonantigem Suffixpronomen bleibt -*w* dagegen erhalten. Eine Endung können zeigen die Kausative, ein Teil der Verben ult[imae] inf[irmae] und gewisse mehr als dreiradikalige starke Verben. Die Endung tritt nicht auf bei einem anderen Teil der Verben ult.inf. und kaum bei den starken zwei- und dreiradikaligen Verben. Einige Verben ult.inf. gehören keiner der beiden Teilklassen eindeutig zu. Der Subjunktiv zeigt Formen mit der Endung -*y* und solche ohne Endung. Die Endung können zeigen die Verben ult.inf. einschließlich der Kausative ult.inf. Alle anderen Verbalklassen zeigen keine Endung.

versions exhibit *rwjj.f* and three exhibit *rw.f.* The early and otherwise overall excellent text of the Papyrus of Nu (ed. Budge 1898: 209, line 2; see now Lapp 1997) instead has the explicit *wḏꜣ.f wḏꜣ.j* 'When he is fine, I am fine', without *ṯs pẖr.*

None of the four verb forms in the first two of the three examples cited above have written characteristics that explicitly identify them as substantival. But in the third example, the affix *jj* () may very well be such a characteristic. *Jj* is then an alternant of *w* before single-consonant suffix pronouns (Schenkel 2000). Then again, *jj* is not an exclusive characteristic of the active future substantival *sḏm.f.* In fact, *jj* more often and more typically marks a *sḏm.f* as active, future, and adverbial. It may be assumed by inference, however, that all six verb forms in the three examples cited above are substantival.

5.1.1.3 A hybrid instance of ṯs pẖr 'and vice versa'

The following example constitutes a special case:

jw pt m rswt ḥms.j m mḥwt ṯs pẖr (CT 4.50f, Coffin B1C [also B2L])

'Heaven will come as a south wind,
I will sit to the north of it, and vice versa'.

jw pt m jmnt ḥms.j m jꜣbt ṯs pẖr (CT 4.50g, Coffin B1C [also B2L])

'Heaven will come as a west wind,
I will sit to the east of it, and vice versa'.

The example has been translated in the future tense. Judging by this example alone, the verb forms could also be interpreted as instances of the general present (aorist) tense. Both verbs are last-weak. Fourth-weak verbs do not often geminate in the general present tense with substantival function, and *ḥms*(*y*) 'sit (down)' appears to be one that does not (cf. Schenkel 1997: 179). The third-weak verb *jw*(*y*) 'come' is irregular and, unlike other third-weak verbs, it appears to geminate only rarely in the general present with substantival function (cf. Gilula 1987). However, there is one small indication that favors the future tense over the present tense for the balanced sentence above. There is a variant, discussed at greater length in §10.5.2 below, that exhibits a form *ḥmsw,* with affix *w.* This affix clearly marks the form as future tense:

jr jw pt m mḥt ḥmsw N tn m rs.s (CT 5.3c–4c, Coffin B4C)

'If heaven comes as a north wind,
this N will sit to the south of it'.

jr jw pt m rsw ḥmsw N [*tn* …]

'If heaven comes as a south wind, this N will sit …'.

Alternative translations of the first clause in CT 4.50f–g are 'Whenever the wind blows from the south …' and 'Whenever the wind blows from the west …'. As equivalents of the translation 'to the north of it', that is, 'to the north of the house, out of the wind', and of the translation 'to the east of it', that is, 'to the east of the house, out of the wind', one expects *ḥr mḥt.s* and *ḥr j3bt.s*. The translation given above is therefore not an exact rendition of the Egyptian original. However, such an inexact rendition is necessary because the text is in fact a hybrid, a mixture of two texts. The full, non-hybrid, version will be spelled out below.

The text exhibits two features that set it apart from other examples of the balanced sentence. First, the inverted statement implied by the simple expression *ṯs pẖr* is not, as might be expected, a full inversion of the two sides of the two balanced sentences. Such a full inversion would produce the following result:

**ḥms.j m mḥwt*	*jw pt m rswt*
**ḥms.j m j3bt*	*jw pt m jmnt*

'I will sit to the north,	heaven will come as a south wind.
I will sit to the east,	heaven will come as a west wind'.

Clearly, such a full inversion is not what is intended. The two sides do not need to be inverted in full. Only the cardinal points need to be switched, but not the verb forms featuring the verbs *ḥms*(*y*) 'sit' and *jw*(*y*) 'come'. Accordingly, the inverted version would look as follows.

**jw pt m mḥwt*	*ḥms.j m rswt*
**jw pt m j3bt*	*ḥms.j m jmnt*

'Heaven comes as a north wind,	I will sit to the south.
Heaven comes as an east wind,	I will sit to the west'.

The second special feature that differentiates CT 4.50f–g from most other instances of the balanced sentence concerns the four cardinal points. The text as it stands is incongruous—because it is hybrid. The cardinal points appear both as a *wind* and as a *place*. In the first member of both balanced sentences, the cardinal point ought to refer to the *wind*. In the second member of both balanced sentences, the cardinal point ought to refer to the *place*. But in fact, as the text stands, the cardinal point refers to the *wind* in *both* members. This is internally inconsistent, but it would be hasty to conclude that the text is in error. The inconsistency is just a secondary outcome of the author's effort to telescope a fuller version containing *four* balanced sentences into *two* balanced sentences. Some loss and distortion of information are to be expected, and in fact, the result is a hybrid or mixture. A full version containing *four* balanced sentences is found at CT 5.3a–6c as well as in chapter 57 of the Book of the Dead (for the version in the Papyrus of Nu, see Budge 1898: 129, lines 3–8). This full version is discussed at greater length in §5.1.6 below.

If CT 4.50f–g had been fully internally consistent, it ought to have sounded something like the following.

**jw pt m rswt ḥms.j ḥr mḥt.s ṯs pẖr*
**jw pt m jmnt ḥms.j ḥr j3bt.s ṯs pẖr*

'Heaven comes as a south wind, I will sit to the north of it, and vice versa'.
'Heaven comes as a west wind, I will sit to the east of it, and vice versa'.

In this modified version, the preposition preceding the cardinal point as a *place* is *ḥr* 'on, at, to' and not *m* 'as', as one would expect. Furthermore, a suffix pronoun has been attached to *ḥr*.

Does CT 4.50f–g still qualify as a balanced sentence with all this telescoping? I believe it does. The essence of the message is still inversion. The essential inversion is more or less as follows: "north is south," therefore "south is north"; "west is east," therefore "east is west." The fact that the cardinal point once refers to a wind and once to a place is of secondary importance. What matters is the equation between the two cardinal points and the fact that the equation is twice reversed.

5.1.1.4 *Ṯs pẖr sp 2 'and twice vice versa'*

Another special case of the balanced sentence is the use of *ṯs pẖr sp 2* 'and twice vice versa':

pr.n.f jm.ṯ ṯs pẖr sp 2 (CT 1.186c, Coffins B10C[a] and B10C[b])

'That he leaves you (fem.), and twice vice versa'

The function of *ṯs pẖr sp 2* is not quite clear. Does it mean that the fully explicit version on two coffins of CT 1.186c consists of *two* balanced sentences, as follows?

**pr.n.f jm.ṯ pr.n.ṯ jm.f pr.n.ṯ jm.f pr.n.f jm.ṯ*

(lit.) 'That he has left you (fem. sg.) means that you have left him. That you (fem. sg.) have left him means that he has left you'.

Another coffin, B10C^c^, has *ṯs pẖr*, without *sp 2* 'twice'. On the face of it, this fully explicit version would presumably be equivalent to just *one* balanced sentence and not *two*, namely **pr.n.f jm.ṯ pr.n.ṯ jm.f.*

5.1.1.5 Ṯs pẖr marking one balanced sentence

In the following examples, *ṯs pẖr* apparently signals the abbreviation of one balanced sentence, not two.

wnn.j jm.k	*ṯs pẖr*	(CT 4.53b, Coffins B3L, B1L)
ḫpr.n.j m jrt Ḥr	*ṯs pẖr*	(CT 4.99g, Coffins S2P, S2C, S1C)
ḫpr.n.j m Rʿ	*ṯs pẖr*	(CT 4.178g, Coffin G1Be)

The explicit versions are presumably as follows.

**wnn.j jm.k wnn.k jm.j*

'That I am (or: will be?) you means that you are (or: will be?) me'.

**ḫpr.n.j m jrt Ḥr ḫpr.n jrt Ḥr jm.j*

'That I have come into existence out of the Eye of Horus means that the Eye of Horus has come into existence out of me'.

**ḫpr.n.j m Rʿ ḫpr.n Rʿ jm.j*

'That I have come into existence out of Re means that Re has come into existence out of me'.

5.1.2 The meaning of paired balanced sentences: Equation (from which reversibility naturally follows)

There are two basic possibilities with pairs of balanced sentences in which the second balanced sentence is the inversion of the first. Either the second balanced sentence is added to convey a different thought from the first, or the second balanced sentence conveys more or less the same thought as the first. It is assumed here that the second possibility is the case. The reason for thinking so is the structure of the balanced sentence. The balanced sentence appears to be nothing but a nominal sentence like *pḥty N pḥty Stš* 'The power of (king) N is the power of Seth' (for this type, see Edel 1955/64: 481 §947). This association between balanced sentence and nominal sentence was first made by Polotsky (1964: 282 [1971: 67]). The function of the nominal sentence is to equate two nouns or noun phrases. Once the full identity of the two entities is declared, the possibility of inversion follows naturally. Indeed, if a certain thing A is declared identical to a certain thing B, then it follows from the very concept of equation that, inversely, thing B is the same as thing A. If A is B, then B is A. The axiom of commutativity is valid in Boolean algebra as much as it is in quantitative mathematics. The same consideration therefore also applies to the balanced sentence, which is nothing but a nominal sentence.

The balanced sentence clearly exhibits the structure of an equation. If the first member is the same as the second, then the second member is also the same as the first. The concept of equation is essential to the balanced sentence. Consequently, in pairs of balanced sentences, the second restates by inversion the general thought of the first balanced sentence, which is to declare an identity. But why the repetition? Presumably, the repetition serves the purpose of emphasizing the veracity and the full impact of the equation.

In spite of the equation, a sentence and its inversion are not one hundred percent the same. They begin and end differently. Then again, speech is by necessity linear. Something has to come first. But does this necessarily mean anything more than just that, namely, that something is first?

5.1.3 Association of the balanced sentences with the logical statement of Type 1 (and of the sentence with *jr* with the logical statement of Type 2)

Balanced sentences signify equation (see §5.1.2). An algebraic expression for equation is $x = y$. This representation is also used in symbolic logic for logical statements of Type 1. It may therefore be concluded that balanced sentences exhibit a close affinity with logical statements of Type 1. By contrast, sentences with *jr* do not denote equation. Inversion therefore produces a sentence with a different meaning (see §5.1.5). There has been a long tradition of identifying sentences with *jr* as conditional sentences. In symbolic logic, conditional sentences are presented as logical statements of Type 2.

5.1.4 Translating balanced sentences

Translations may now be proposed for the balanced sentence that better bring out the concept of equation. Consider the well-known example cited in chap. 1 above, along with the translation given above.

prr.ṯn r pt m nrwt prr.j ḥr tpt ḏnḥwy.ṯn (CT 3.61f–g)

'Whenever you depart for heaven as vultures,
I depart on top of your wings'.

This translation is identical to that of a sentence with *jr*. More distinctive translations are as follows:

'You depart for heaven as vultures,
I depart on top of your wings'.

'I depart on top of your wings only and always
when you depart for heaven as vultures',

Both translations better bring out the equation $x = y$, from which the statement $y = x$ naturally follows. In conclusion, inversion often occurs with balanced sentences because the inverted statement means basically the same thing. The aim of explicitly mentioning the inverted statement is to emphasize the fact.

5.1.5 The absence of inversion with sentences with *jr*

To my knowledge, no examples exist of inversion involving sentences with *jr*. Thus, the expression *ṯs pẖr* 'and vice versa' never seems to follow sentences with *jr*. The reason would appear to be that the inverted statement has a different meaning. For example, (a) "When it rains, I stay inside" differs markedly in meaning from its inversion, (b) "When I stay inside, it rains." With (a), though not with (b), it can happen that it does not rain ($\bar{r}$) and I stay inside (i). In this case, $r\bar{i} = 0$ but $\bar{r}i \neq 0$; the technical details are discussed in §4.3 above. With (b), though not with (a), it can happen that it rains (r) and I do not stay inside ($\bar{i}$). That is because $\bar{r}\bar{i} = 0$ whereas $r\bar{i} \neq 0$ (see §4.3). With both (a) and (b), it can happen that it rains and I stay inside and it can also happen that it does not rain and I do not stay inside. That is because with both (a) and (b), neither ri nor $\bar{r}\bar{i}$ equals zero (see §4.3). In sum, the structure of sentences with *jr* is not that of an equation.

In theory, inversion is not impossible with sentences with *jr*. Indeed, it is a fact of logic that two logical statements of Type 2 (such as sentences with *jr*) produce a single statement of Type 1 (such as balanced sentences). Thus, if one combines "When it rains, I stay inside" (Type 2) with its inversion "When I stay inside, it rains" (Type 2), the result is "I stay inside only and always when it rains" (Type 1). In practice, however, this combination seems like an unusually cumbersome thought. That may be why there are no examples of it to be found.

5.1.6 A single thought expressed both as a balanced sentence and as a sentence with *jr*, confirming the postulated difference between the two

In order to confirm the difference between the balanced sentence and the sentence with *jr* as defined above, it would be convenient to have a single general thought that is expressed both ways. In fact, there is such an example. On the one hand, the balanced sentence can be followed by *ṯs pẖr* 'and vice versa'. On the other hand, the sentence with *jr* is not. The contrast is striking. Moreover, the meaning of the thought is fully transparent and both sentence types are preserved in more than one copy. More such examples remain desirable.

The general thought is of a person sitting on the lee side of a house in order to stay out of the wind.

BALANCED SENTENCE (repeated from §5.1.1.3)

jw pt m rswt ḥms.j m mḥwt ṯs pẖr
jw pt m jmnt ḥms.j m j3bt ṯs pẖr

'Heaven comes as a south wind,
I will sit to the north, and vice versa'.
"Heaven comes as a west wind,
I will sit to the east, and vice versa'.

SENTENCE WITH *jr*

jr jw pt tn m mḥwt ḥms⟨.j⟩ ḥr rs.s
jr jw pt tn m rswt ḥms.j ḥr mḥt.s
jr jw pt m jmnt ḥms.j ḥr j3bt.s
jr jw pt tn m j3bt ḥms.j ḥr jmnt.s

'When this heaven comes as a north wind,
I will sit to the south of it (the house).
When this heaven comes as a south wind,
I will sit to the north of it.
When this heaven comes as a west wind,
I will sit to the east of it.
When this heaven comes as an east wind,
I will sit to the west of it'.

(CT 5.3c–4a, 4b–c, 5a–b, 5c–6a, Coffin B2Bo, and several other coffins; cf. also Book of the Dead, chap. 57, Papyrus of Nu [Budge 1898: 129, lines 3–8])

5.2 BALANCED NEGATION AS A FACT OF BALANCED SENTENCES

Compare the second logical property of Type 1, described in §4.2.

There are at least two instances in which a balanced sentence is accompanied by a balanced negation. In a balanced negation, both members of the balanced sentence are negated in the same way. Two examples is not much, but they are at least fully transparent and therefore suitable for supporting a grammatical point. In the first example, which is often cited in works on grammar, the negation is grammatical:

jw.k r.j *ḏd.j r.k*
tm.k jw r.j *tm.j ḏd r.k*

'You come against me, I say against you.
You do not come against me, I do not say against you'.

(Book of the Dead, chap. 90, Papyrus of Nu [Budge 1898: 192, lines 10–12]; see also CT 5.326g–h, Coffin B2L [cited on p. 71])

A literal translation is: 'That you come against me means (the same as) that I say against you. That you do not come against me means (the same as) that I do not say against you'. Future tense cannot be excluded, however: 'That you will come …'

In the second example, the negation is lexical:

mrw.f mt.ṯn *mt.ṯn*
mrw.f ꜥnḫ.ṯn *ꜥnḫ.ṯn* (Pyr. §153c)

'He wants you to die, you will die.
He wants you to live, you will live'.

'Die' and 'live' are opposites—'die' is the negation of 'live'. 'Die' is the equivalent of 'not-live', and vice versa.

In both examples, two balanced sentences seem to convey the same message, once affirmatively and once negatively. In other words, making both members of the balanced sentence negative does not produce a different message. Such synonymity is not the case when both members of a conditional sentence with *jr* are negated. For example, it does not follow from "When it rains, I stay inside" that "When it does not rain, I do not stay inside." I may be either inside or outside when it does not rain.

Balanced negation is also a logical property of logical statements of Type 1, therefore associating the balanced sentence and logical statements of Type 1 with each other. By contrast, no examples of balanced negation with conditional sentences with *jr* are known. Sentences with *jr* are therefore apparently affiliated with logical statements of Type 2. An example is *jr swrj.f mw stp.ḫr.f* 'When he drinks, he chokes' (Papyrus Smith 9,19–20). This sentence tells us what happens if the man in question drinks. It does not tell us whether he chokes or does not choke when he does *not* drink. All that is stated is that when he drinks (d), he chokes (c). Put differently, in the formulation preferred by Venn, it does not happen that he drinks (d) and does not choke ($\bar{c}$) ($d\bar{c} = 0$). The total truth (1) is therefore that he either drinks and chokes (dc), does not drink and yet chokes ($\bar{d}c$), or neither drinks nor chokes ($\bar{d}\bar{c}$): $d\bar{c} + \bar{d}c + \bar{d}\bar{c} = 1$.

5.3 RESTRICTION TO CERTAIN COMBINATIONS AS A FACT OF BALANCED SENTENCES

Compare the third logical property of Type 1 described in §4.3.

There are at least two instances in which a balanced sentence is accompanied by a parallel balanced sentence exhibiting nonbalanced negation. In nonbalanced negation, both members of the balanced sentence are negated, but in different ways. Two examples is not much, but both examples are fully transparent and so are suitable for drawing grammatical conclusions.

mrr.f jrr.f msḏḏ.f n jr.n.f (Pyr. §412b)

'He likes (something), he does (it).
He hates (something), he does not do (it)'.

The first negation, *msḏḏ.f*, is lexical. *Msḏ(y)* 'hate' is the lexical opposite or the negation of *mr(y)* 'love'. The second negation, *n jr.n.f*, is grammatical. A balanced negation would have been **tm.f mr tm.f jr* (lit.) 'That he does not like (something) means that he does not do (something)'. It is not clear why this kind of negation was not chosen. One might perhaps speculate that, once the writer had somehow chosen the lexical negation *msḏḏ.f* over the grammatical negation *tm.f mr*, it was not possible to obtain the stylistically balanced *tm.f mr tm.f jr* anyhow. *Tm.f jr* was therefore not used either.

The second example appears in the Coffin texts:

jw.k r.j ḏd.j r.k   *tm.k jw r.j nn ḏd.j r.k*

'You come against me, I say against you.
You do not come against me, I will not say against you'.

(CT 5.323h–i, Coffin B2L; *nn* is absent in coffin B17C, perhaps in error)

The example has been translated here in the present tense. It may be a future tense.

The negation with *nn* in *nn ḏd.j* seems to suggest that the verb form *nn ḏd.j* is future tense, as translated above. Neither *msḏḏ.f n jr.n.f* nor *tm.k jw r.j nn ḏd.j r.k* is a balanced sentence, because neither *n jr.n.f* nor *nn ḏd.j r.k* is substantival. *N jr.n.f* is general present; *nn ḏd.j* ought to be future. By contrast, in the parallel passage later in the same spell (at 326g–h), the same coffin B2L exhibits balance in the negated version: *jw.k r.j ḏd.j r.k tmm.k* (probably for *tm.k*) *jw r.j tm.j ḏd r.k* (cf. Loprieno 1995: 216). Another parallel version found in chapter 90 of the Book of the Dead (see §5.2) also exhibits balanced negation.

What, then, is the sentence structure of the two balanced sentences *jw.k r.j ḏd.j r.k* and *tm.k jw r.j nn ḏd.j r.k*? First, there is no denying that verb forms like *n jr.n.f* and *nn ḏd.j r.k* could be adverbial. They often function as the verb forms of adverbial clauses. Second, nothing speaks against interpreting the sentences as adverbial sentences, consisting of a substantival verb form followed by an adverbial verb form. The effect is emphasis on the latter. The literal translation is then something like 'It is while (*or:* by) not doing (something) that you (show that you) hate (it)' and 'It is while (*or:* by) not being about to say (something) against you that I will come against you'. In other words, it is specifically in the circumstances described in the second members of the two sentences that the events described in the first members occur.

Likewise, Allen (1984: 295–96 §437) identifies *n jr.n.f* in *msḏḏ.f n jr.n.f* as "predicate in the 'emphatic' sentence … whereby positive and negative elements are 'balanced' through the use of the 'emphatic' sentence: thus, 'that-he-likes is that he acts' but 'that-he-hates is without him acting'." Also similarly, Wente (1977: 312) writes about exceptional cases in which *wnn.f ḥr sḏm* of Late Egyptian is *not* followed by *jw.f* (*ḥr*) *sḏm*.

> The existence of such exceptions need not vitiate the supposition that the relationship between the *wnn*- and *ỉw*- clauses is that between a nominal (emphatic) form and a circumstantial clause any more than does the puzzling *n ỉr.n.f* of Pyr. 412b in the context of a 'Wechselsatz' nullify the thrust of Polotsky's remarks in *Orientalia* 33 (1964): 282 [= 1971: 67].

Polotsky (ibid.) too originally favored the interpretation of *msḏḏ.f n jr.n.f* as a single sentence, against interpreting *n jr.n.f* as a sentence in its own right.

> Instead of assuming it to be self-evident that *n ỉr.n.f* must under all circumstances be an independent sentence, one must rather reckon with the possibility that [*n ỉr.n.f*] contracts a syntactic relation with [*msḏḏ.f*], in the same way that *ỉrr.f* does with *mrr.f*, and is conditioned by *msḏḏ.f*."*

Then again, Polotsky does not explicitly identify that single sentence as an adverbial sentence. Rather, he confesses ignorance in this respect (ibid.):

> I do not know why [*n sḏm.n.f*] is used and how [*n sḏm.n.f*] relates to *tm.f sḏm*, which one expects judging from the example cited above.†

In fact, Polotsky later (1969: 471) showed preference for interpreting *n jr.n.f* as the main clause.

> [*N*] *ỉr.n.f* is to be recognized as an independent main clause, just like *nn ḏd.j*. However, the construction must be characterized in both cases as unharmonious. In other words, it represents the transition to a "conditional clause without introduction."‡

"Konditionalsatz ohne Einleitung" is Edel's term (Edel 1955/64: 534 §1033). "Ohne Einleitung" is here synonymous with "not introduced by *jr*." The "conditional clause not introduced by *jr*" consists of a substantival verb form placed at the head of the sentence. I suspect that this type of *jr*-less conditional clause does not exist, even though I have in the past non-committally made reference to it (Depuydt 1993a: 147 n. 40). Upon closer inspection, alternative interpretations always seem possible for passages

* Statt als selbstverständlich vorauszusetzen dass *n ỉr.n.f* unter allen Umständen ein selbständiger Satz sein müsse, ist vielmehr mit der Möglichkeit zu rechnen, dass es, ebenso wie *ỉrr.f* mit *mrr.f*, in syntaktischem Verband mit *msḏḏ.f* steht und durch dieses bedingt ist.

† Ich weiss nicht warum [*n sḏm.n.f*] gebraucht ist und wie sie sich zu dem nach dem oben zitierten Beispiel zu erwartenden *tm.f sḏm* verhält.

‡ [*N*] *jr.n.f* ist ebenso wie *nn ḏd.j* als selbständiger Hauptsatz anzuerkennen, die Konstruktion muss aber wohl in beiden Fällen als inkonzinn bezeichnet werden, bzw. sie stellt den Übergang zum "Konditionalsatz ohne Einleitung" dar.

adduced as instances of this structure (see also chap. 10 below). The following example is related. It also consists of two negated clauses, but unlike the examples cited above, it is not the second member of a pair of which the first member is affirmative. Still, it seems to provide some additional evidence of how balanced sentences can be negated.

tm.f wnn jm n wn.n.j [or *wnn.j*?] *jm*

'He is not [*or:* will not be?] there,
I am not [*or:* will not be?] there'.

(Book of the Dead, chap. 8, Papyrus of Ani [Budge 1898: 30, lines 7–8]; cf. Gunn 1924: 104, who cites the Papyrus of Amenemhat)

I consider the most probable interpretation of the sentence to be as an adverbial sentence. Presumably, *n wn.n.j* would be general present tense and *n wnn.j* would be future tense. *Tm.f wnn* could be either general present (aorist) or future. In general, there is still much uncertainty as to how the various verb forms of *wnn* are written.

The interpretation of the pattern *msḏḏ.f n jr.n.f* as an adverbial sentence ('That he hates is while he does not do') matches what we know about related expressions. Three structures are well attested (see e.g. Niccacci 1980).

(a) *prr.f* *prr.j* geminating + geminating

(b) *prr.f* *pr.j* geminating + non-geminating

(c) *pr.f* *pr.j* non-geminating + non geminating

The following does not seem to occur.

(d) *pr.f* *prr.j* non-geminating + geminating

Structure (a) is no doubt a balanced sentence. Structure (c) is presumably a balanced sentence in the future tense. It is difficult to see what else it could be. As regards structure (b), two interpretations are possible. First, (b) is a balanced sentence consisting of a general present (aorist) substantival verb form and a substantival future verb form, with the meaning 'That he departs means that I will depart'. The difference in tense presents a certain unbalance. I would therefore consider this possibility less likely. Alternatively, (b) is an adverbial sentence with a present (aorist) substantival verb form and a present or future adverbial verb form: 'It is while I (too) depart (*or:* will depart) that he departs'.

So far, the three structures (a), (b), and (c) can be interpreted without too much difficulty as either a balanced sentence or an adverbial sentence. There is therefore no need for more than two options. Therefore, when a structure is *not* a balanced sentence, the attention readily turns to the adverbial sentence as the obvious candidate for identification. It is the only other option.

The fourth structure, (d), would be much more difficult to interpret as either a balanced sentence or an adverbial sentence. On the one hand, the fact that the second verb form, *prr.j*, is geminating eliminates the interpretation of this structure as an adverbial sentence. The fact that the first verb form, *pr.j*, does not geminate whereas the second verb form does eliminates the possibility of interpreting this structure as a balanced sentence; one expects the same verb form in both members of a balanced sentence. It is then not surprising that pattern (d) seems absent from the surviving evidence. It probably never existed.

The balanced sentences with two negated clauses cited above (Pyr. §412b; CT 5.323h–i; Book of the Dead, chap. 8) are interpreted here as adverbial sentences. Other related expressions support this interpretation. Some of them are also easily interpreted as adverbial sentences. The following related expression is almost certainly an adverbial sentence.

prr.sn r pt m bjkw	*jw.j ḥr ḏnḥw.sn*	(CT 3.100h–101b,
hꜣꜣ.sn r tꜣ m ḥfꜣw	*jw[.j] ḥr ḳꜣbw.sn*	Coffins S1C, S2C)

'That they depart for heaven as falcons
is while I am on their wings.
That they descend to earth as snakes
is while I am on their coils'.

Niccacci (1980: 203–4) interprets the example differently, namely as a substantival verb form (*prr.sn* … and *hꜣꜣ.sn* …) functioning as a part of a conditional clause without *jr* plus a main clause (*jw.j* …). He describes it as a "personal substantival verb form … used in extraposition (protasis) with respect to an independent construct (apodosis)" (ibid. 222).*

If *msḏḏ.f n jr.n.f* (Pyr. §412b) and *tm.k jw r.j nn ḏd.j r.k* (CT 5.323h–i) are both adverbial sentences, then *n jr.n.f* and *nn ḏd.j r.k* are adverbial clauses. Adverbial clauses denote concomitant circumstances. Furthermore, the initial verb forms *msḏḏ.f* and *tm.k jw* are substantival. The concomitant circumstances expressed by *n jr.n.f* and *nn ḏd.j r.k* as adverbial clauses ought

* Forma verbale sostantiva personale … usata in estraposizione [protasi] rispetto a un costrutto independente [apodosi].

therefore to be *emphasized*. The literal meaning of the sentences would therefore need to be more or less as follows: it is in this or that circumstance that the event expressed in the first, substantival, clause takes place.

It appears that grammatical structure can be aligned with logical structure as follows. Consider the well-known expression *mrr.f jrr.f msḏḏ.f n jr.n.f* (Pyr. §412b) 'He likes (something), he does (it). He hates (something), he does not do (it)'. The first sentence is a balanced sentence. The second sentence is interpreted here as an adverbial sentence. The temptation may arise to interpret this difference with regard to type of sentence as a mismatch. That perceived mismatch might in turn undermine the interpretation of the two sentences as a balanced sentence and an adverbial sentence. However, the difference with regard to type of sentence can easily be explained in logical terms.

The first sentence is a balanced sentence. Its structure may be presented as $x = y$, Type 1. Now consider $\bar{x}$, the negation of x. The crucial question arises: By which of the two options y and $\bar{y}$ can $\bar{x}$ be accompanied as a *circumstance*, if one first assumes that $x = y$? It follows necessarily from the equation $x = y$ that $\bar{x} = \bar{y}$. The answer to the question therefore is: only by $\bar{y}$ can $\bar{x}$ be accompanied. Only xy and $\bar{x}\bar{y}$ exist; $x\bar{y} = 0$ and $\bar{x}y = 0$. Or, *it is* by circumstance $\bar{y}$ *that* $\bar{x}$ is accompanied. Note the emphasis. It is exactly such contrastive emphasis that the second sentence, *msḏḏ.f n jr.n.f*, would express if it were interpreted as an adverbial sentence, meaning 'It is while he does not do (it) that he hates (it)'. The juxtaposition of a balanced sentence and an adverbial sentence, in that order, is therefore altogether explainable in logical terms, in spite of the fact that the types differ. By contrast, the logical considerations just presented do not apply to Type 2. Logic therefore serves to confirm that balanced sentences are affiliated with logical statements of Type 1.

5.4 TWO SENTENCES WITH *JR* CORRESPOND TO ONE BALANCED SENTENCE

Compare the fourth logical property described in §4.4.

There is an instance in the Coffin texts in which two sentences with *jr* are telescoped into a single balanced sentence. The sentences with *jr* are found in CT 5.3c–4a, 4b–c, 5a–b, 5c–6a. The corresponding balanced sentences are found in CT 4.50f–g (cf. §§5.1.1, 5.1.6).

TWO SENTENCES WITH *JR*

jr jw pt tn m mḥwt ḥms.j ḥr rs.s
jr jw pt tn m rswt ḥms.j ḥr mḥt.s

'When heaven comes as a north wind,
I will sit to the south of it (the house).
When heaven comes as a south wind,
I will sit to the north of it'.

TELESCOPED INTO ONE BALANCED SENTENCE

jw pt m rswt ḥms.j m mḥwt ṯs pẖr

'Heaven comes as a south wind, I will sit as a northwind [*sic*], and vice versa'.

TWO SENTENCES WITH *jr*

jr jw pt m jmnt ḥms.j ḥr j3bt.s
jr jw pt tn m j3bt ḥms.j ḥr jmnt.s

'When heaven comes as a west wind,
I will sit to the east of it.
When heaven comes as an east wind,
I will sit to the west of it'.

TELESCOPED INTO ONE BALANCED SENTENCE

jw pt m jmnt ḥms.j m j3bt ṯs pẖr

'Heaven comes as a west wind, I will sit to the east, and vice versa'.

Two sentences with *jr* are telescoped into a single balanced sentence. This does not come without some distortion, however. In the sentences with *jr*, the four cardinal points appear both as a *wind* and as a *place*. The writing of the names of the four winds—*mḥwt*, *rswt*, *jmnt*, and *j3bt*—exhibits the hieroglyph depicting a sail as a determinative. In the *second* member of both balanced sentences, one expects the cardinal point as a place. What one finds is the cardinal point as a wind. The result is that the balanced sentences are hybrids. The matter has already been discussed in §5.1.6.

It is assumed here that *ṯs pẖr* 'and vice versa' refers to the complete reversibility that characterizes the relationship between $x = y$ and $y = x$. Indeed, the structure of the balanced sentence is that of a nominal sentence.

The nominal sentence denotes an equation. The purpose of *ṯs pẖr* appears to be to emphasize the fact of the equation, the fact that the two halves of the balanced sentence are the same. The two balanced sentences together stress the full identity between the two halves in rhetorical fashion. Of course, exactly the opposite might be argued, namely, that *ṯs pẖr* rather indicates that $x = y$ and $y = x$ are in fact different in language even if they are the same in mathematics. The repetition by means of *ṯs pẖr* would then be necessary because $x = y$ differs from $y = x$. The question remains, then, how a full equation running one way ($x = y$) can be different from an equation running the other way ($y = x$). It seems difficult to accept that the addition *ṯs pẖr* could mark an entirely different message.

6 From Condition to Premise and Back

> Reasoning about causes and effects is rather difficult. I believe that only God can be the judge. We have so much difficulty in establishing a relation between such an obvious effect as a scorched tree and the lightening that has set it on fire that pursuing sometimes long chains of causes and effects in reverse order seems to me as foolish as trying to construct a tower that reaches up to the sky.*
>
> Umberto Eco, *Il nome della rosa* (1980), 38 ("William of Baskerville" speaking)

6.1 Condition and Premise

6.1.1 Definition

In my opinion, a clear understanding of the difference between condition and premise is fundamental to comprehending how we think and what the limitations of our thinking are. Such an understanding also makes it possible to understand why causality is not a relevant concept when it comes to thought. Moreover, the ancient Egyptian language sharply distinguishes between the denotation of a condition and the denotation of a premise. Egyptian ideally illustrates the difference.

At the outset, it will be useful to stress once more how sharp the difference between condition and premise is. The statement "When(ever) it rains, I stay inside" implies neither that it rains nor that I stay inside. That makes "when(ever) it rains" a condition (or protasis) and "I stay inside" a contingent statement (or apodosis). By contrast, the statement "If it is in fact raining, I am staying inside" does imply that I assume it true that it is raining and that I am, as a necessary consequence, staying inside. That makes "if it is raining" a premise and "I am staying inside" a consequence or result. A premise is *an assumption that something is true.* Once such an assumption is made, a consequence necessarily follows.

* Ragionare sulle cause e sugli effetti è cosa assai difficile, di cui credo che l'unico giudice possa essere Dio. Noi già fatichiamo molto a porre un rapporto tra un effetto così evidente come un albero bruciato e la folgore che lo ha incendiato, che il risalire catene talora lunghissime di cause ed effetti mi pare altrettanto folle che cercare di costruire una torre che arrivi sino al cielo.

Condition and premise are related. Consider someone adopting the policy, "Whenever it rains, I stay inside," which consists of a condition and a contingent statement. Exactly four premises can be derived from this conditional sentence. Two of them trigger necessary consequences. First, if (it is the case that) it is raining, then it follows necessarily by the stated policy that I am staying inside. Second, if (it is the case that) I am not staying inside, it necessarily follows that it is not raining. The two other premises are inconclusive. First, if (it is the case that) it is not raining, I may or may not stay inside. There is no way to know. Second, if (it is the case that) I am staying inside, it may or may not be raining. There is no way to know.

Condition is often associated with causality, but the two have nothing to do with one another. In some sense, no one really knows what causality is anyhow. The sentence "When it rains, I stay inside" may seem to state that the rain is *causing* me to stay inside. But this impression is false. Even if the rain causes me to stay inside, that is not what the sentence structure of "When it rains, I stay inside" says. The proof is in the logically equivalent statement, "When I do not stay inside, then it does not rain." Obviously, my being outside does not *cause* rain to stay away. Therefore, the sentence structure by itself does not express causality, whatever causality may be.

I believe the distinction between condition and premise to be of fundamental importance to the study of modern scientific logic. Yet I have not found the distinction treated anywhere ever, from De Morgan and Boole up to the present time. There is one chapter in Venn's *Symbolic Logic* (1894: 209–44) that comes somewhat close, entitled "Hypotheticals." The chapter's topic is located somewhere in the vicinity of the (unrecognized) distinction between condition and premise. As a result of the absence of the distinction, Venn clearly struggles. First he recants (232 n. 1) an earlier version of some parts of the same chapter in the first edition (1881). Furthermore, the chapter is full of inconclusive statements. One is that hypotheticals "are to a considerable extent either material or psychological, and therefore the necessity of forcing them into our system will crush much of the life out of them." Rather than crushing the life out of the system of modern scientific logic, the distinction between condition and premise would rather seem to add life to it, and add to modern scientific logic what seems a cornerstone of its edifice.

6.1.2 Definite and indefinite processes

I have elsewhere described how premises can be derived unfailingly from conditions by a rigorous method (Depuydt 1999a, with references to earlier discussions). But what is the nature of the reverse process, deriving conditions from premises?

There can be no doubt that the process of deriving premises from conditions is *definite*, in the sense described by Venn (1894: 74–77) and others. Each derivation of a premise from a condition has just one solution. Likewise, the mathematical operation of addition is definite. Adding 5 to 15 has one solution only, namely 20. The inverse process, subtraction, is also definite. Subtracting 5 from 20 has one solution only, namely 15. Squaring is also a definite operation. The result of squaring 2 is 4, and 4 only, but in this case the inverse operation is indefinite. The square root of 4 can be either +2 or –2.

6.1.3 The problem: Is deriving conditions from premises a definite process?

Deriving premises from conditions is no doubt definite. But what about the inverse process? I have elsewhere explicitly left the question open (Depuydt 1997: 109). Additional investigation now reveals that deriving conditions from premises is also unequivocally definite. To prove this, no step in the argument will be left unexpressed. At stake is the notion that we always inevitably think in certain ways in spite of ourselves. Once the afore-mentioned proposition has been proven, the relationship between the two following variants of a single general thought can be described.

CONDITION:

ⲉⲣϣⲁⲛ ⲡⲉⲕⲃⲁⲗ ⲥⲕⲁⲛⲇⲁⲗⲓⲍⲉ ⲙ̄ⲙⲟⲕ ⲡⲟⲣⲕ̄ϥ
eršan (Gk. ἐὰν) *pe=k-bal skandalize mmo=k pork=f* (Mark 9:47)

'When(ever) your eye causes you to sin, pluck it out'.

PREMISE:

ⲉϣϫⲉ ⲡⲉⲕⲃⲁⲗ ⲥⲕⲁⲛⲇⲁⲗⲓⲍⲉ ⲙ̄ⲙⲟⲕ ⲡⲟⲣⲕ̄ϥ
ešje (Gk. εἰ) *pe=k bal skandalize mmo=k pork=f* (Matthew 18:9)

'If (it is the case that) your eye is causing you to sin, pluck it out'.

The difference between the two seems tantalizingly subtle, yet it can be defined in the most exact of terms. In what follows, the steps leading from condition to premise and back again are laid out in full.

6.2 FROM CONDITION TO PREMISE

6.2.1 Two events correspond to eight sequences of protasis and apodosis

In order to guarantee complete transparency about the conversions of condition into premise and premise into condition, it will be necessary to lay out the two conversions in fullest detail by means of a generic example. The example concerns two events: "It rains" (*r*) and "I stay inside" (*i*). These two events combine in eight different ways into a sequence consisting of a condition, or "if/when(ever)" clause (protasis), and a contingent statement, or "then" clause (apodosis). This division into eight is obtained by two criteria each of which exhibits two options.

THE TWO CRITERIA

(1) The event is either affirmative (r "It rains"; i "I stay inside") or negative ($\bar{r}$ "It does not rain"; $\bar{i}$ "I do not stay inside").

(2) The event appears either in the "if/when" clause or in the "then" clause.

The resulting eight combinations are as follows.

THE EIGHT SEQUENCES

(1) When(ever) it rains, I stay inside.
(2) When(ever) it does not rain, I stay inside.
(3) When(ever) it rains, I do not stay inside.
(4) When(ever) it does not rain, I do not stay inside.
(5) When(ever) I stay inside, it rains.
(6) When(ever) I do not stay inside, it rains.
(7) When(ever) I stay inside, it does not rain.
(8) When(ever) I do not stay inside, it does not rain.

All eight sequences are thinkable or conceivable, but perhaps only two, (1) and (8), refer to more or less plausible situations in reality. The other six are implausible but still conceivable. Cases (2) and (6) describe someone who always stays inside when it does not rain; (3) and (7) describe someone who is always outside when it rains; (4) and (5) describe someone who is always outside when it does not rain.

Boole (1854) would present (1) as $r = \frac{0}{0}\, i$ "The time that it rains (= *r*) coincides with an indefinite amount (none, some, or all) (= $\frac{0}{0}$) of the time that I stay inside (= *i*).” Accordingly, (2) would correspond to $\bar{r} = \frac{0}{0}\ i$. And so on for the other six sequences.

6.2.2 Eight sequences correspond to four denials of a combination’s existence, making four pairs

Modern scientific logic, or symbolic logic, is about how *things* and *events* relate to one another. It is sometimes called mathematical logic, but this term is in danger of obscuring the exact relationship between what is normally called logic and what is normally called mathematics. As the present work’s title suggests, both are branches of what may be called mathematics in the large sense. In other words, modern scientific logic is nothing but another kind of mathematics. The familiar mathematics is quantitative and describes that which is capable of increase or diminution. Logic is another, non-quantitative type of mathematics (§0.7).

A fundamental principle of this other mathematics is the partition of the universe of discourse (that is, anything one could possibly think about) in terms of classes and their contradictories. Consider the two classes “blue” (*b*) and “car” (*c*). The two contradictories are “non-blue things” ($\bar{b}$) and “all things that are not cars” ($\bar{c}$). As far as these two classes of *things* are concerned, the universe is subdivided into exactly four combinations of classes, which are themselves also classes.

FOUR COMBINATIONS OF CLASSES OF THINGS, THEMSELVES ALSO CLASSES

(a)	bc	blue cars
(b)	$\bar{b}c$	non-blue cars
(c)	$b\bar{c}$	blue objects that are not cars
(d)	$\bar{b}\bar{c}$	things that are neither blue nor cars

This writer belongs to combination (d). There are of course other possible combinations of the two classes and their contradictories, but none of these other combinations produce additional combination classes. Three types may be distinguished among the other combinations. First, the combinations above may be inverted, as in *cb*; but *cb* is the same combination class as *bc*. Second, a class symbol may be combined with the same symbol, as in *bb*. However, *bb* is the same class as just *b*. Third, a class symbol may be combined with its contradictory, as in $b\bar{b}$. However, the class of things that are both blue (*b*) and not blue ($\bar{b}$) at the same time does not exist. That is because, according to the fundamental axiom of logic, a thing cannot have a certain attribute and not have that same attribute at the same time.

Four combination classes have been obtained. But the existence of each of the four combinations can be denied, as in $bc = 0$ "There are no blue cars."

The same partition into combination classes applies to events. Consider the two events "It rains" (r) and "I stay inside" (i). As far as these two *events* are concerned, the universe of thought is subdivided into exactly four combinations and each of these combinations can be absolutely denied.

FOUR COMBINATIONS OF CLASSES OF EVENTS, THEMSELVES ALSO CLASSES

(a) $ri = ir = 0$ It never happens that it rains and I stay inside.
(b) $r\bar{i} = \bar{i}r = 0$ It never happens that it rains and I do not stay inside.
(c) $\bar{r}i = i\bar{r} = 0$ It never happens that it does not rain and I stay inside.
(d) $\bar{r}\bar{i} = \bar{i}\bar{r} = 0$ It never happens that it does not rain and I do not stay inside.

What is said above about other possible combinations in relation to things also applies here in relation to events. Each of the eight sequences in §6.2.1 can be associated with one absolute denial. For example, (1), "When(ever) it rains, I stay inside," absolutely denies that it ever rains while I am not inside. Absolute denials add great sharpness to the sequences in question. Venn (1894: 159) praises the advantages of absolute denials:

> The proposition "All x is y" may be written, and for the purposes in hand is better written, "No x is not-y" ($x\,(1 - y) = 0$, or more briefly $x\bar{y} = 0$, ...) that is, it empties the compartment $x\bar{y}$, or declares the non-existence of a certain combination, viz. of things which are the combination of x and not-y.

The eight sequences in §6.2.1 can be associated with four denials of combinations. Each denial pertains to two sequences. Each of the two sequences of a pair corresponds to the same denial. The two sequences are therefore *identical in logical purport*; they convey the same logical message. Both members deny the existence of the same combination. Sequences and denials correspond to one another:

EIGHT SEQUENCES AND FOUR DENIALS

(1) When(ever) it rains, I stay inside. *Corresponds to*

(b) $r\bar{i} = 0$ It never happens that it rains and I do not stay inside.

(2) When(ever) it does not rain, I stay inside. *Corresponds to*
(d) $\bar{r}\bar{i} = 0$ It never happens that it does not rain and I do not stay inside.

(3) When(ever) it rains, I do not stay inside. *Corresponds to*
(a) $ri = 0$ It never happens that it rains and I stay inside.

(4) When(ever) it does not rain, I do not stay inside. *Corresponds to*
(c) $\bar{r}i = 0$ It never happens that it does not rain and I stay inside.

(5) When(ever) I stay inside, it rains. *Corresponds to*
(c) $i\bar{r} = 0$ It never happens that I stay inside and it does not rain.

(6) When(ever) I do not stay inside, it rains. *Corresponds to*
(d) $\bar{i}\bar{r} = 0$ It never happens that I do not stay inside and it does not rain.

(7) When(ever) I stay inside, it does not rain. *Corresponds to*
(a) $ir = 0$ It never happens that I stay inside and it rains.

(8) When(ever) I do not stay inside, it does not rain. *Corresponds to*
(b) $\bar{i}r = 0$ It never happens that I do not stay inside and it rains.

In terms of logical purport, these eight sequences represent four pairs whose two members are identical.

FOUR PAIRS OF SEQUENCES,
EACH PAIR EXHIBITING BOTH MEMBERS OF IDENTICAL PURPORT

I.	(1) = (8) $r\bar{i}$	=	$\bar{i}r$	$= 0$
II.	(2) = (6) $\bar{r}\bar{i}$	=	$\bar{i}\bar{r}$	$= 0$
III.	(3) = (7) ri	=	ir	$= 0$
IV.	(4) = (5) $i\bar{r}$	=	$\bar{r}i$	$= 0$

6.2.3 Conclusive and inconclusive derivations of premises from conditions

Deriving premises from conditions is achieved by making assumptions about a condition. There are two possible assumptions: a condition is either true or false—that is, it equals either 1 or 0. Each of the eight sequences listed in §6.2.1 contains a condition. Each condition can be assumed to be either true (1) or false (0). There are therefore sixteen ways of deriving premises from conditions, but eight of the sixteen derivations are inconclusive.

Examples of a conclusive and an inconclusive derivation now follow. Consider sequence (1), "When(ever) it rains, I stay inside." Two derivations in terms of premises can be achieved by taking the condition "It rains" (r) to be either true (1) or false (0). To make matters simpler, the algebraic formula of absolute denial favored by Venn will be chosen here for the substitution of 1 and 0. "When(ever) it rains, I stay inside" is logically equivalent to the absolute denial $r\bar{i} = 0$. Let us first assume, then, that it is true that it rains. "It rains" (r) is replaced by 1 if r is true and by 0 if r is false. Substituting 1 for r in $r\,(1 - i)$, we obtain:

$$1\,(1 - i) = 0;$$

$$1 - i = 0;$$

$$1 = i;$$

$$i = 1.$$

The interpretation is as follows. If it is true that it rains ($r = 1$), then it is also true that I stay inside ($i = 1$). That is a necessary consequence, in contrast to a contingent statement. This specific derivation of premises from conditions is therefore conclusive.

An example of an inconclusive derivation is as follows. Consider again "When(ever) it rains, I stay inside." The corresponding denial is again $r\bar{i} = 0$, or $r\,(1 - i) = 0$, that is, "It never happens that it rains (r) and I do not stay inside ($\bar{i}$)." Let us assume this time that it is *not* true that it rains: $r = 0$. Substituting 0 for r in $r\,(1 - i)$, we obtain:

$$0 \times (1 - i) = 0;$$

$$0 = 0.$$

Zero equals zero ($0 = 0$) is logically inconclusive. Indeed, anyone heeding the directive "Whenever it rains, I stay inside" can choose to be either inside or outside if it is a fact that it is not raining. There is no necessary consequence in this case. This specific derivation of premises from conditions is therefore inconclusive.

6.2.4 Sixteen derivations of premises from conditions

To serve the cause of transparency, the sixteen derivations may now be listed in full. Completeness should reveal the exact nature of the derivations in a manner that is "unembarrassed by exception or failure" (Boole 1854: 92). The eight sequences of condition and contingent statement are:

EIGHT SEQUENCES OF CONDITION AND CONTINGENT STATEMENT

(1) When(ever) it rains, I stay inside.
(2) When(ever) it does not rain, I stay inside.
(3) When(ever) it rains, I do not stay inside.
(4) When(ever) it does not rain, I do not stay inside.
(5) When(ever) I stay inside, it rains.
(6) When(ever) I do not stay inside, it rains.
(7) When(ever) I stay inside, it does not rain.
(8) When(ever) I do not stay inside, it does not rain.

In each sequence, the condition can be considered to be either true (1) or false (0). In sequences (1)–(4), *r* is the condition; in (5)–(8), *i* is the condition. For the sake of transparency, the replacements of *r* and *i* by either 1 (true) or 0 (false) will again be made by using the corresponding denials. That means that the eight sequences will be rearranged in four pairs whose two members correspond to the same denial:

I	=	(1) + (8)
II	=	(2) + (7)
III	=	(3) + (6)
IV	=	(4) + (5)

The two members of each pair have identical logical purport. The sixteen derivations are shown in the table at the top of p. 88; bold type marks the eight conclusive derivations.

6.2.5 The eight conclusive derivations of premises from conditions

Of the sixteen derivations in §6.2.4, the eight conclusive ones are shown in the table at the bottom of p. 88.

This is the interpretation of the first line: Given sequence (1), "When(ever) it rains, I stay inside," if it is assumed that it is in fact now raining ($r = 1$), then it follows that I am now staying inside ($i = 1$). The other seven lines can be interpreted in the same way.

§6.2.4	I. Sequences (1) and (8), corresponding to the denial $r(1 - i) = 0$			
I.1	$\boldsymbol{r = 1}$	therefore $1(1 - i) = 0$	or $1 - i = 0$	or $\boldsymbol{i = 1}$
I.2	$i = 1$	therefore $r(1 - 1) = 0$	or $r0 = 0$	or $0 = 0$
I.3	$r = 0$	therefore $0(1 - i) = 0$		or $0 = 0$
I.4	$\boldsymbol{i = 0}$	therefore $r(1 - 0) = 0$	or $r1 = 0$	or $\boldsymbol{r = 0}$
	II. Sequences (2) and (6), corresponding to the denial $(1 - r)(1 - i) = 0$			
II.1	$r = 1$	therefore $(1 - 1)(1 - i) = 0$	or $0(1 - i) = 0$	or $0 = 0$
II.2	$i = 1$	therefore $(1 - r)(1 - 1) = 0$	or $(1 - r)0 = 0$	or $0 = 0$
II.3	$\boldsymbol{r = 0}$	therefore $(1 - 0)(1 - i) = 0$	or $1 - i = 0$	or $\boldsymbol{i = 1}$
II.4	$\boldsymbol{i = 0}$	therefore $(1 - r)(1 - 0) = 0$	or $1 - r = 0$	or $\boldsymbol{r = 1}$
	III. Sequences (3) and (7), corresponding to the denial $ri = 0$			
III.1	$\boldsymbol{r = 1}$	therefore $1i = 0$		or $\boldsymbol{i = 0}$
III.2	$\boldsymbol{i = 1}$	therefore $r1 = 0$		or $\boldsymbol{r = 0}$
III.3	$r = 0$	therefore $0i = 0$		or $0 = 0$
III.4	$i = 0$	therefore $r0 = 0$		or $0 = 0$
	IV. Sequences (4) and (5), corresponding to the denial $(1 - r)i = 0$			
IV.1	$r = 1$	therefore $(1 - r)i = 0$; or $(1 - 1)i = 0$		or $0 = 0$
IV.2	$\boldsymbol{i = 1}$	therefore $(1 - r)1 = 0$; or $1 - r = 0$		or $\boldsymbol{r = 1}$
IV.3	$\boldsymbol{r = 0}$	therefore $(1 - 0)i = 0$; or $1i = 0$		or $\boldsymbol{i = 0}$
IV.4	$i = 0$	therefore $(1 - r)i = 0$; or $(1 - r)0 = 0$		or $0 = 0$

§6.2.5	The sequence	The premise	The necessary conclusion
I.1	If in sequence (1)	$\boldsymbol{r = 1}$	then $\boldsymbol{i = 1}$
II.3	If in sequence (2)	$\boldsymbol{r = 0}$	then $\boldsymbol{i = 1}$
III.1	If in sequence (3)	$\boldsymbol{r = 1}$	then $\boldsymbol{i = 0}$
IV.3	If in sequence (4)	$\boldsymbol{r = 0}$	then $\boldsymbol{i = 0}$
IV.2	If in sequence (5)	$\boldsymbol{i = 1}$	then $\boldsymbol{r = 1}$
II.4	If in sequence (6)	$\boldsymbol{i = 0}$	then $\boldsymbol{r = 1}$
III.2	If in sequence (7)	$\boldsymbol{i = 1}$	then $\boldsymbol{r = 0}$
I.4	If in sequence (8)	$\boldsymbol{i = 0}$	then $\boldsymbol{r = 0}$

6.2.6 Conclusion: Derivation of premises from conditions is a definite process

Closer inspection of the equivalences in §6.2.5 shows that, first of all, each of the eight sequences of condition and contingent statement corresponds to one specific sequence of premise and result. But second, importantly, each sequence also corresponds to a *different* sequence of premise and result. The fact that each derivation of premises from conditions has a *different* result makes the process of deriving premises from conditions into a *definite* process. There are in fact sixteen derivations of premises from conditions, but eight of them, the ones producing the equation 0 = 0, are inconclusive. Thus eight derivations are conclusive, and they all differ from one another; each relates to one and only one sequence of condition or protasis and contingent statement or apodosis. Again, the process of derivations is fully definite.

The present line of argument now reaches a crucial stage. The time has come to invert the process, namely to derive conditions from premises instead of premises from conditions. The question arises: Is the inverse process of deriving conditions from premises also definite? Does each possible premise lead back to one and only one condition? It will be shown that it does and that the process of deriving conditions from premises is therefore definite.

6.3 FROM PREMISE TO CONDITION

6.3.1 Two events correspond to eight sequences of premise and result or consequence

Two events can be combined into eight possible sequences of condition, or protasis, plus contingent statement, or apodosis, as listed in §§6.2.1 and 6.2.4. There are also eight possible sequences of premise plus consequence. A premise is an assumption that something is true or real. Therefore, an event serving as an assumption must involve reference to a certain tense, namely past, present, or future. The present tense is chosen here.

EIGHT SEQUENCES OF PREMISE PLUS CONSEQUENCE

(1) If (it is the case that) it is raining,
then I am (in fact) staying inside.

(2) If (it is the case that) it is not raining,
then I am (in fact) staying inside.

(3) If (it is the case that) it is raining,
then I am (in fact) not staying inside.

(4) If (it is the case that) it is not raining,
then I am (in fact) not staying inside.
(5) If (it is the case that) I am staying inside,
then it is (in fact) raining.
(6) If (it is the case that) I am not staying inside,
then it is (in fact) raining.
(7) If (it is the case that) I am staying inside,
then it is (in fact) not raining.
(8) If (it is the case that) I am not staying inside,
then it is (in fact) not raining.

6.3.2 Derivation of premises from conditions: A definite process

If the derivation of condition from premise is definite, one and only one sequence of condition and contingent statement should correspond to each sequence of premise and consequence. It appears that the eight sequences of premise and result listed in §6.3.1 are the same as those listed in the table of conclusive derivations of premises from conditions in §6.2.5, in one-to-one correspondence. Importantly, the table in §6.2.5 clearly reveals that each sequence of premise and consequence corresponds to *only one* sequence of condition and contingent statement. *It may therefore be considered proven that deriving conditions from premises is just as definite as deriving premises from conditions.*

6.4 FROM CONDITION TO PREMISE AND BACK: DERIVATIONS CHANNELED THROUGH FOUR PAIRS

There are exactly eight sequences of condition plus contingent statement, and exactly eight sequences of premise plus consequence. It has been stated so far that each sequence of the first kind corresponds to one and only one sequence of the second kind, but that statement is not yet precise enough. It needs to be further refined.

Upon closer inspection, it appears that each sequence of the first kind corresponds to *two* sequences of the second kind, and vice versa. In other words, each *pair* of sequences of condition and contingent statement corresponds to one and only one sequence of premise and result. And each *pair* of sequences of premise and result corresponds to one and only one sequence of condition and contingent statement. Does this mean that the processes of deriving conditions from premises and premises from conditions are not definite after all? No, it does not—because the two members of all the pairs concerned are in fact logically equivalent.

Take, for example, two §6.2.1 sequences of condition and contingent statement, (1) "When(ever) it rains, I stay inside" and (8) "When(ever) I do not stay inside, it does not rain," which are logically equivalent and convey the same message. Both sequences correspond to the same absolute denial, "It never happens that it rains and I do not stay inside" (see §6.2.2).

Furthermore, both of these sequences lead to the two §6.3.1 sequences of premise and result (1) "If (it is the case that) it is raining, then I am (in fact) staying inside" and (8) "If (it is the case that) I am not staying inside, then it is (in fact) not raining." In the other direction, the two sequences of premise and result can produce two sequences of condition and contingent statement. However, the two sequences of condition and contingent statement are fully equivalent in purport. The definiteness of the derivations of premises from conditions is therefore in no way affected.

In conclusion, as a result of the above refinement, the definiteness of the two processes of deriving premises from conditions and of deriving conditions from premises is now better described as follows. Each *pair* of §6.2.1 sequences of condition and contingent statement corresponds to one and only one *pair* of §6.3.1 sequences of premise and consequence, and vice versa.

Four absolute denials serve as the *interface* between the two sets of four pairs: (a) $ri = 0$, (b) $r\bar{i} = 0$, (c) $\bar{r}i = 0$, and (d) $\bar{r}\bar{i} = 0$ (§6.2.2). The derivations in both directions, from conditions to premises and from premises to conditions, are channeled through these four options, as it were. In one direction, eight conditions shrink to four denials, and these four denials expand into sixteen derivations involving premises. But only eight of these sixteen derivations are conclusive. In the other direction, each sequence of premise and result corresponds to two sequences of condition and contingent statement, but the latter two sequences are fully equivalent; in that sense, the two sequences are just a single sequence.

In conclusion, it has been shown how the derivations from condition to premise and from premise to condition are strictly controlled by the invariable laws of thought. Both are one hundred percent definite.

6.5 CONCRETE APPLICATION: THE DIFFERENCE BETWEEN MARK 9:47 AND MATTHEW 18:9

We repeat the following from §6.1.3.

CONDITION:

ⲉⲣϣⲁⲛ ⲡⲉⲕⲃⲁⲗ ⲥⲕⲁⲛⲇⲁⲗⲓⲍⲉ ⲙ̄ⲙⲟⲕ ⲡⲟⲣⲕ̄ϥ
eršan (Gk. ἐὰν) *pe=k-bal skandalize mmo=k pork=f* (Mark 9:47)

'When(ever) your eye causes you to sin, pluck it out'.

PREMISE:

ⲉϣϫⲉ ⲡⲉⲕⲃⲁⲗ ⲥⲕⲁⲛⲇⲁⲗⲓⲍⲉ ⲙ̄ⲙⲟⲕ ⲡⲟⲣⲕ̄ϥ
ešje (Gk. εἰ) *pe=k–bal skandalize mmo=k pork=f* (Matthew 18:9)

'If (it is the case that) your eye is causing you to sin,
pluck it out'.

The two are tantalizingly close, yet, they are separated by a well-defined distance. The steps needed to bridge this distance are described above. By taking these steps, the first sentence can be converted into the second and the second back into the first. Both processes are definite. The first sentence inevitably leads to the second and the second inevitably leads to the first. The definiteness of these processes creates a bond between the two sentences, which goes a long way toward explaining why the two alternate as expressions of the same thought. However, this bond is not one of synonymity. There is a semantic gap between the two. This gap can be defined in the sharpest of terms. In modern Bible translations, the difference between the two is not expressed. For example, in the New Revised Standard Version, the translation is uniformly "If your eye causes you to stumble, tear it out." "If" seems to denote both conditions and premises in English (see §0.11). This is a problem of English grammar, however; but "when(ever)" unambiguously denotes a condition, "if it is the case that" unambiguously denotes a premise.

7 THE *CONDITIO SINE QUA NON*

Conditio sine qua non is a familiar concept, but it is difficult to find a definition. The literal translation is 'condition without which not'; it is also known as a "necessary" condition. How exactly does the *conditio sine qua non* differ from the regular condition? Symbolic logic allows an answer to this question.

I have previously (Depuydt 1993a: 226 n. 28) quoted a definition of "condition" by John Wesley (*Works*, 10: 308) from the *Oxford English Dictionary* s.v., sense I 1 [a]:

> The word *condition* means neither more nor less than something *sine quâ non*, without which something else is not done.

I now no longer subscribe to the notion that condition and *conditio sine qua non* are the same. They are in fact rather opposites, as will be shown. Perhaps the most important statement about the relation between the necessary condition and the regular condition is that the two exhibit the same logical structure. Conditional sentences involve two events, allowing a four-way division of the universe. Both the necessary condition and the regular condition involve the nonexistence of just one of these four combination classes.

What, then, is the difference between the necessary condition and the regular condition? Both involve a relation between two events or occasions. On the surface, the difference between the two relations is one of linguistic expression. Compare the regular condition in "When it is sunny, I lie at the beach" with the necessary condition in "Only when it is sunny do I lie on the beach" (or: "I lie on the beach only when it is sunny"). Where the two part ways is in that the nonexistent class is different.

(1) "When it is sunny, I lie on the beach" corresponds to $s\bar{b} = 0$: It does not happen that it is sunny (s) and I do not lie at the beach ($\bar{b}$).

(2) "Only when it is sunny do I lie on the beach" corresponds to $\bar{s}b = 0$: It does not happen that it is not sunny ($\bar{s}$) and I lie at the beach (b).

In (2), it needs to be sunny for me to lie on the beach. That is a necessary condition. I cannot lie on the beach when it is *not* sunny. By contrast, in (1), lying on the beach when it is not sunny is possible. Again, in (2), if I am going to lie on the beach, the sun has to shine. That does not mean that I *have to* lie on the beach when it is sunny. By contrast, in (1), I do have to lie on the beach when it is sunny.

The three existing combination classes in the two cases are:

(1) $sb + \bar{s}b + \bar{s}\bar{b} = 1$. That is, either sb or $\bar{s}b$ or $\bar{s}\bar{b}$ is the case.

(2) $sb + s\bar{b} + \bar{s}\bar{b} = 1$. That is, either sb or $s\bar{b}$ or $\bar{s}\bar{b}$ is the case.

The following diagrams allow optimal visualization of the difference between (1) and (2).

(1) *s* as a regular condition (2) *s* as a necessary condition

$s\bar{b} = 0$ $\bar{s}b = 0$

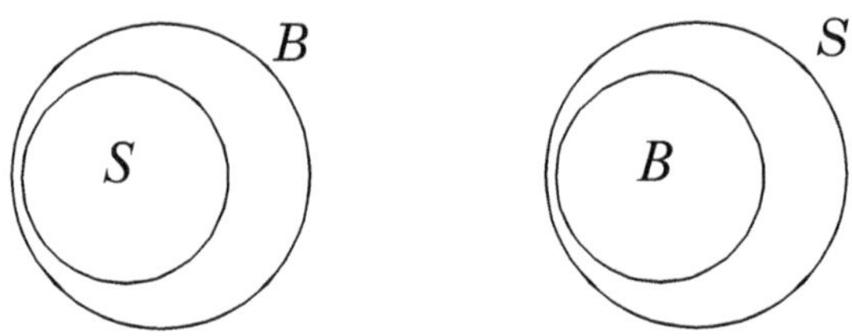

These diagrams reveal that (1) is in a sense the opposite of (2). In (2), the circle *B*, which encompasses all the occasions when I am at the beach, is entirely encompassed by the circle *S*, which encompasses all the occasions when the sun shines. In other words, *b* does not exist outside *s*. That makes *s* into the necessary condition of *b*.

It is now also possible to convert (1) into its logical equivalent in (2) and (2) into its equivalent in (1). Logical equivalence means that the same combination class is equated to zero (= 0). The logical equivalent of "When it is sunny, I lie on the beach" in (1) is "Only when I lie on the beach is it sunny" in (2). In both cases, $s\bar{b} = 0$. The equivalent of "Only when it is sunny do I lie at the beach" in (2) is "When I lie on the beach it is sunny" in (1). In both cases, $\bar{s}b = 0$.

If "Only when it is sunny do I lie on the beach" contains a necessary condition, then how does one express the equivalent of a balanced sentence? "It is sunny, I lie on the beach" has been suggested above. Another possible formulation is "I lie at the beach only and always when it is sunny."

Klenk (1983: 36) uses the term "biconditional" to describe the sentence

> "Mary will be elected student body president
> if and only if Fred is elected treasurer."

I am not sure, however, that this is a "two-way conditional" in the sense that, "If Mary is elected student body president, then Fred is elected, *and* that if Fred is elected, then Mary will be elected." Rather, it seems to me that Fred's election is only a necessary condition. That means that Fred can be elected without Mary being elected. A biconditional might more unambiguously be expressed as "Mary will be elected only when and always when Fred is elected," adding the specification "and always."

Finally, "Only when the sun shines do I lie at the beach" conveys the same logical purport as "Unless the sun shines, I do not lie at the beach." Obviously, lying at the beach is presented affirmatively in the first sentence and negatively in the second sentence. But what is the difference in final terms? A look at the corresponding diagram on the opposite page in which circle *B* is entirely encompassed by circle *S* readily reveals that there are two ways of describing my lying at the beach. On the one hand, all the occasions on which I lie at the beach fall inside circle *S*. This corresponds to "Only when the sun shines do I lie at the beach." On the other hand, all the occasions on which I do *not* lie at the beach lie outside circle *S*. This corresponds to "Unless the sun shines, I do not lie at the beach." "Unless" derives from "upon less than." "Upon less than" *S* can be understood as "outside" or "except" *S*.

8 THE BALANCED SENTENCE: COLLECTION OF EXAMPLES

8.0 PURPOSE OF THE COLLECTION AND CRITERIA OF CLASSIFICATION OF THE EXAMPLES

8.0.1 Exclusive writings and distinctive writings of verb forms

This chapter provides as many examples of the balanced sentence in Egyptian as I have been able to find, most of them already well known and none of them new to the literature. Most examples are followed by some "references to discussions," which are not exhaustive but include the more detailed and more significant ones.

The main principle of classification of the examples is the degree of distinctiveness of the verb forms they contain. That distinctiveness pertains to the substantival character of the two verbs paired in a balanced sentence. It is the bane of grammarians of Old and Middle Egyptian that the writings of verb forms are mostly ambiguous. An unambiguous writing of a verb form is one that points to one and only one form. Old and Middle Egyptian do exhibit such unambiguous writings, many of them belonging to compound verb forms, such as that transcribed as *ḫr.f sḏm.f.* Others concern verb forms with affixes, such as *sḏm.k3.f* and *sḏm.n.tw.f*, the latter verb form exhibiting *two* affixes, *n* and *tw*.

The writing of most Old and Middle Egyptian verbs is not unambiguous, however. Among the countless ambiguous writings of Old and Middle Egyptian verbs, a distinction is possible between more distinctive and less distinctive writings. All writings are distinctive to some degree, even if only minimally so. Consider *sḏm.n.tw.f*, the unambiguous character of which was established only as recently as 1957 (Polotsky 1957 [1971: 43–51]), when *sḏm.n.tw.f* was identified as passive, past, and substantival. It follows that every other verb form of Old and Middle Egyptian is distinctive at the very least in *not* being *sḏm.n.tw.f.* Complete absence of distinctiveness therefore does not exist. Even the absence of explicit written characteristics is distinctive: absence of features is a characteristic. In that capacity, absence is called zero in linguistics. As opposed to nothing, zero is a *significant* absence that contributes to the recognition of a verb form.

Still, there are marked differences in the degrees to which verb forms are distinctive. Some writings are highly distinctive, offering as few as two options. Moreover, criteria other than the writing of the verb may narrow the choice to just a single option. In that sense, the identification of a highly distinctive writing may on occasion be nearly as secure as the identification of an unambiguous writing.

Below, the term "distinctive writing" denotes a *highly* distinctive writing, one that along with other criteria leaves little or no doubt about the identification of a verb form.

8.0.2 *Sḏmm.f* is always (1) passive, (2) future, and (3) substantival

8.0.2.1 Sḏmm.f and sḏm.n.tw.f as the only two exclusive writings of noncompound substantival verb forms

There are quite a few writings of substantival verb forms that are highly distinctive in balanced sentences. These writings together with their context leave little or no doubt that the verb forms in question are substantival (§§8.2–8.4). But are there any writings that, all by themselves, without considering the context at all, *exclusively* mark verb forms as substantival (whatever their syntactic context)? Among noncompound verb forms, I can think of only two features that are exclusively substantival, each of them marking the writing of only one verb form: the combination of the affixes *n* and *tw*, which marks the verb form *sḏm.n.tw.f* as substantival, and gemination in strong verbs, which marks the verb form *sḏmm.f* as substantival.

Sḏm.n.tw.f is characterized by two affixes, *n* and *tw*. There are many *sḏm.f* forms but there appears to be only one *sḏm.n.tw.f* form. It exhibits the three coordinates (1) passive, (2) past, and (3) substantival (for a full catalogue of all the coordinates of the verb, see Depuydt 1996a; for a tightened version, see Depuydt 1999d: 331–564). This does not mean that other forms could not also have exhibited the same three coordinates (for the possible predecessor of *sḏm.n.tw.f* in earlier Egyptian, see Polotsky 1976: 19 §2.6.3). However, no occurrence of *sḏm.n.tw.f* in a balanced sentence has come to my attention, so it is not relevant to the present investigation.

The other form that is exclusively substantival is *sḏmm.f*. As regards other coordinates, *sḏmm.f* is always both passive and future. A writing such as *jr.tw.f*, characterized by absence of gemination and the affix *tw*, might also be passive, future, and substantival, but not exclusively so—it could also be passive, aorist, and adverbial. There are many *sḏm.f* forms, but apparently only one *sḏmm.f* form. The form *sḏmm.f* is characterized uniquely by the combination of two features, (1) gemination in strong verbs and (2) suffix conjugation.

The following statements by Edel and Allen regarding the form classes in which *sḏmm.f* is attested are worth noting. Edel (1955/64: 264 §557) on Old Egyptian in general:

> The repetition of the last, strong, consonant is attested with certainty only with biliteral and triliteral verbs and with fourth-weak and causative verbs.*

And also:

> It is not certain and in fact not very probable that second-doubling verbs can build a geminating passive.†

As regards the Pyramid texts specifically, Allen (1984: 349 §515.B) writes:

> The geminating passive (*sḏmm.f*) occurs in conjugations with strong or geminating final radicals and in the conjugation of quart[ae]-inf[irmae] prefixing verbs.

There are two additional types of gemination in Old and Middle Egyptian (Depuydt 1996a: 94–98 §§129–35, 1999d: 440–42 §§5.185–89). By far the most frequent is the well-known type that differentiates the active aorist substantival *mrr.f* 'that he likes' from the active aorist adverbial *mr.f* 'while he likes'. The last type occurs in the past passive participle of biliteral verbs, as in the form *ḏdḏt* 'that which has been said'.

A problem arises, however. No one has ever explicitly stated, as far as I know, that *sḏmm.f* is just one verb form and that that form is always passive, future, and substantival. Until recently, I have myself not gone any further than describing *sḏmm.f* as "mostly, if not always, substantival" (Depuydt 1996a: 98 §133, 1999d: 442 §5.188). But looking at all the evidence again, I can find nothing to suggest that it is ever anything but substantival. It is the passive counterpart of the active future *sḏmw.f*. In fact, Schenkel (1997: 195 §7.3.1.2.4) concludes much the same when he declares *sḏmm.f* to be the "passive counterpart" (*passivisches Pendant*) of the *sḏmw.f*. Schenkel otherwise considers *sḏmm.f* always to denote future tense (see most recently Schenkel 2004/5: 174). In earlier Egyptology, there was a widespread belief that

* Die Wiederholung des letzten starken Konsonanten ist mit Sicherheit nur bei den 2– und 3–rad[ikaligen] Verben, sowie bei den Verben IV. inf[irmae] und Kausativen belegt.

† Ob auch Verba II. gem[inatae] ein geminierendes Passiv bilden könnten, ist unsicher und wenig wahrscheinlich.

sḏmm.f can also denote past tense. However, Allen (1984: 358 §529) draws the following conclusion.

> In all its uses in the Pyramid texts, the *sḏmm.f* seems to have clear or at least applicable future tense, as Edel [1955/64: 266 §559] has noted.

More recently, Allen (2002: 91) has made reference to the "prospective *sḏm.f* ('*sḏmw.f*') and its passive counterpart (*sḏmw.f*/*sḏmm.f*)." It is my firm belief that the active future *sḏmw.f* also always exhibits the coordinate "substantival" in the verbal dimension of function, alongside the coordinate "passive" in the verbal dimension of voice and the coordinate "future" in the verbal dimension of time. But as far as I know, unequivocal declarations to this effect in works of grammar are rare. Perhaps this reflects a reluctance to generalize; in this connection, the following principle may be invoked.

8.0.2.2 *A principle of inference for establishing that* sḏmm.f *is always substantival: The search for a grammatical fingerprint*

What grounds are there for concluding that *sḏmm.f* is always substantival? The following logical consideration seems paramount. Consider the set of all the verb forms in all of Egyptian that have been found to mark emphasis on an adverbial phrase. The range of usage of the Old and Middle Egyptian members of the set is wider than the range of usage of the Late Egyptian, Demotic, and Coptic members. The former appear in most positions in which a substantive can be found and can therefore be called substantival. By contrast, the range of use of their descendants in Late Egyptian, Demotic, and Coptic has shrunk mainly to the role of marking emphasis on an adverbial phrase. The Late Egyptian and Demotic members are therefore usually called emphatic verb forms. "Emphatic" may be understood as "marking emphasis on an adverbial phrase." The Coptic members of this same set are usually called Second tenses in Coptic grammar, but they might just as well be called "emphatic."

Every verb form that has been securely determined to be a member of this set is assumed to *always* be substantival or emphatic. Indeed, the function of being substantival or emphatic is so specific that it is difficult to assume that a verb form could be substantival or emphatic and nonsubstantival or nonemphatic at the same time. An analogy may be adduced in support of this notion. It seems as if most everyone would accept that a verb form that expresses a certain tense *always* expresses that tense. Nothing is more natural to assume that, once it is established that a verb form denotes future tense, it *always* denotes future tense. Or, with regard to syntactic function, is there anyone who thinks that a verb form that is relative (adjectival) is not always relative? Substantival and adjectival verb forms have much in common, including gemination in third-weak forms in the aorist tense and the *sḏm.n.f* of verbs of motion. Why would substantival verb forms be any different from adjectival verb forms? Therefore, everyone agrees: *sometimes* adjectival, *always* adjectival. Then why not likewise: *sometimes* substantival, *always* substantival? What is the difference? How can a linguistic system function in which a verb form can express a certain meaning and its absence or opposite at the same time?

Yet, when it comes to a property such as substantival or emphatic, one observes much hesitancy among grammarians of Egyptian to generalize a conclusion that is obvious for many instances of a verb form to all the instances of that form. Such generalizations obviously remain logical inferences, rather than straight observations of fact. Principles of inference are, however, tacitly accepted by every student of Old and Middle Egyptian grammar; one such is the principle of parallelism (for a formulation and references, see §8.2.4; for this principle as an essential characteristic of Polotsky's method in Egyptian grammar, see Depuydt 2006b).

By the principle of parallelism, everyone accepts that the verb *sḏm* 'hear' exhibits both an active past participle ('who has heard') and an active present participle ('who is hearing'), because verbs such as *jr(y)* 'do, make' evidently exhibit two participles, namely *jr* 'who has done' and *jrr* 'who does'. Yet no one has ever seen the two participles of *sḏm* distinguished in writing. Via the principle of parallelism, a distinction that is observable in a sufficient number of verbs is expanded to *all* verbs. The logic is as follows. There are basically two theoretical choices. Either the distinction applies to all the other verbs or it does not. The two choices need to be carefully considered in light of all else that is known about Egyptian and about language in general. It seems everyone has always agreed that it makes little sense for a distinction such as active past versus active present participle to exist with some verbs but not with others. Absolute proof that *sḏm* 'hear' does indeed have both a past participle and a present participle may seem to be lacking. But then, in inductive and empirical reasoning, absolute proof is *never* available anyhow.

By a similar principle of inference, a certain distinction applying to some instances of a form may be expanded to *all* instances of a form. Thus, if a significant number of facts line up in favor of *sḏmm.f* being substantival and no facts contradict the proposition, then there are two choices: either *sḏmm.f* is *always* substantival, or it is sometimes substantival and sometimes not. One possible attitude is to avoid the choice altogether, but making no choice seems to favor the second option. Failing to choose leaves open the possibility that *sḏmm.f* is sometimes *not* substantival. Failing to choose is therefore not really failing to choose. Failing to choose also dodges the requirement to provide positive proof. It is by no means self-evident that a verb form can be both substantival and not substantival at the same time. This proposition requires proof in its own right. Then again, it takes a certain kind of confidence in the nature of systems, linguistic and otherwise, to accept that *sḏmm.f* is always substantival on the basis of an incomplete picture of the verb form.

In choosing between the two options, one needs to rely on circumstantial evidence. In evaluating the evidence, the following reasoning seems paramount. Late Egyptian, Demotic, and Coptic emphatic forms are descendants of Old and Middle Egyptian substantival forms. The later emphatic forms derive their function from the earlier substantival forms. The range of usages has shrunk, however. Emphatic function is just one of the many uses of Old and Middle Egyptian substantival verb forms, but Late Egyptian, Demotic, and Coptic emphatic forms are *as a rule* emphatic. They *as a rule* exercise the function that they derive from Old and Middle Egyptian substantival forms. In Late Egyptian, Demotic, and Coptic, the matter is clear. The grammar of those stages of the language is fairly transparent. Can one therefore with equal justification conclude that certain Old and Middle Egyptian verb forms are *always* substantival if they are clearly substantival in a large set of examples? It is difficult to see why the Old and Middle Egyptian substantival verb forms would differ in this respect from their emphatic descendants in Late Egyptian, Demotic, and Coptic. One observes a certain agnosticism in Old and Middle Egyptian grammar. There is a reluctance to draw the same inference that is made in Late Egyptian, Demotic, and Coptic grammar. Yet agnosticism is not as noncommittal as it may seem. Absence of a commitment is in itself a tacit admission of the possibility that the verb forms in question are sometimes not substantival.

Still, three explicitly written elements are widely interpreted as marking verb forms as substantival. (1) The combination of the affixes *n* and *tw* in the verb form *sḏm.n.tw.f* '(the fact) that he has been heard', a form that is always passive, past, and substantival. Most everyone who accepts that this form is substantival also accepts that it is *always* substantival. Engsheden (2003: 336) has recently described the consensus on this matter:

> The form *sḏm.n.tw=f* exhibits the characteristic, rare among verb forms of classical Egyptian, of not being the subject of controversy among grammarians. Since the demonstrations of Polotsky [1957 (1971: 43–51)], everyone agrees to regard it as the passive counterpart of the substantival form *sḏm.n=f*, which consolidates by itself the notion of [substantival *sḏm.n=f*] as an entity in its own right.*

Engsheden (ibid. n. 1274) also notes that only a single discordant voice is on record as of now, that of Chris Reintges.

(2) Gemination in third-weak verbs, as in the forms *mrr.f* and *jrr.f*. It is generally assumed that these writings always mark either substantival or adjectival (relative) forms. By contrast, *sḏm.n.tw.f* can *only* be substantival. That is because there are no passive adjectival (relative) forms. In the 1940s, 1950s, and 1960s, balanced sentences constituted an obstacle on the road to interpreting writings such as *mrr.f* as always nominal, that is, substantival or adjectival. However, Polotsky's explanation of the balanced sentence as a nominal sentence consisting of two substantival components each containing a substantival verb form removed this obstacle. (3) The affix *n* with verbs of motion, as in *pr.n.f*, also either substantival or adjectival.

No one has seen every instance of these three features. An impressive list of indications can be compiled that all point in the same direction, namely that they mark verb forms as substantival. Furthermore, no compelling argument seems to exist that falsifies this proposition. The only remaining obstacle to preventing anyone from accepting that the written forms in question are always substantival is the theoretical possibility that a single verb form could sometimes be substantival and sometimes not substantival. However, it seems that the burden of proof rests on those who claim that there is such a thing as a verb form that is both substantival and not substantival at the same time. Everything else we know about Egyptian grammar of all the stages of the language seems to argue against such an assumption.

An alternative to the conclusion just proposed is the assumption that there are *two* forms written *sḏm.n.tw.f*, *two* non-adjectival (nonrelative) forms written *mrr.f*, or *two* non-adjectival forms written *pr.n.f*. It is undeniable that, in hieroglyphic writing, different forms are often written alike, but the proposition that any of the three written forms denotes *two* verb forms is by no means self-evident. It would need to be proven in its own right.

This alternative conclusion is only possible because hieroglyphic writing is defective, which may be relevant to the widespread unwillingness that I perceive among grammarians of Egyptian to accept that the three features listed above *always* mark verb forms as substantival (whenever the verb forms are not adjectival). Grammarians of Old and Middle Egyptian know that much of the language cannot be seen in the script. That fact encourages

* La forme *sḏm.n.tw=f* a la qualité, rare chez les formes de l'égyptien classique, de ne pas être un sujet de controverse entre les grammairiens. Depuis les démonstrations de Polotsky [1957 (1971: 43–51)], on s'accorde à y voir la contrepartie passive de la forme substantive *sḏm.n=f*, renforçant en soi l'idée de cette dernière comme entité indépendante.

an attitude of agnosticism. However, the verbal system is limited in size. Each form must have a well-defined place for the system to function. In my opinion, every Old and Middle Egyptian verb form that there is to know is now known. No new generic ones are likely to be discovered. Assuming that there are *two* non-adjectival verb forms written *sḏm.n.tw.f*, *mrr.f*, or *pr.n.f* would necessitate a considerable expansion of the master-list of generic verb forms.

Quite a few grammars of Middle Egyptian have appeared in recent years. Explicit statements to the effect that writings such as *sḏm.n.tw.f*, *mrr.f*, and *pr.n.f* are always substantival (when they are not adjectival, which *sḏm.n.tw.f* cannot be because it is passive) are difficult to find. As a result, beginning students receive little encouragement to accept that the verb forms in question are always substantival. I myself do not doubt that the three writings mark substantivalness. I would also without hesitation adopt the active future *sḏmw.f* into the substantival family, with Polotsky.

In this connection, Allen (1984: 2 §3) writes:

> Syntactically, the *sḏmw.f seems* to supply the nominal/"emphatic" counterpart of the "prospective" *sḏm.f*. [emphasis added]

But Allen stops short of accepting that *sḏmw.f* is *always* substantival. Instead, he assumes that his "prospective" *sḏm.f* "also has nominal [that is, substantival] uses, [namely] as the object of verbs and prepositions." In other words, the notion that another future *sḏm.f* might also have substantival function somehow undermines the notion that the future *sḏmw.f* is *always* substantival. How could there be two substantival verb forms of the future tense? First of all, it needs to be noted that Allen's view contains no arguments *against* *sḏmw.f* being always substantival. What about Allen's belief that the other future *sḏm.f* (the one exhibiting characteristic forms such as *jwt.f*, *jnt.f*, and *m3n.f*) can sometimes be substantival? The grammatical fingerprint of the other future *sḏm.f* form is not obviously substantival. There is one verb of which it is generally thought to be the direct object, *rd(y)* 'give'. *Rd(y)* is then translated in English as 'cause, let' despite its literal meaning (cf. Depuydt 1993b: 28), but *rd(y) sḏm.f* is therefore more likely to mean, at least in origin, 'give *so that* he might hear' (for an argument from Akkadian, see Depuydt 1996c). That interpretation makes the future *sḏm.f* in question adverbial rather than substantival. In my opinion, that other future *sḏm.f* form is *always* adverbial. Historically speaking, adverbial verb forms presumably evolved from independent verb forms (Depuydt 1995c).

8.0.2.3 Three features evidencing the substantivalness of sḏmm.f

The purpose of this section is to expand the substantival family with another member, *sḏmm.f*. *Sḏmm.f* occurs only in Pyramid texts, which represent the oldest attested stage of Egyptian, and in Coffin texts derived from Pyramid texts (Edel 1955/64: 264 §556). *Sḏmm.f* has been described in detail by Edel (§§556–69) and especially by Allen (1984: 348–58 §§515–29). I refer to their accounts for examples in addition to those cited below.

The grammatical fingerprint that *sḏmm.f* has left in the texts is unmistakably substantival. Substantivalness is so specific as a property that it seems difficult to suppose that a single verb form could both possess it and not possess it at the same time. Being partly substantival would be like being partly pregnant. The only alternative would be to suppose that there are *two* *sḏmm.f* forms, but such an assumption would itself not be without considerable consequences.

The time seems ripe to induct *sḏmm.f* into the company of substantival verb forms as the passive future substantival verb form. Gemination in the suffix conjugation of strong verbs is then an exclusive writing of this substantival verb form. To my knowledge, the only other written feature that exclusively marks noncompound verb forms as substantival is the combination of the affixes *n* and *tw* in the form *sḏm.n.tw.f*. (I leave compound substantival forms such as *wnn.f ḥr sḏm* out of consideration.) Other written features that are typical of substantival verb forms such as gemination in third-weak verbs, as in *mrr.f*, mark forms as either substantival or adjectival (relative). That fact makes these features less than exclusively substantival.

There are at least three indications that *sḏmm.f* is substantival, and indications to the contrary seem absent. Two of them mark substantivalness *indirectly*, which may help explain why no one has proposed that *sḏmm.f* is *always* substantival, and which may also obscure the substantivalness to casual observers. Of the three features that identify *sḏmm.f* as a substantival form, only one does so directly. One direct feature is not much. The substantival character of *sḏmm.f* does not impose itself with overwhelming immediacy, but closer inspection reveals that every indication points in the direction of substantivalness. In his recent study of the traditional Egyptian of texts written in 400–30 B.C.E., Engsheden (2003: 306) intriguingly classifies *sḏmm.f* as a passive "temps second" in a table, though without any further comment.

8.0.2.3a The first feature: Sḏmm.f as a substantival form in the balanced sentence

The first feature, the direct one, concerns the appearance of *sḏmm.f* in balanced sentences (§8.1). One example (Edel 1955/64: 269 §569; Allen 1984: 352 §520) may suffice at present.

šnṯṯ N šnṯṯ Tm (Pyr. §492b)

'When N will be reviled, Atum will be reviled'.

Šnṯ 'shun (or the like)' is a triliteral verb. *Šnṯṯ* is therefore an instance of the form *sḏmm.f*. Translation as a passive is obvious—the verb is transitive and there is no direct object. It seems widely accepted that verb forms appearing in balanced sentences are *always* substantival. *Sḏmm.f* appears in balanced sentences. This *sḏmm.f* is therefore substantival. If a verb form cannot be substantival *and* not substantival, this example alone suffices to declare *sḏmm.f* always substantival. Other examples of the balanced sentence (§8.1) then only strengthen this conclusion. However, it will be good to cite additional examples and also to look beyond the balanced sentence for additional features; the two following point indirectly to the substantivalness of *sḏmm.f*.

8.0.2.3b The second feature: Sḏmm.f as a substantival verb form in the cleft sentence

The first of the two additional, indirect, features is an occurrence of the *sḏmm.f* in what looks like a cleft sentence.

nṯrw nb rd.t(y).sn nfr rwḏ mr pn … (Pyr. §§1650a, 1651c N;
jntjsn stpp n.sn stpwt.sn … M has *ntjsn*)

'As for all gods who will cause this pyramid to be good and strong …
it is they for whom (*or:* it is for them that) their choice cuts will be chosen'.

Edel, in his grammar of Old Egyptian (1955/64), does not classify the cleft sentence as a structure in its own right. Consequently, isolating contrast ('It is I who …') is not sharply distinguished from distinctive contrast ('As for me, I …') in his grammar.

When the emphasized element is a noun, the cleft sentence begins with the particle *jn*. Such a sentence is unambiguously identifiable as a cleft sentence, which is often called a "participial statement" in Egyptian grammar. When the emphasized element is a personal pronoun, however, cleft sentences begin with an independent pronoun (*jnk*, *ntk* [or earlier *ṯwt*], *ntṯ* [or earlier *ṯmt*], *ntf* [or earlier *swt*], and so on), as in the generic example *jnk sḏm.j* 'It is I who will hear'.

It will be useful to consider an alternative interpretation of *jnk sḏm.j*, according to which *jnk* is topicalized; I call this "distinctive contrast," as opposed to the "isolating contrast" of cleft sentences (see Depuydt 1991a). According to this alternative interpretation, *jnk* is "topicalized" or placed in "anticipatory emphasis." Such emphasis can be expressed in English by "as for" or "as regards." The sentence above would then not be a cleft sentence at all, and the translation of *jntsn stpp n.sn stpwt.sn* would be something like 'As for them (on their part, as opposed to someone else to whom something else has happened), choice cuts are chosen for them'. But there are two arguments against interpreting *jntsn* as being "topicalized."

(1) *Jntsn* is preceded by a phrase that is surely "topicalized," namely *nṯrw nb* ... '(As for) every god ...'. If *jntsn* is "topicalized," then the sentence exhibits a kind of "double topicalization," with both topicalizations referring to the same entity. If *nṯrw* 'gods' is already "topicalized," why "topicalize" it yet once more by means of *jntjsn* 'they'? This would be awkward. Double "topicalization" is otherwise unattested in Egyptian. One would therefore need to postulate and prove the existence of the phenomenon just for the purpose of the present passage. By contrast, a cleft sentence does not "topicalize." The cleft sentence conveys a meaning of its own, namely contrastive emphasis ('It is they that ...'). This meaning does not clash with that of the "topicalization" of *nṯrw* ... '(As for) the gods ...'.

But even if one does not consider the presence of the initial phrase *nṯrw nb rd.t(y).sn* ..., it is still not certain that *jnk* in the pattern *jnk sḏm.j* can ever be interpreted as "topicalized," with a translation such as 'As for me (on my part, as opposed to someone else who does or does not), I will hear', as opposed to 'It is I who will hear'.

Thus, it has been widely assumed that the independent pronoun *jnk* is topicalized in examples such as *Urkunden* 1, *jnk wn jr* (102.9), *jnk wn sḥtp* (217.7), and *jnk wn d(y)* (217.9). Accordingly, *jnk* would need to be translated as 'as for me'. However, *jnk* is in all probability not "topicalized"; rather, the three sentences make more sense if they are interpreted as cleft sentences (cf. Depuydt 1991a: 39–40). The translations then are 'It is I who used to make', 'It is I who used to appease', and 'It is I who used to give'. The three sentences can then be read as *jnk wn* (a participle!) *jr(.j)*, *jnk wn* (a participle!) *sḥtp(.j)*, and *jnk wn* (a participle!) *d(y)(.j)*. Importantly, *wn* is in my opinion a participle. It is probably not a relative verb form; if it were, it would be a writing for *wn(.j)*. The first person suffix pronoun (*j*) remains as a rule unwritten in Old Kingdom texts. What is known otherwise about the cleft sentence or "participial statement" seems to require an interpretation of *wn* as a participle. The pattern of the cleft sentence is as follows:

past	*jnk jr* (past participle)	'It is I who have made'
present	*jnk jrr* (present participle)	'It is I who make'
future	*jnk jr.j* (*sḏmw.f*)	'It is I who will make'

The suffix conjugation appears only in the future but there is no need for the future in the three examples of *jnk wn* cited above. *Wn* is therefore probably a participle. There is no doubt that *wn* can on occasion be a participle in the sequence *wn sḏm.f*, as in *rmṯ wnw jr.sn* 'people who used to make' (*Urkunden* 1 50.3) and *rmṯ wnw m33.sn* 'people who used to see' (cited by Edel 1955/64: 441 §873).

In sum, in the *Urkunden* 1 passages cited above, *jnk* is part of a cleft sentence and therefore expresses isolating contrast ('It is I …'). How, then, is distinctive contrast expressed? One means is the preposition *r* + suffix pronoun (see Shisha-Halevy 1986). A clear example is *r.k* in the following passage.

> *jr gm.k ḥˁbw.sn ḥmsjj ḥms.k3.k r.k jmt.sn* (CT 1.273d–g, Coffin T2C)
> *jr gm.k nṯrw ḥmsjj ḥms.k3.k r.k ḥnˁ.sn*
>
> 'If you find their *ḥˁbw* seated, then you on your part (*or:* you too) will sit among them.
> If you find the gods seated, then you on your part (*or:* you too) will sit with them'.

(2) The second argument against interpreting *jntsn* as being "topicalized" in Pyr. §§1650a, 1651c, cited above, concerns the writing  *jntjsn*, with reed leaf, in version N (pyramid of Pepi II) of Pyr. §1651c. *Jntjsn* is opposed to *ntjsn* in version M (pyramid of Merenre). *Jntsn* is also distinct from writings that became the norm, such as 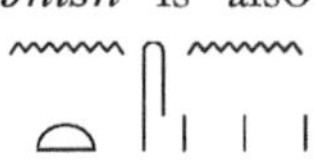 *ntsn*. It is tempting to interpret the form *jntjsn* as a combination of the *jn* of cleft sentences and *ntsn*. Edel (1955/64: 80 §173) cites one more example of  *jntsn* from version N of Pyr. §711b (see also §711d). The other versions of Pyr. §711b,d exhibit  *ntsn*. In

this case, *(j)ntsn* no doubt heads a cleft sentence. In fact, in §711b, d, *(j)ntsn* is even twice parallel to the sequence *jn* + Noun that heads unequivocal cleft sentences.

jn ḥm jr-js ẖnnw Rᶜ	*jntsn ẖn.sn N*	(Pyr. §711a–d)
jn ḥm jr-js pšrw Rᶜ ḥ3 3ẖt	*jntsn pšr.sn N ḥ3 3ẖt*	

'It is the *ḥm jr-js*(?) that row Re. It is they that will row N.
It is the *ḥm jr-js*(?) that accompany Re around the horizon. It is they that will accompany N around the horizon'.

Allen (1984: 430 §615 with n. 416) comments on this passage.

In sum, *jntjsn stpp n.sn stpwt.sn* (Pyr. §1651c N) would appear to be a cleft sentence, meaning 'It is for them that their choice cuts will be chosen'. And I am convinced that suffix conjugations in the cleft sentence are *substantival* (Depuydt 1993b: 25–26). The view that they are substantival may not be widely held. Yet, all indications clearly point to the substantival character of the *sḏm.f* in the cleft sentence. Among those indications are the following.

First, in Old and Middle Egyptian, the *sḏm.f* form appears to be the *sḏmw.f* form. The evidence has recently been surveyed by Schenkel (2000). *Sḏmw.f* is in my opinion always substantival. Polotsky suggested as much, but not as forcefully as those in favor of it might like. Second, in Late Egyptian, *i.ir.f sḏm* appears in the cleft sentence. *I.jr.f sḏm* descends from Old and Middle Egyptian substantival verb forms. By Late Egyptian, the usage range of substantival verb forms had shrunk mainly to the function of emphasizing adverbial phrases. The forms are therefore better called emphatic verb forms in Late Egyptian. Why then do emphatic forms not emphasize adverbial phrases in cleft sentences? Late Egyptian cleft sentences are in large part fossilized holdovers from Old and Middle Egyptian structures. The slot in which verb forms need to fit in cleft sentences seems clearly nominal. Nominal verb forms consist of substantival verb forms and adjectival verb forms. It comes as no surprise that verbs enter into the cleft sentence structure as either substantival or adjectival forms. True, there is something elusive about the use of substantival *sḏmw.f* in *jnk sḏmw.f* 'It is I that will hear'. But so there is in English about the use of "that" in "It is I that will hear." Some kind of verb form had to enter the cleft sentence structure. The cleft sentence exhibits the structure of a nominal sentence. The slot in which a verb form needs to be entered is nominal. The appearance of a substantival verb form in that slot would therefore appear to be altogether normal and expected.

8.0.2.3c The third feature: Sḏmm.f *as a substantival verb form in the negation* n sḏmm.f *'he will not hear'*

Other substantival verb forms following the negation n.—There are quite a few examples in which *sḏmm.f* follows the negation 𓈖 *n* (Allen 1984: 355–57 §527). There are about forty instances of *sḏmm.f* in the Pyramid texts; in about a third of them, *sḏmm.f* is negated. A survey of these negated instances follows.

The combination *n sḏmm.f* may be interpreted as an indication that *sḏmm.f* is substantival. It is not clear *why* a verb form following the negation *n* would be substantival, because the etymology or original meaning of the negation *n* is no longer transparent.

One form that often follows the negation *n* is *sḏm.n.tw.f*, which is substantival *par excellence.* Another is the *sḏm.n.f* of verbs of motion. This form too is generally assumed to be substantival (when it is not adjectival). More recently, it has become clear that active future *sḏmw.f* also frequently follows *n*. Many now accept that *sḏmw.f* is at least sometimes substantival; I believe, with Polotsky, that it always is. To be sure, one typical substantival verb form, the geminating aorist, as in *mrr.f* and *jrr.f*, does *not* follow *n*. But the negation of the aorist *jrr.f* is in fact *n jr.n.f*, with the affix *n*, which otherwise denotes past tense, which is called Gunn's rule. Geminating verb forms such as *mrr.f* and *jrr.f* are therefore not expected after *n* in the general present tense.

The special case of the verb mꜣꜣ 'see'.—As regards the negated verb form *n sḏmw.f* 'he will not hear', second-doubling verbs such as *mꜣꜣ* 'see' are a special case. First, in the *sḏmw.f* in general, be it affirmative or negated, *mꜣꜣ* 'see' exhibits the stem *mꜣꜣ*, with gemination. *Mꜣꜣ* never exhibits the *w* that gives the form *sḏmw.f* its name. To call *mꜣꜣ* the stem of the *sḏmw.f* form of *mꜣꜣ* 'see' may seem incongruous. In the case of *mꜣꜣ*, it is perhaps more suitable to speak of the active future substantival *sḏm.f*, the label for the *sḏmw.f*.

At one time, Edel (1984: 30–31) thought that he might have located a possible instance of *mꜣw* in Hatshepsut's temple at Deir el-Bahri. The text is *Urkunden* 4 245.13, but the editor, Kurt Sethe, hesitated between reading *mꜣw* and *mꜣꜣ*. Jürgen Osing collated the inscription and read *mꜣꜣ* (Edel, ibid. n. 13).

In his foundational account of the *sḏmw.f*, Edel (1955/64: 238–48 §§511–31) did not discuss *mꜣꜣ* 'see'. Edel always took a guarded attitude toward the so-called principle of parallelism (§8.2.4 below), which asks grammarians to accept something that they cannot see, such as the distinction between the active past participle *sḏm* 'who has heard' and the active aorist (or present) participle *sḏm* 'who hears'. Edel's early account of the *sḏmw.f* is completely agnostic regarding the verb *m33* 'see'; it is not clear from his account whether *m33* does or does not have a form corresponding to the *sḏmw.f* of other verbs.

In the 1980s, it became clear that the geminating form *m33* is found in positions in which undeniable *sḏmw.f* forms are found. Allen's (1984) study of the Pyramid texts was instrumental in this development. The active aorist adverbial *sḏm.f* of *m33* "see" also exhibits gemination, and as recently as 1976, Polotsky (1976: 46 n. 73) still assumed that *n m33.f* 'he will not see' contains the adverbial or circumstantial *sḏm.f.* That assumption is now obsolete. *M33.f* in *n m33.f* must be the active future substantival *sḏm.f* of *m33* 'see'.

By the same token, the expected negated verb form *n sḏmw.f* of the verb *m33* 'see' must be *n m33.f* 'he will not see'. As examples, Polotsky cites the following: *n m33.f sẖdw* 'he will not see people turned upside down' (Pyr. §323b); *n m33.k wj* 'you will not see me' (CT 5.324c, Coffin B2L); *n m33 N tn sẖdw* 'this N will not see people turned upside down' (CT 6.189f, Coffin B3Bo); and *n m33.ṯn* … 'you will not see …' (CT 6.389i, Coffin B1C).

Polotsky also observes that *n m33.k wj* is replaced later in time by *nn m3n.k wj* 'you will not see me' in a parallel in chapter 90 of the Book of the Dead (Papyrus of Nu; Budge 1898: 191, lines 9–10).

8.0.2.3d The presumed coalescence of two future sḏm.f forms

As Schenkel (1992: 379 top) has rightly pointed out, no one has done more to separate the following two synonymous constructions from one another than Vernus (1990: 117–42).

EARLIER	*n*	+	(substantival future)	*sḏmw.f*
LATER	*nn*	+	(adverbial future)	*sḏm.f*

The earlier form *n jw.f* 'he will not come' is therefore synonymous with the later form *nn jwt.f* 'he will not come'. Earlier *n m33.f* 'he will not see' is synonymous with later *nn m3n.f* 'he will not see'.

Vernus supposes that the two future non-adjectival (nonrelative) *sḏm.f* forms gradually merged into a single form in the course of the history of the Egyptian language. Allen's most recent statements (2002: 91–96) about future *sḏm.f* verb forms clearly imply such a coalescence, e.g.

> Otherwise, [in Middle Egyptian] the prospective has been replaced by the subjunctive *sḏm.f* and the analytic future *jw.f r sḏm*, a process whose beginnings are already evident in Old Egyptian. (91)

(His "prospective," the *sḏmw.f*, is my "substantival future *sḏm.f*"; his "subjunctive" is my "adverbial.") In my opinion, no such replacement or coalescence ever occurred.

> The [active] prospective *sḏm.f* ("*sḏmw.f*") and its passive counterpart (*sḏmw.f*/*sḏmm.f*) are productive forms of the suffix conjugation in the language of the Pyramid Texts and their Middle Kingdom descendant, the Coffin Texts, but have largely disappeared elsewhere by Middle Egyptian. (ibid.)

I again see matters slightly differently. I am not at all prepared to accept that the prospective *sḏmw.f* has disappeared. What seems to have decreased is explicit writings of the characteristic marker *w*.

A distinction is necessary between the active and the passive. First, the passive *sḏmw.f*/*sḏmm.f*. The passive future *sḏmm.f*, which I believe is always substantival, has indeed "largely disappeared by Middle Egyptian." Of course, this does not mean that there no longer existed a passive future substantival verb form. The grammatical category itself cannot have disappeared. Instead, the affix *tw* came to be used to denote this combination of coordinates. The resulting verb form is *sḏm.tw.f* 'that he will be heard', which does not exhibit gemination in the third-weak verb form, as in *jr.tw.f*. Second, the active *sḏmw.f*. It does seem to be case that the affix *w* has to an extent "disappeared" from Middle Egyptian, but perhaps not as "largely" as Allen assumes. In fact, already in Old Egyptian the form *sḏmw.f* is mostly written *without* the characteristic affix *w*. In other words, *jr.f* is often the writing of the very same form that is elsewhere written *jrw.f*. What "largely disappeared by Middle Egyptian," then, is explicit writings with *w* such as *jrw.f*. This does not mean, however, that the form itself has disappeared. My impression is that the form in question occurs as frequently in Middle Egyptian as it does in Old Egyptian. After all, in Middle Egyptian, there must have been as much of a need to denote the characteristic combination of the three characteristics (1) active, (2) future, and (3) substantival as there was in Old Egyptian. How could the language have changed in this way? I also believe that the active future substantival *sḏm.f* only truly began to disappear when it was being replaced by *i.jr.f sḏm* in Late Egyptian.

Again, considering all the recent arguments, I do not see any support for the notion that the two future *sḏm.f* forms have coalesced to any degree. Certain factors that may have mistakenly led to the assumption that a coalescence has occurred can be identified. One concerns the characteristic affix ending 𓇌 *jj*. Schenkel's recent detailed account (2000) identifies nonrelative future *sḏm.f* forms exhibiting *jj*, showing that both future non-adjectival *sḏm.f* forms can exhibit that ending. Schenkel confirms that the two future verb forms in question overlap in writing. Both can exhibit 𓇌 *jj*. This overlap in writing may have contributed to creating the false impression that the two future verb forms gradually merged.

The existence of two nonrelative future *sḏm.f* forms, rather than one, was not widely accepted until the 1980s. But no sooner did the existence of two future non-adjectival *sḏm.f* forms become widely accepted in the 1980s than theories about a merger between the two began springing up.

It is now generally agreed that there are two future nonrelative *sḏm.f* forms. For example, Schenkel's earlier uncertainty (e.g. 1981) was resolved by Allen's work (Schenkel 1985: 485), and he now (e.g. 2000) accepts the existence of two such forms. Polotsky had postulated two distinct forms as early as 1944 and had observed the two future *sḏm.f* forms of *jn*(*y*) 'bring', *jn.f* and *jnt.f* (1969: 473; see Allen 1979a: 25 n. 19). Edel's *sḏmw.f* clarified the issue, but he himself did not interpret *sḏmw.f* as a future tense. Wolfhart Westendorf was the first to do so, and H.J. Polotsky (1976: 23–25 §2.7, with references) first clearly interpreted it as a substantival future.

Schenkel (2000) makes it clear that the theory that two nonrelative future forms merged into one is not valid, but despite his considerable detail regarding the *sḏmw.f*, there is hardly a reference to *sdmw.f* even sometimes being substantival, let alone always. In fact, Schenkel clearly implies that *sḏmw.f* can sometimes *not* be substantival, as in the following statement (Schenkel 2000: 28).

> When a *sḏm.f* occupies initial position in the main clause, which often happens, then it is difficult to decide whether it is a future tense (prospective [my "active future substantival"]) or a wish expressed by the subjunctive [my "active future adverbial"], that is, whether the *sḏm.f* in question is to be interpreted as 'he will hear' or as 'may he hear'.*

Schenkel states that the difference between an initial "prospective" *sḏm.f* and an initial "subjunctive" *sḏm.f* is the same as the difference between 'he

* Steht [der Prospektiv oder der Subjunktiv], was häufig vorkommt, in initialer Position im Hauptsatz, so läßt sich oft schwer entscheiden, ob ein Futur (Prospektiv) oder ein durch den Subjunktiv ausgedrückter Wunsch vorliegt, ob ein *sḏm=f* also eher als "er soll/wird hören" aufzufassen ist oder als "er soll/möge hören."

will hear' and 'may he hear', but he implies that the difference between the two is not between substantival and nonsubstantival function. I believe that the "prospective" *sḏm.f* (the form that exhibits writings such as *jw.f*, *jn.f*, and *m33.f*) is always substantival and that the "subjunctive" *sḏm.f* (the form that exhibits writings such as *jwt.f*, *jnt.f*, and *m3n.f*) is always adverbial. When the adverbial future *sḏm.f* stands at the beginning of a sentence and denotes a wish or the like, there are two theoretical solutions. Either the form is an elliptical construction, or the usage is a vestige from when the adverbial forms of the suffix conjugation were independent. Therefore, whenever an initial "prospective" *sḏm.f* is not part of a balanced sentence, I would consider it to emphasize an adverbial phrase.

8.0.2.3e A recent proposal to positively verify two nonrelative* sḏm.n.f *forms

Many grammarians now accept that transitive verbs typically exhibit *two* past tense *sḏm.n.f* forms besides the relative *sḏm*(*t*).*n.f*, sometimes called the "substantival" and the "adverbial" *sḏm.n.f*, while many intransitive verbs exhibit just *one* additional *sḏm.n.f* form; transitive verbs overall differ from intransitives with regard to past tense verb forms.

The main problem is that this particular difference is not visible in hieroglyphic writing, which does not represent vowels, so the belief that transitive verbs can exhibit *two sḏm.n.f* forms where intransitive verbs exhibit *one* has been on the decline. This decline works in two directions. In one direction, some now assume the existence of two forms in cases in which formerly only one was assumed to exist. In the other direction, some now assume the existence of only one in cases in which the existence of two was assumed. Because these are all assumptions that are not based on observable fact, distinguishing various *sḏm.n.f* forms has been called "a game of grammarians" (Winand 1992: 190). In addition, a view that was once apparently accepted as an unshakeable tenet, namely that the *sḏm.n.f* of verbs of motion is as a rule substantival, has now lost much support.

The matter would be resolved if the various *sḏm.n.f* forms were not merely postulated but appeared differently in writing. Schenkel (2006a) now claims to have detected visible evidence for such differentiation. Two questions arise:

(1) What is the source of the theory that most intransitive verbs exhibit *one* nonrelative *sḏm.n.f* form where transitive verbs exhibit *two*?

(2) Which orthographic features, according to Schenkel, positively evidence the distinction between two nonrelative *sḏm.n.f* forms in transitive verbs?

(1) One main line of reasoning is as follows. Consider the following expressions including *jw/jj* 'come' reported in grammars.

**jw jj.n.k*	'You have come'.
jw.k jj.tj	'You have come'.
jj.n.k ṯn	'Where have you come from?'
**jw.k jj.tj ṯn*	'Where have you come from?'

First, the expressions marked with the asterisk do not exist. Second, the two expressions that do exist differ in meaning by only one element, *ṯn* 'from where?'. "It is a matter of some importance … ," writes Polotsky (1965: 12 §24 [1971: 82 §24]), "that *jj.n.k* must necessarily be used before an adverbial 'adjunct' consisting of a preposition plus interrogative pronoun. … An interrogative pronoun … is by its very nature [emphasized]." The addition of *ṯn* 'from where?' to the first expression triggers a concurrent shift from the verb form *jw.k jj.tj* to the verb form *jj.n.k*. A parallel relation exists in Coptic. The addition of **ΤⲰΝ** 'from where?' to **ⲀⲔⲈⲒ** 'you have come' produces the expression **ⲚⲦⲀⲔⲈⲒ ΤⲰΝ** 'Where have you come from?'

Third, *ṯn* (Coptic **ΤⲰΝ**) by definition expresses "emphasis" or "isolating contrast." Thus (nonrelative) *jj.n.k* marks isolating contrast in a way that *jw.k jj.tj* does not. There are still two solutions at this point. Either there is *only one* (nonrelative) form *jj.n.f* and it specifically serves to express emphasis—as opposed to *jw.f jw* 'he has come', which does not—or there are *two or more* (nonrelative) forms *jj.n.f*. Many—including myself—believe, following Polotsky, that there is *just one* (nonrelative) form *jj.n.f* and that this one (nonrelative) *sḏm.n.f* is always substantival. It is not clear what the function of any other postulated (nonrelative) *jj.n.f* forms might be.

One indication in favor of the assumption that (nonrelative) *jj.n.f* is always substantival is that sentence-initial *jj.n.f* is *as a rule* accompanied by an adverbial phrase. The adverbial phrase would appear to be emphasized, but not always. An instance in which it is not emphasized is found in lines 2–3 of a passage inscribed on the east wall, south of the entrance, in the tomb of Qereri (for a full discussion, see §9.10.3.5[4]).

jj.n(.j) ḥm r ẖrt-nṯr š3d.n(.j) |³ *š n mḥ 100 ḥr m3ʿ.f nb nhwt 10 jm.f*

'When I came to the necropolis, I dug a pool of 100 cubits on each side and ten sycamore trees in it'.

By contrast, the *sḏm.n.f* form of transitive verbs does appear after *jw*. Many—including myself—believe, following Polotsky, that transitive verbs typically exhibit *two* (nonrelative) *sḏm.n.f* forms, one following *jw* and the other substantival.

One facet of these beliefs is that, in the past tense, transitive verbs typically exhibit a *nonsubstantival, nonrelative suffix conjugation* form whereas intransitive verbs typically do not. That is what is found in Late Egyptian. The nonrelative *sḏm.f*, which is a suffix conjugation, denotes the past tense in Late Egyptian and is restricted to transitive verbs. The past *sḏm.f* of intransitive verbs does occur, but only in certain types of texts. That is also what is found in Middle Egyptian. The past (nonrelative) *sḏm.n.f* of intransitive verbs is typically restricted to substantival function.

In sum, many now accept two tenets regarding Middle Egyptian grammar, following Polotsky:

(a) transitive verbs exhibit a nonsubstantival, nonrelative *sḏm.n.f*;
(b) the nonrelative *sḏm.n.f* of verbs of motion is always substantival ("emphatic").

The same tenets apply in Late Egyptian if "past *sḏm.f*" is substituted for "nonrelative *sḏm.n.f*." Indeed, since *sḏm.n.f* evolved into *sḏm.f* via loss of *n*, the Late Egyptian verb forms are descendants of these Middle Egyptian ones. All this favors the notion that in Late Egyptian we can *see* only one (nonrelative) *sḏm.n.f* with verbs of motion and there *is* only one; whereas we can *see* only one (nonrelative) *sḏm.n.f* with transitive verbs but there *are* two.

The above tenets are still widely held. So many facts, not all of which can be reviewed here, fall into place in the system that includes them. Still, it is my impression that their acceptance has been declining since the 1970s. This decline may result from a number of factors. First, the distinctions are postulated, not explicit in the data. (But Polotsky and others would deem the reasoning that leads to their postulation to be quite sufficient.) The complexity of any proposed reasoning may also lead to distrust. (But the Middle Egyptian verbal system undeniably exhibited a certain structure. The refusal to try and combine the facts into a coherent framework implies that the distinctions within it did not matter.)

(2) The need for *postulating* distinctions inside the *sḏm.n.f by inference* would dissipate if the distinctions in question were in fact explicit in the data. Schenkel (2005: 186–92, 2006a) has recently claimed to have detected such distinctions. What has he seen? The focus is on last-doubling verbs, such as *jšš* 'spit out', *nẖẖ* 'become old', *ngg* 'cry out', and *ḳbb* 'become cool'. The Coffin texts include not only forms with gemination, such as *nẖẖ.n*, but also forms without gemination, such as *ḳb.n*. Schenkel has collected all the

sḏm.n.f forms of last-doubling verbs in the Coffin texts and makes two observations: instances following *jw* as a rule do not exhibit gemination; and instances in which emphasis presumably rests on an adverbial phrase do exhibit gemination. These facts suggest that last-doubling verbs exhibit a distinction in writing between substantival *sḏm.n.f* (with gemination) and another *sḏm.n.f* (without gemination). A similar distinction is already well known elsewhere in the verbal system: gemination is associated with substantival function in the *sḏm.f* of third-weak verbs, as in *mrr.f* and *jrr.f*.

Schenkel's evidence is not as persuasive as it could be. Of the last-doubling verbs in question, only one occurs in both positions. Otherwise, *either* a last-doubling verb is attested in the form *jw sḏm.n.f* exhibiting absence of gemination, *or* it is attested in emphatic function exhibiting presence of gemination (see Schenkel's table, 2006a: 45). That leaves open the possibility that the former verbs also exhibit absence of gemination in emphatic function and that the latter verbs exhibit presence of gemination after *jw*. One verb, attested in both positions, contradicts this possibility: *ngg* 'cry out':

ngg.n N pn ⟨m⟩ smn (CT 4.23b, Coffin T1L; *m* is secure from the parallel versions Sq6C and Sq1Sq)

'It is as a goose that this N cried out'.

The context strongly supports the notion that *ngg.n* is emphatic. The sentence is preceded at 23a by the parallel *pꜣ.n N pn m wr* 'This N has flown away as a great one'. The verb *pꜣ(y)* is intransitive and a verb of motion, its *sḏm.n.f* hence probably substantival.

In the same coffin, one finds:

jw ng.{n}n.k ⟨m⟩ smn (CT 1.74b, Coffin T1L)

'You have cried out as a goose'.

The second *n* follows the determinative and is therefore probably spurious. The contrast between presence of gemination in emphatic function at 4.23b and absence of gemination at 1.74b may well be significant.

The writing of the verb form in CT 1.74b in versions on other coffins otherwise varies. Like T1L, Coffin BH5C exhibits *ng.nn.k*, presumably to be interpreted as in T1L. Coffin B1P has *ng.n.k*, expected from the theory. B3Bo and B4C show *ngn.n.ṯ*, with the second *n* preceding the determinative, as if the verb is *ngn*. T9C and T1C have *jw ngnn.k* without determinative. But what is most troublesome is the writing of Coffin B6C, *ngg.n.k*, *with* gemination. Schenkel does not list this coffin. It is unfortunate that the only last-doubling verb that is attested after *jw without* gemination and in emphatic function *with* gemination is also attested *with* gemination after *jw*. The opposition between presence of gemination and absence of gemination is obscured.

Schenkel's proposal has two further weaknesses. First, some of the last-doubling verbs in question are intransitive, such as *wrr* 'become great' and *nḫḫ* 'become old'. The (nonrelative) *sḏm.n.f* of many intransitive verbs is always substantival. Any differentiation in the writing of the *sḏm.n.f* form is therefore unlikely to be significant because there is only one form.

A second difficulty is that, according to Schenkel's own observations, it is the nongeminating forms that are found after the negation *n*. These are otherwise the forms that do not have emphatic function and are present after *jw*. Yet elsewhere in the verbal system, forms found after the negation *n* do have emphatic function and are absent after the particle *jw*, as Schenkel (2006a: 48) realizes. Three examples of verb forms found both in emphatic function and after the negation *n* are (1) the *sḏm.n.f* of verbs of motion, (2) the *sḏm.n.tw.f* passive form, and (3) the future *sḏmw.f*. Accordingly, writings with gemination of the *sḏm.n.f*—the substantival verb form—of last-doubling verbs after the negation *n* would be expected. This expectation is contradicted by the data concerning the *sḏm.n.f* of last-doubling verbs. All is not clear.

Schenkel (2002a) had earlier made a similar observation regarding the *sḏm.n.f* of *wnn* 'be' after the negation *n*, suggesting that the affirmative may be written differently from the form following the negation *n*. He claimed (92) that the form *wn.n.f* is written *with* gemination when it follows the negation *n* and *without* gemination when the verb form is affirmative and has emphatic function. There would therefore be two forms of the *sḏm.n.f*, one negated and one substantival. Again, what is known about the Middle Egyptian verbal system would predict the same form to be found in both uses.

One potential problem is that Schenkel's two examples of negated  *n wnn.f* at CT 4.18f and 7.501c might be instances of the future *n sḏmw.f*. Judging from *mꜣꜣ* 'see', second-doubling verbs geminate in the *sḏmw.f* form. If *n wnn.f* is future, then  should be transcribed as *n wnn.f* and not as *n wn.n.f*. The translation would be 'he will be' instead of 'he can be' or 'he is'. The adjectival (or relative) forms 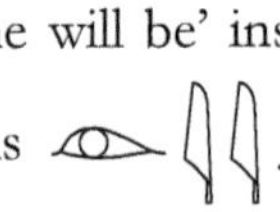 *jry* and  *jr* appear nearby in CT 4.18e and 7.501b. Absence of gemination marks these two forms as future tenses (as opposed to the aorist or general present tense *jrr.f* and the past tense *jr.n.f*).

8.0.2.3f The nature of sḏm.f in past tense n sḏm.f 'he has not heard'

Three negated verb forms have been discussed so far: (1) *n sḏm.n.tw.f*, (2) *n sḏm.n.f* when the verb is a verb of motion, and (3) *n sḏmw.f*. The indications are that substantival forms are normal after the negation *n*. This leaves a fourth form, the negated past *n sḏm.f* 'he has not heard'. The identity of this *sḏm.f* form remains obscure. Polotsky (1976: 46) notes that it exhibits the characteristic combination of writings of the stem *rd(y)* with *r* and writings of the stem *m33* without gemination. There are just a couple of examples, but they seem reliable. Moreover, he adds the remarkable observation that *n m3n.f* 'he has not seen' may on occasion be a variant of *n m3.f* 'ditto'. The first of his two examples is CT 1.334/335c: two out of sixteen versions have *n m3n.f* instead of *n m3.f*; the second is Hammamat 191.6, in which *n m3n s(j) jrt nbt* 'no eye has seen her' appears to be parallel to *n ḫr ḥr n rmṯw ḥr.s* 'no one's face (that is, looks) has fallen on her'. A possible third example is *n m3n.k (sw?)* in Papyrus Ramesseum (Papyrus British Museum 10755 [Barns 1956: plates 1–6]), at A4, if the verb form is past tense. The stem *m3n* is well known as a writing of the adverbial future *sḏm.f*, also called the subjunctive. Polotsky's observation raises the possibility that the writing *m3n.f*, with *n*, may on occasion also denote past tense. That possibility complicates the problem of identifying the *sḏm.f* that follows the negation *n* in the negated past tense verb form *n sḏm.f*. This matter too requires further investigation.

8.0.2.3g A survey of negated sḏmm.f 'he will not hear' in the Pyramid texts

A survey of negated *sḏmm.f* forms in the Pyramid texts follows. It is assumed, by the principle of parallelism, that all verbs occur in all forms. Therefore, the lack of distinctive writings of the *sḏmm.f* form for many verbs does not mean that these verbs do not possess a form fully equivalent to it. About a third of the Pyramid texts' forty or so instances of gemination of this type are negated; all are given here.

The biliteral verb *wn* 'open' appears at Pyr. §485b P, *n wnn n.f ꜥ3wy msktt n wnn n.f ꜥ3wy mꜥnḏt* 'The doors of the *msktt*-bark will not be opened for him, the doors of the *mꜥnḏt*-bark will not be opened for him'. That *wnn* is an instance of *sḏmm.f* is clear from the fact that *n wnn* here is followed at §485c by *n wḏꜥꜥ mdw.f* 'He will not be judged' in both P and N. Version N twice exhibits *n wn* ..., without gemination. As often in Egyptian grammar, one assumes that something in the pronunciation of *n wn* in version N made *n wn* recognizable as the same form as *n wnn* in P. Gemination also appears in *n wnn.sn n j3btyw* 'They will not be opened for the Easterners' (Pyr. §1590b N), which is immediately preceded by absence of gemination in *n wn.sn n jmntyw* 'They will not be opened for the Westerners'.

Two other biliteral verbs appearing in the form *n sḏmm.f* are *dm* 'penetrate(?)' (so Allen 1984: 547 §728) in *n dmm.s* (Pyr. §392a WT, §603a TN) and *dr* 'suppress, remove' (so Allen 1984: 547 §728) in *n drr.s* (Pyr. §*1902d; see Allen 1984: 687). As regards triliteral verbs, *wḏ*ʿ (Pyr. §485c PN) has already been cited. There is also *nḥm* 'take away' in *n nḥmm sʿḥw N m-ʿ.f* 'The dignities of N will not be taken from him' (Pyr. §411c WT). At Pyr. §748d T, *n nḥmm ḥ3ty.f* (M: *ḥ3ty.k*) 'His heart (M: your heart) will not be taken away' is parallel to *n jṯw jb n N* (M: *jb.k*) 'N's heart (M: your heart) will not be taken'. The parallelism of *nḥmm* and *jṯw* illustrates that the single passive, future, and substantival form can be written either with gemination or with *w*. Another example of the same characteristic combination of gemination in a triliteral verb and affix *w* in a third-weak verb is found in a spell inscribed in cursive script in a margin of the back-side of the unpublished coffin of *Ppjj-jm3* found in tomb N 4003 at Nag ed-Dêr and probably dating to Dynasty 9. Brovarski (1989: fig. 63) provides a facsimile of the coffin, cited by Hussein (2004). The text is as follows.

n mdss.f n rdw ʿ ḥr nḥbt.f jn jṯnww

'He will not be cut down. No arm will be placed on his neck by the adversaries'.

The same parallelism also occurs when two *affirmative* passive, future, and substantival verb forms are parallel. Examples are cited in §8.0.2.4.

A third triliteral verb is *ẖbs* 'hack up'. Negated *n ẖbss n.f t3* 'The earth will not be hacked up for him' appears in close vicinity to affirmative *ẖbss n.f t3* in Pyramid spell 539 at Pyr. §1323c P and §1326a P. Both exhibit gemination. Negated *n ẖbss n.f t3* is also found at Pyr. §978c N and affirmative *ẖbss n.f t3* at §1138a PM. If *sḏmm.f* is always substantival, then *ẖbss n.f t3* ought to be translated as 'It is for him that the earth will be hacked up'. *N.f* is the only adverbial phrase in the sentence. On the other hand, a negated *n sḏmm.f* form such as *n ẖbss* as a unit is not known to be emphatic. If *n.f* was somehow emphasized anyway in *n ẖbss n.f t3*, then that would have to be by some other means, for instance by intonation.

Emphatic affirmative *ẖbss n.f t3* 'It is for him that the earth will be hacked up' is apparently opposed to non-emphatic negated *n ẖbss n.f t3* 'The earth will not be hacked up for him'. Does this opposition between absence of contrast and presence of contrast pose a conflict? I believe not. The two can be reconciled. The affirmative world is not a reflection of the negated world, as it were. An affirmative statement is not necessarily a mirror image of its corresponding negation. The statement "It is for him that the earth will be hacked up" can be negated in two ways: as "It is not for him that the earth will be hacked up," or "It is for him that the earth will not be hacked up." The first manner of negation is expressed elsewhere in Old and Middle Egyptian by the discontinuous two-part negation *n* ... *js*, the second by the negation verb *tm*. I know of no instances in which either of these expressions negates a passive future substantival verb form.

The contrastive emphasis in affirmative *ẖbss n.f tꜣ* seems to come naturally: "It is for him that the earth will be hacked up" (and not for someone else). The implied contrast is with a non-event—the earth will *not* be hacked up for someone else. In other words, no positive event is implied by contrast. Assuming that the negated counterpart is also emphasized, contrastive emphasis comes with necessary implications. "It is for him that the earth will not be hacked up" must imply by contrastive emphasis that it is *not* for someone else that the earth will not be hacked up. This is a possible but rather convoluted thought. It is difficult to imagine that *n ẖbss n.f tꜣ* has this meaning. The other negated counterpart, "It is not for him that the earth will be hacked up," implies that it is for someone else that the earth will be hacked up. There is the implication of another event: a certain other person will not receive the benefit of having the earth hacked up. There is no compelling reason to assume that *n ẖbss n.f tꜣ* conveys this implication, either.

Both *ẖbss n.f tꜣ* and *n ẖbss n.f tꜣ* concern a person as well as a benefit The benefit is having the earth hacked up. The person either receives or does not receive the benefit. In affirmative *ẖbss n.f tꜣ*, he receives it. In this case, contrastive emphasis increases the focus on the person in question by singling him out as special and unique. No one else receives the benefit. No other event is taking place. The contrast is with a non-event. If negated *n ẖbss n.f tꜣ* exhibited contrastive emphasis with the meaning 'It is not for him (but for someone else) that the earth will be hacked up', the existence of other people who do receive the benefit is implied by contrast. But it is not obvious that this is the meaning of the text.

In sum, *n ẖbss* is probably not overtly marked as emphatic by some empirical feature, written or not. It may therefore be preferable to assume that it *is* not emphatic. In the passages cited above, both affirmative *ẖbss n.f tꜣ* and *n ẖbss n.f tꜣ* feature the substantival form *ẖbss*. In the case of negated *n ẖbss*, *ẖbss* is substantival because substantival forms for some unknown reason follow the negation *n*. But *n ẖbss as a whole* is not emphatic. In the case of affirmative *ẖbss*, *ẖbss* is substantival because the adverbial phrase *n.f* is emphasized. It is then only a coincidence that *non-emphatic n ẖbss* and *emphatic ẖbss* both exhibit *substantival ẖbss*. In light of these assumptions, there is now no longer a need to suppose that affirmative *ẖbss* can sometimes be nonsubstantival (independent) just because it can be parallel to non-emphatic *n ḫbss*.

A similar pair of sentences that seems to exhibit a contrast similar to the one just discussed is as follows. On the one hand, *n wnn.sn n jꜣbtyw* 'They will not be opened for the Westerners' (Pyr. §1593b N; cf. also Pyr. §485b P, cited above) presumably exhibits absence of contrastive emphasis. On the other hand, *wnn.sn n Ḥr* 'It is for Horus that they will be opened' (Pyr. §1594a N) is followed by parallel cleft sentences and therefore presumably also exhibits contrastive emphasis (see below).

The other examples of *n sḏmm.f* in the Pyramid texts involve the triliteral verbs *nḏr* 'grasp (or the like)' (Pyr. §658d T), *ḫsf* 'drive away (or the like)' (Pyr. §658e T, §1441b PMN), and *šꜣs* 'trample (or the like)' (Pyr. §722c TN), as well as the fourth-weak verb *swꜣ(y)* 'cut (or the like)' (Pyr. §722c TN).

Negated examples are also attested in the Coffin texts: *n ḫfꜥꜥ bꜣ.j jn ꜣkrw n ꜣmm bꜣ.j jn ḥkꜣ* (CT 1.398a–b, Coffin S1C); *n ḥꜣmm.j m jꜣdt* (CT 3.296f, Coffin G1T); and *n šnꜥꜥ.j ḥr sbꜣw* (CT 3.303i).

8.0.2.4 Inference: Sḏmm.f is a substantival form everywhere else

8.0.2.4a A logical principle of inference: Sometimes substantival, always substantival

From the three features listed in §8.0.2.3, it may be inferred that *sḏmm.f* is always substantival. The principle of inference (§8.0.2.2) is deductive, involving logical reasoning. The logic is that a verb form that leaves a grammatical fingerprint of substantivalness is always substantival. Indeed, it seems difficult to imagine that the same verb form could be both substantival and not substantival at the same time. No one would ever assume, it seems to me, that the same verb form could be both adjectival (relative) and not adjectival. Then why adopt a different approach in the case of substantival verb forms?

8.0.2.4b That sḏmm.f is always substantival as a probable conclusion

But there is also an empirical, inductive component to the notion that *sḏmm.f* is always substantival. On the one hand, there is a need for positive indications that *sḏmm.f* is at least sometimes clearly substantival. Instances of *sḏmm.f* in the balanced sentence are unequivocal in this respect. On the other hand, there is a need for the absence of any indication to the contrary, namely any indication to the effect that *sḏmm.f* is sometimes *not* substantival. In this sense, it is reassuring that Allen has noted that there are no examples of *sḏmm.f* "in clearly adverbial function" (1984: 353 §522) and that the "independent status of … the *sḏmm.f* cannot be demonstrated with certainty"

(§525). Then again, Allen (ibid.) remains noncommittal when he observes that "the possibility of independent use cannot be dismissed."

The value of inductive statements is by necessity always limited. No grammarian will ever have access to *all* instances of *sḏmm.f* ever produced. To the extent that they are inductive, the laws of grammar are always probable laws. In this respect, they are like the laws of nature. Boole (1854: 4) writes the following about the laws of nature.

> The general laws of Nature are not, for the most part, immediate objects of perception. They are either inductive inferences from a large body of facts, the common truth in which they express, or, in their origin at least, physical hypotheses of a causal nature serving to explain phænomena with undeviating precision, and to enable us to predict new combinations of them. They are in all cases, and in the strictest sense of the term, *probable* conclusions, approaching, indeed, ever and ever nearer to certainty, as they receive more and more of the confirmation of experience. But of the character of probability, in the strict and proper sense of that term, they are never wholly divested.

8.0.2.4c The interpretation of inconclusive instances of sḏmm.f as substantival

Once it has been accepted by inference that *sḏmm.f* is always substantival, it is possible to approach any ambiguous instances of *sḏmm.f* and interpret them too as substantival. One requirement is obviously that the interpretation of *sḏmm.f* as substantival is at least possible.

Consider the use of *sḏmm.f* in the Pyramid texts. Instances of three types have already been cited because they suggest the presence of substantival function—in the balanced sentence, in a cleft sentence, and following the negation *n*. The other instances of *sḏmm.f* typically stand at the beginning of a sentence. In that position, substantival verb forms typically function as first component of an adverbial sentence. The effect is to apply contrastive emphasis to the second component of the adverbial sentence, an adverbial phrase. Verifying the presence of isolating contrast in a passage is to a certain extent always a subjective undertaking. But if it is accepted that *sḏmm.f* is always substantival, then diagnosing contrastive emphasis imposes itself. Examination of all the remaining instances of *sḏmm.f* found in the Pyramid texts reveals that identifying an emphasized adverbial phrase is at least possible in every instance.

First is a set of related instances involving the verb *wn* 'open'. The translations, here as below, reflect the assumption that emphasis rests on the adverbial phrase.

wnn ꜥ3wy pt n N pn jsnjj ꜥ3wy ḳbḥ n N pn (Pyr. §1408c P)

'It is for this N that the doors of the sky will be opened. It is for this N that the doors of heaven will be drawn aside'.

The significance of *jj* (double reed leaf) in *jsnjj* is not clear (cf. Allen 1984: 30 §53). *Jj* is otherwise absent in similar examples (ibid.). Allen (§736) classifies *jsn* as a triliteral verb, in which *j* is presumably a strong consonant because it is often written out, "although it is absent in 11 of 87 examples." Well-known words in which *j* is always written out include *jwr* 'conceive' and *jḳr* 'become excellent'.

wnn ꜥ3wy pt n N pn jsn n.f ꜥ3wy sḥdw (Pyr. §1474c P)

'It is for this N that the doors of heaven will be opened. It is for him that the doors of the firmament will be opened'.

wnn n.k sb3 pt jr 3ḫt (Pyr. §1720a M, P fr. Or. D4 [for the latter, see Allen 1984: 746 and Leclant 1968: plate 23, fig. 25b])

'It is for you that the door of heaven will be opened up to(?) the horizon'.

With a verb like *wn* 'open,' the only other possible interpretation of the second *n* is as an affix of the past *sḏm.n.f.* However, *wn.n.f* would then be active, and *ꜥ3wy* 'the doors' ought to be the actor. It is clear, however, that the doors undergo the event of opening and do not perform it. As a result, it is in this somewhat exotic corner of Old and Middle Egyptian that one finds unambiguous identifications of a verb form such as *sḏmm.f*, at least according to a certain theoretical framework.

Version M of Pyr. §1408c exhibits *wn ꜥ3wy pt jsnjj ꜥ3wy ḳbḥ n N*, which presumably means 'It is for N that the doors of the sky will be opened and the doors of heaven drawn aside'. It is assumed that M's *wn*, with absence of gemination, is the same form as the one written *wnn* in P. In version N, *n N* 'for N' is resumed by *n.f* 'for him': *wn ꜥ3wy pt n N jsn n.f ꜥ3wy ḳbḥ* 'It is for N that the doors of the sky will be opened. It is for him that the doors of heaven will be drawn aside'. In *wn ꜥ3wy pt jsnjj ḳbḥ n Ḥrw nṯrw* (Pyr. §1408a PMN), the phrase *n Ḥrw nṯrw* 'for Horus of the gods' perhaps modifies both *wn* and *jsnjj*, as presumably does *n N* 'for N' in version M of Pyr. §1408c, just cited.

The expression at hand involving *wn* 'open' is repeated several times in Pyramid spell 563, in the section Pyr. §§1408–10. P is the only version exhibiting instances of *wnn* (§§1408c P, 1410c P, and 1411c P); it also has instances of the writing *wnj* () (§1409c P, §1411a P). The significance of *j* as part of this writing is not fully clear to me.

Allen (1984: 604) classifies *wnn* as a *sḏmm.f* passive. But he also (15 §23) considers the possibility that *wnn*, with its gemination, is a "misreading." It might reflect an original **wn n*(.*j*), with *n*(.*j*) 'for me'. Presumably *n* was held over erroneously when an original *n*(.*j*) 'for me' was changed to *n N* 'for N'. The multiple occurrences of *wn* 'open' in Pyramid spell 563 otherwise await additional analysis (see meanwhile Allen 1984: 345–46 §512).

Emphasis on the beneficiary of the act of opening seems supported by the parallel cleft sentences in *wnn.sn n Ḥr swt jr sn swt sꜥḥꜥ sn swt* ... 'It is for Horus that they will be opened. It is he who has made them. It is he who has erected them. It is he ...' (Pyr. §1594a–b N, §1595a–c PN). In Pyr. §1595a–c P, the cleft sentences are introduced by *nswt* () 'It is he ...'.

In addition to biliteral *wn* 'open', nine more verbs need to be considered. All are triliteral (glosses from Allen 1984: 552–67 §736): *ꜣbḫ* 'mix', *jwr* 'conceive', *npḏ* 'slaughter', *rḫs* 'slaughter', *ḥsḳ* 'decapitate', *ḫbs* 'hack up', *stp* 'choose', *šsp* 'receive', and *ṯs* 'tie' (often classified as biliteral, but Edel [1955/64: 14 §29] derives it from triliteral *ṯꜣs* [cf. Allen 1984: 565 §736, 615 §772]).

Before citing the texts themselves, a few defining traits may be noted. Each of the nine passages exhibits an adverbial phrase. The presence of contrastive emphasis is plausible in each instance, if not certain. Contrastive emphasis can be diagnosed with certainty only with cleft sentences and verb forms that are unambiguously substantival. In six of the nine passages, the presumed emphasized adverbial phrase is the beneficiary: *n.j* 'for me' with *ḫbs* 'hack up'; *n.k* 'for you' with *ꜣbḫ* 'link', *npḏ* 'slaughter', and *ḥsḳ* 'decapitate'; and *n.f* 'for him' with *stp* 'choose' and *ṯs* (from *ṯꜣs*) 'tie'. The beneficiary is often called, in imitation of Latin grammar, the indirect object. In Egyptian, of course, prepositional phrases are adverbial phrases. There is no doubt that prepositional phrases introduced by *n* 'to, for' can be emphasized by substantival verb forms, just like any other adverbial phrase. The following examples leave little doubt that the adverbial phrase headed by the preposition *n* is emphasized. The examples exhibit two characteristics: (1) unambiguous substantival verb forms; (2) parallelism with emphatic constructions. The fourth example exhibits only (1), however.

nnk ḏfꜣw jr.n.t(w) n.j ḥtpt (CT 4.134b–c; cited by Polotsky 1957: 111 [1971: 45])

'Mine are the provisions. It is *for me* that offerings have been made'.

dd.tw n ḫnms[w.f] jn ḫnmsw.f dd n.f st (Eloquent Peasant R 18.6–7 [old numbering: 130–31])

'It is *to his friend* that one gives (it). It is his friends that give it to him'.

ntf grt jnn ḳnbtyw nw w	'It is he who brings the district-assessors.
ntf h(ꜣ)b sn	It is he who sends them.
smj.sn n.f ḫrt ww.sn	It is *to him* that they report the affairs of their districts.
jnn.t(w) n.f jmt-pr nb	It is *to him* that every will is brought.
ntf ḫtm st	It is he who seals them'.

(From Rekhmire's tomb; cited from Gardiner 1957: 362.)

njs.n.tw n wꜥ jm jst wj ꜥḥꜥ.kw

'It is *to one of them* (that is, to one of the princes) that a call went out, and that while I was standing (there)'.

(Story of Sinuhe B1 = R24). Sinuhe seems to have expected that it was to someone else (perhaps himself) that a call would go out.

As regards the last example, I had earlier considered it possible (Depuydt 1993a: 181 ex. 26) that *jst wj ꜥḥꜥ.kw* '…, and that while I was standing there' (perhaps meaning: '…, and that while I was available') might be emphasized. Others have also considered this possibility, including Polotsky (1957: 113 [1971: 47] ex. 18). But after subsequent study of the particle *jst/jsṯ* (Depuydt 1993c), I am convinced that clauses introduced by *jst* are never emphasized by explicit substantival verb forms in Old and Middle Egyptian. Rather, I believe that *jst* denotes sentence anaphora ('and that …'). The only candidate left as emphasized adverbial phrase is *n wꜥ jm* 'to one of them'. The contrast suits the message conveyed by the sentence. Sinuhe appears to have expected the call to go out to someone else, but against his expectation, it is to one of them, one of the princes, that the call went out.

The adverbial phrases in the other three passages are as follows: *jm* 'there' with *jwr* 'conceive', *m wꜣg* 'on the Wag feast' with *rḫs* 'slaughter', and *jn Rꜥ* 'by Re' and *jn ꜣst* 'by Isis' with *šsp* 'receive, take'.

One argument in support of the presence of contrastive emphasis in the nine examples given below is that the presumed emphasized phrase is repeated for rhetorical effect. Such repetition lends itself easily to contrastive emphasis, as in "It is for you that .... And it is for you that ..." and so on. Eight of the nine examples involve repetition. In the immediate context of *ꜣbẖ* 'link', *n.k* 'for you' appears three times; with *jwr* 'conceive', *jm* 'there' appears twice; with 'hack up', *n.j* 'for me' appears twice; *npḏḏ* and *ḥsḳḳ* appear side by side, both followed by *n.k* 'for you', which is therefore also repeated; with *stp* 'chose', *n.f* 'for him' is repeated; with *šsp* 'take', *jn* 'by' appears four times, each time followed by the name of a different god; and with *ṯs* 'tie', *n.f* 'for him' appears twice.

The texts are as follows. The translations involve the assumption that the adverbial phrase is indeed emphasized.

ꜣbẖ 'link'

ꜣbẖẖ n.k ꜥwy(.j?) 'It is for you that (my?) arms will be linked' (Pyr. §743d TM).

The verb form that immediately follows in *rww n.k rdwy(.j?)* 'It is for you that (my?) feet will move', namely *rww,* exhibits the affix *w.* The affix *w* denotes many things. The passive future substantival would seem to be one of them. In this case, *rww* therefore presumably does not exhibit gemination but rather the affix *w.*

jwr 'conceive'

jwrr N jm 'It is there that N will be conceived' (Pyr. §1960b N).

Msjw in parallel *msjw N jm* 'It is there that N will be born' is an argument in favor of the notion that *w* can serve as a distinctive marker as much in passive future substantival *msjw* as in active future substantival *msjw.*

npḏ 'slaughter'

npḏḏ n.k ṯrp 'It is for you that the *ṯrp*-bird will be slaughtered' (Pyr. §746a TM).

The example with *ḥsḳ* (see just below) immediately precedes.

rẖs 'slaughter'

rẖss sbnwt m wꜣg 'It is on the Wag feast that the cows in suck will be slaughtered' (Pyr. §716c P).

Version T slightly differs: *rẖss sbnwt ḥmwt n N ⟨m⟩ Wꜣg* 'It is on the Wag feast that the female cows in suck will be slaughtered for N'. This is followed in version T by *n-ntt N js pw kꜣ Jwnw* '..., and that because N is the bull of Heliopolis'. I believe that *js* marks afterthoughts that are emphasized in their own right (for details, see Depuydt 1993c, 2008).

ḥsḳ 'decapitate'

ḥsḳḳ n.k smn 'It is for you that the goose will be slaughtered' (Pyr. §746a TM).

The example with *npḏ* (see just above) immediately follows.

ẖbs 'hack up'

ẖbss n.j tꜣ 'It is for me that the land will be hacked up' (Pyr. §1138a P; the younger text in version P has *n.f*; version M has *n N*).

stp 'choose'

('As for any god who will take this king N to the sky …,) *stpp n.f ẖpšw* it is for him that forelegs will be selected' (Pyr. §1026a–b P).

šsp 'receive, take'

šspp ꜥ.f jn Rꜥ … *šspp ꜥ ḥꜣt.f jn ꜣst* 'It is by Re that his hand will be taken … It is by Isis that his bow-warp will be seized' (Pyr. §1347a–b P).

The verb form that immediately follows, *spjw* (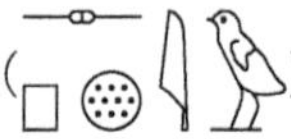), exhibits the affix *w*, which denotes many things. The combination of the features passive, future, and substantival would seem to be one of them.

ṯs (from *ṯꜣs*) 'tie'

ṯss n.f ḥmꜣṯt 'It is for him that the *ḥmꜣṯt*-rope will be tied' (Pyr. §373c W; fragmentary in N).

The verb form that precedes, *jrw* (), exhibits the affix *w*.

Additional instances of what appears to be the affirmative *sḏmm.f* in the Pyramid texts are as follows: *dbḥḥ* (Pyr. §*1888b N; Allen 1984: 687), *sẖnn* (Pyr. §2059a N), *snfẖẖ* (Pyr. §852e N), *sfẖẖ* (Pyr. §1078d P), and *sfẖẖ* (Pyr. §*1923d Nt, completed with PN^{b}Ou; Allen 1984: 690). Interpreting these instances, let alone diagnosing contrastive emphasis in them, is difficult for various reasons, including the fragmentary character of some examples.

8.0.2.5 *Sḏmm.f* and the future *sḏm.f* in the history of Egyptian grammar

The path of the study of the three prospective verb forms has been long and tortuous (for details and bibliographical references, see Depuydt 1993b). First of all, there is no reason not to call the forms in question just simply "future *sḏm.f*" rather than "prospective *sḏm.f*." The term "prospective" needlessly suggests some exotic nuance; the term "future" should therefore do just fine.

The champion of the future *sḏm.f* is Battiscombe Gunn, who in 1924 distinguished *two* future *sḏm.f* forms, the relative one he discovered and Erman's "subjunctive". In 1927, in the first edition of his influential grammar, Alan H. Gardiner accepted only the relative future *sḏm.f*, but in the second and third editions (1950, 1957), Gardiner rejected all future *sḏm.f* forms; this became the dominant view in Egyptology into the 1980s.

Meanwhile, in 1944, H. J. Polotsky had suggested that there were *three* future *sḏm.f* forms: (1) abstract relative (later: substantival); (2) concrete relative (later: adjectival); and (3) predicative (later: adverbial). An opportunity for additional articulation of the future *sḏm.f* arose in the mid 1950s with Edel's proof that *sḏmw.f*, a *sḏm.f* form sometimes exhibiting *w*, is a distinct form in its own right. As for passive *sḏmw.f*, Edel only tentatively identified it as a future tense (cf. Allen 1984: 358 §529). He writes as follows (Edel 1955/64: 266 §559).

> The passive [verb form] that is marked by the repetition of the last root consonant is attested more rarely [in Old Egyptian] because it is used only in the Pyramid texts and in certain verbal forms. Outside of the future and the cases listed in §568, the existence of [passive *sḏmm.f*] cannot be verified with certainty.*

The "cases in §568" are in fact balanced sentences, found in Pyr. §492a–d and cited in §8.1 below. Edel does translate them in the future tense. It was Westendorf (1962) who first definitively classified active *sḏmw.f* as a future tense; but he accepted only a single future nonrelative (non-adjectival) *sḏm.f*. For example, *sḏmw.f* was considered the same form as the future *sḏm.f* that follows *rd*(*y*) when *rd*(*y*) means 'cause'. The evidence shows that it is not. It is now clear that there are definitely *two* future nonrelative (non-adjectival) *sḏm.f* forms, as Polotsky had already surmised in 1944. To this day, the tacit assumption that there is only one future nonrelative *sḏm.f* seems widespread. The complex nature of the evidence and the tortuous path of the history of research on the future *sḏm.f* forms are to an extent responsible.

In the late 1960s and the 1970s, Polotsky first proposed that the *sḏmw.f* is always substantival. He also tentatively suggested that the other nonrelative future *sḏm.f* is adverbial. The "splitting" of the future nonrelative (non-

* Das Passiv mit Wiederholung des letzten Radikals ist [im Altägyptischen] infolge seiner Beschränkung auf die [Pyramidentexte] und auf bestimmte Verbalformen seltener belegt und scheint ausserhalb des Futurs und der Fälle des §568 nicht mit Sicherheit nachgewiesen werden zu können.

adjectival) *sḏm.f* into two distinct forms gained definitive articulation in works by James P. Allen and Pascal Vernus in the 1980s and 1990s. Incidentally, Allen (1984) calls the *sḏmw.f* the "prospective" and the other future *sḏm.f* the "subjunctive." Allen had earlier (1979a) called the *sḏmw.f* the "future" and the other one the "prospective." Vernus calls the *sḏmw.f* the "prospective *sḏmw.f*" and the other one the "prospective" *tout court.* I would myself call the *sḏmw.f* the "substantival" future *sḏm.f* and the other one the "adverbial" future *sḏm.f.*

Allen's and Vernus's *sḏmw.f* is not always substantival. The inevitable conclusion is that there is both a substantival *sḏmw.f* form and a nonsubstantival *sḏmw.f* form. The alternative is the implausible assumption that a single form can be both substantival and not substantival at the same time. A further inevitable conclusion from Allen's and Vernus's shared views is that there are then *more than three* future *sḏm.f* forms: (1) the relative future *sḏm.f*; (2) the substantival *sḏmw.f*; (3) the nonsubstantival *sḏmw.f*; and (4) the future *sḏm.f* that follows *rd*(*y*) 'cause'. There would be *five* future *sḏm.f* forms if (4) is also split into a substantival form and a nonsubstantival form. However, these necessary logical consequences were never followed through.

In light of all the evidence, my view is that there are three, and no more, future *sḏm.f* forms: (1) A future substantival form, sometimes exhibiting *w* and therefore called *sḏmw.f.* (2) A future adverbial form, probably originally independent since this appears to be the origin of adverbial past *sḏm.n.f* and adverbial aorist *sḏm.f.* It follows *rd*(*y*) 'cause' (literally probably: 'give in order that', cf. Depuydt 1996c) and exhibits exclusive forms such as *jwt.f*, *jnt.f*, and *mꜣn.f.* (3) A future adjectival (relative) form, as in *jrt.f* 'that which he will do', as opposed to *jrt.n.f* 'that which he has done' and *jrrt.f* 'that which he does'.

This section §8.0.2 has shown that the time has come to add *sḏmm.f* to the substantival family.

8.0.3 Balanced sentences with the second versus the first verb form written distinctively as substantival

Certain strings of words contain two verb forms denoting two events that seem to be equated with one another. These are likely balanced sentences. When both verb forms are written *exclusively* as substantival, their identification as balanced sentences is secure (§8.1). Exclusive writings of substantival verb forms are writings that cannot be interpreted as anything else than substantival on the basis of visible empirical features alone. In other words, *both* verb forms are *distinctively* written as substantival (§8.2). Distinctive writings mark verb forms as substantival in contrast with many other verb forms but not with all. In some cases, the writing of *one* verb form is distinctively substantival; when it is the second of the two verb forms (§8.3), the diagnostic value is greater than when it is the first.

Why is this so? To answer this question, we begin with the fact that only three of the four possible pairings of forms with gemination (seen in third-weak verbs) in balanced sentences occur:

		Model	Attested
gemination	+ gemination	(1) *prr.f prr.j*	yes
gemination	+ no gemination	(2) *prr.f pr.j*	yes
no gemination	+ no gemination	(3) *pr.f pr.j*	yes
no gemination	+ gemination	(4) **pr.f prr.j*	no!

There are two cases. First, the *second* verb form is written distinctively as substantival (see *prr.j* in [1]). Second, the *first* verb form is written distinctively as substantival (see *prr.f* in [1] and [2]). Within the confines of the pairings listed above, the following applies. First, when the second verb form is written distinctively as substantival (see *prr.j* in [1]), the first is also substantival (see *prr.f* in [1]). The substantivalness of the second verb form implies the substantivalness of the first. Second, when the first verb form is substantival (see *prr.f* in [1] and [2]), the second is either distinctively substantival (see *prr.j* in [1]) or not (see *prr.j* in [2]). The substantivalness of the second verb form allows no conclusion about the first.

This scenario is narrow, being limited to certain types of verbs; but third-weak verbs are sufficiently common to suggest that the situation revealed by them applies by analogy (§8.2.4 below) also to all instances in which two events are juxtaposed and equated. When a second verb form is substantival, the first is too (§8.3).

8.0.4 A notable absence in balanced sentences as formulas

In an article that has generated much reaction, Vernus (1981: 75) proposed that a subordinate clause commonly *precedes* its main clause in Old and Middle Egyptian (cf. §§9.8.3, 9.9.2.1 below). He calls this configuration the "Second" scheme. By contrast, in the "First" scheme, a subordinate clause *follows* its main clause.

> In the "first scheme," the process conveyed by the second clause (= circumstantial) conditions the process conveyed by the first clause.*

* Dans le "premier schème," le procès représenté par la seconde proposition (= circonstant) conditionne le procès représenté par la première proposition.

A construction that belongs to the First scheme is the adverbial sentence consisting of a substantival and an adverbial verb form. An example is *jj.n.j ḫr.k … jnt.k wj m3n.j nfrw.k* (lit.) 'That I have come to you is in order that I might bring you back and that I might see your beauty' (Introduction to chap. 25 of the Book of the Dead; Budge 1898: 249, lines 5–6).

(A possible alternative analysis of this type of sentence is that both the substantival and the adverbial verb forms are subordinate to the overarching structure of the adverbial sentence.) More than one construction meets the criterion for the Second scheme. Two constructions are as follows: an *adverbial* or *circumstantial* clause precedes its main clause; or, a *substantival* verb form *used adverbially* precedes its main clause. In Vernus's formulation (1981: 77), the latter consists of a "substantival form with circumstantial function plus nonsubordinated verbal form":*

> The substantival form of the first member assumes circumstantial function, just as a substantive can assume this function.†

Niccacci (1980) accepts the existence of a construction of the same form, but he does not give it the same name. Furthermore, unlike Vernus, Niccacci (1980: 207) interprets the initial substantival verb form as extraposed ("As for the fact that …") rather than adverbial. He describes it as "a nominal form placed at the head of the sentence in extraposition."‡

I do not believe that the Second scheme exists, so it is irrelevant whether the second member of the Second scheme is interpreted *à la* Vernus or *à la* Niccacci. Most of their examples can be interpreted instead as adverbial sentences. The structure of some others is obscure.

Two general observations about the Second scheme are in order. The establishment of a new grammatical category requires absolutely unambiguous examples. In the second construction, substantival verb form preceding main clause containing an independent verb form, it does not matter whether the initial substantival verb form is interpreted adverbially, with Vernus, or as extraposed in anticipatory emphasis, with Niccacci. I would accept the existence of this construction if types such as **prr.sn jw.j pr.j* 'When they ascend, I ascend' and **prr.sn pr.ḫr.j* 'When they ascend, then I ascend' were attested. This assumes that *prr.f* begins a new sentence. There is no doubt that the forms *jw.f sḏm.f* and *sḏm.ḫr.f* are independent, but no examples of those two types are attested anywhere.

* Forme substantive en fonction circonstantielle + forme verbale non subordonnée.

† La forme substantive du premier membre assum[e] la *fonction* circonstantielle, puisqu'aussi bien un substantif peut assumer cette fonction.

‡ Una forma nominale messa in testa alla frase in estraposizione.

Instead, Niccacci (1980: 204) quotes examples like *prr.sn r pt m bjkw jw.j ḥr d̲nḥw.sn* 'When they ascend to heaven as falcons, I am on their wings' (CT 3.100h–i, Coffin S1C). Niccacci believes that *jw.j ḥr …* is an independent adverbial sentence serving as a main clause. However, *jw* plus suffix pronoun should be interpreted as marking a subordinate adverbial or circumstantial clause, as so often. The statement as a whole is then an adverbial sentence, 'It is while I am on their wings that they ascend to heaven as falcons'. The free translation would still be the same as Niccacci's ('When they ascend to heaven as falcons, I am on their wings'). This alternative interpretation is not certain, but it is distinctly possible. This possibility renders the example unsuitable for establishing a new grammatical construction as proposed by Niccacci.

Another of Niccacci's examples (ibid.) can also be translated as an adverbial sentence.

n-ntt p3.n.sn js r pt m bjkw *jw N pn ḥr d̲nḥw.sn*
n-ntt h3⟨.n⟩.sn js r t3 m ḥf3w *jw N pn ḥr ḳ3bw.sn*
(CT 6.283f–i, Coffin B1Bo)

'And that because it is while this N is on their wings
that they have flown to heaven.
And that because it is while this N is on their coils
that they have descended to earth as snakes'.

Allen and Niccacci both propose translations different from the above. Allen (1979b: 8) translates so that *m bjkw* is emphasized, 'For it is as falcons that they have flown to the sky'. Niccacci (1980: 201) interprets *jw N pn …* as the main clause: 'Because, when they have climbed to heaven as falcons, this N was on their wings; because, when they have descended to earth, this N was on their coils'.*

As for *js*, I have proposed that it is directly associated with adverbial *n-ntt* 'because'. I believe that *js* marks sentence anaphora, reference to a preceding thought (see Depuydt 1993c, 2008 [with bibliography]). Allen (1979b: 8), however, following Mordechai Gilula, believes that *js* marks its sentence as a nominal clause. Since *ntt* also marks what follows as a nominal clause, Allen follows Gilula in interpreting the present example as "'hypercharacterization' of the noun-clause"—with both *ntt* and *js* denoting nominalization. I, however, believe that unlike *n-ntt*, *ntt* alone cannot trigger *js*.

* Poiché, quando essi sono saliti al cielo come falconi, questo N era sulle loro ali; poiché, quando essi sono scesi a terra come serpenti, questo N era sulle loro spire.

Niccacci proposes a third example.

p3.n N pn m bjk *sbk pw N pn* (CT 6.295s–
ngg.n N pn m bjk *sbk pw N pn* 296c, Coffin B1Bo)
sp3.n N pn m gbg3 *jnpw pw N pn nb ḳrst*

Once again, Niccacci's interpretation is not binding. The text might contain six independent sentences.

'It is as a falcon that this N has flown away. This N is Sobek.
It is as a falcon that this N has cried out. This N is Sobek.
It is as a vulture that this N has flown away.
This N is Anubis, lord of the funeral'.

Instances of balanced sentences found in the Coffin texts are discussed by Parker (1996: 47–52, 112–15). Many of the texts are preserved in good copies and parallel versions, but the esoteric content of the texts often makes it difficult to understand them, which makes it difficult to choose between rival interpretations.

The following consideration also argues against the existence of the Second scheme (see also §§9.9.3, 9.10.2.1 below). Adverbial or circumstantial *jw* never precedes its main clause in Late Egyptian or Demotic, so why would its Old and Middle Egyptian equivalents do so?

Also, consider the following formula from the Coffin texts.

prr.ṯn r pt m nrwt prr.j ḥr tpt ḏnḥwy.ṯn (CT 3.61f–g, Coffins B1C, B2L; B2L has *ḏnhw.ṯn*)

'When you (pl.) ascend to heaven as vultures, I ascend on top of your wings'.

Niccacci (1980) has studied all the variations of this formula. By his account, only the following three combinations of absence and presence of gemination in third-weak verbs are attested (§8.0.3).

(1) *prr.f* *prr.j*
(2) *prr.f* *pr.j*
(3) *pr.f* *pr.j*

What is striking is that none of these three combinations lends itself to interpretation as the Second scheme, adverbial verb form preceding main clause. The patterns of the adverbial sentence and of the balanced sentence, which is a nominal sentence, suffice to interpret these examples. If the Second scheme was characteristic of Old and Middle Egyptian, why is the formula not unambiguously attested?

8.1 BOTH VERB FORMS EXCLUSIVELY WRITTEN AS SUBSTANTIVAL

šnṯṯ N *šnṯṯ Tm*
ḫsbb N m w3t tn (M: *w3t*) *ḫsbb Tm* (Pyr. §492b,d WPMN)

'When N is (lit., will be) reviled, Atum will be reviled.
When N is barred on this (M: the) road, Atum will be barred'.

The interpretation of this example is clear and its identification as a balanced sentence is secure. Before the concept of balanced sentence was introduced into Old and Middle Egyptian grammar, Edel (1955/64: 269 §568) classified this passage as the use of *sḏmm.f* "in the initial subordinate clause and the main clause of a condition,"* citing only this passage. In fact, Pyr. §492b,d may well contain the only securely established examples of balanced sentences featuring *sḏmm.f* anywhere.

Other examples have been adduced (Allen 1984: 351 §520). Two geminating forms are concerned: *tmm* (Pyr. §230c W, §*1075a–b; see Allen 1984: 677) and *bšš* (Pyr. §1460c PM). The interpretation of these additional instances poses various problems, one of them concerning the verb classes of the verbs involved. These classes are otherwise not known to occur as *sḏmm.f*. The following verb classes are attested for *sḏmm.f*. According to Edel about Old Egyptian (1955/64: 264 §557), "Repetition of the final strong consonant is securely attested only in bi- and triliterals and in fourth-weaks and causatives."† According to Allen about Pyramid texts (1984: 349 §515), "The geminated passive (*sḏmm.f*) occurs in conjugations with strong or geminated final radicals and in the conjugation of 4ae-inf[irmae] prefixing verbs." The difference between Edel's list and Allen's list is the inclusion of geminating verbs in the latter and the exclusion of causatives from it.

The problems with the verb-class membership of the presumed additional instances of *sḏmm.f* in balanced sentences are as follows. First, as regards *bšš*, the verb *bš*(*y*) 'spit' is a well-known third-weak verb, but no instances of the *sḏmm.f* of third-weak verbs are otherwise known. In order to treat *bšš* nevertheless as an instance of *sḏmm.f*, one solution is to classify the verb in question exceptionally as a biliteral verb. This is what Allen (1984: 545 §728) does when he concludes that "the passive has a geminated

* Im Vorder- und Hauptsatz bei einer Bedingung.

† Die Wiederholung des letzten starken Konsonanten ist mit Sicherheit nur bei den 2- und 3-rad[ikaligen], sowie bei den Verben IV. inf. und Kausativen belegt.

stem, which rules out the M[iddle] K[ingdom] *bšj*." The problem remains that *bš*(*y*) is then considered biliteral only to accommodate this one example, even though this verb is elsewhere clearly third-weak. The hypothesis that *bš* is biliteral therefore rests on a very narrow base.

Other problems in addition to those pertaining to the verb-class affect the interpretation of Pyr. §1460c PM. First, its meaning is not entirely clear. Second, version P exhibits the negation *n* (𓂜), whereas version M does not. The reliability of the transmitted text is therefore called into doubt. At Pyr. §1460b–c, version P's *n wg.s bšš.s n wgw.f bšš.f* contrasts with version M's *wgjt bššt wgjw.f bšš.f*. The verb *wg*(*y*) means 'chew' or the like. All in all, Pyr. §1460b–c is beset with too many problems to serve as evidence for *sḏmm.f* in the balanced sentence.

The other presumed examples of balanced sentences involve a verb whose first two consonants are *tm*.

tmm r.k jn šsmt tmm r n šsmt jn mꜣfdt (Pyr. §230c W)

tmm(*.j*) *tmm* [*tꜣ*] *tmm ṯpḥwt tmm jmt.sn* (Pyr. §*1075a–b; see Allen 1984: 677)

At first sight, without even attempting any translations, the second example suits the pattern of the balanced sentence much better than the first. The first sentence could more easily be interpreted as a balanced sentence if it exhibited some such reciprocity as **tmm r n mꜣfdt jn šsmt tmm r n šsmt jn mꜣfdt*. As it stands, the sentence more resembles two juxtaposed emphatic sentences, with emphasis resting on the two agents introduced by *jn* 'by': 'It is by Shesmet that your mouth … and it is by Mafdet that the mouth of Shesmet …'. Neither the meaning nor the verb class of *tm* is fully clear. Clearly, *tm* is not the negation verb *tm*, which is biliteral. This other *tm* appears to have the meaning 'close (off)' in the Pyramid texts. The notion of closing might fit the mention of mouths in the first example and of caves in the second example. Allen (1984: 351 §520) classifies *tm* in the two examples cited above as a second-doubling verb. These two examples are apparently the origin of his belief (Allen 1984: 349 §515) that the *sḏmm.f* is also attested in last-doubling verbs, but there appear to be no other examples. It is therefore premature to accept that the form existed on the basis of examples whose interpretation is problematic.

8.2 BOTH VERB FORMS DISTINCTIVELY WRITTEN AS SUBSTANTIVAL

8.2.1 Book of the Dead, chapter 90, Papyrus of Nu (Budge 1898: 192, lines 10–12)

jw.k r.j *ḏd.j r.k*
tm.k jw r.j *tm.j ḏd r.k*

'You come against me, I say (something) against you.
You do not come against me, I do not say (something) against you'.

This example deserves pride of place. Polotsky (1964: 282 [1971: 67]) considered the negation with *tm* as positive proof of the existence of the balanced sentence:

> Instances in which both *sḏm.f*-forms are negated must constitute hard proof [of the substantival nature of verb forms in balanced sentences]: one would expect *tm.f* in both members.*

It is a fact that *tm* negates both substantival and adjectival verb forms. In the present case, however, *tm.k jw r.j* could not be an adjectival or relative clause—but **tm.k jw r.f* could be; it would mean something like 'the one against whom you do not come'.

Tm.k jw is, at least, a highly distinctive writing of a substantival verb form. The question arises: Is it an exclusive writing? It is widely believed that *tm* negates *only* substantival and adjectival verb forms. If that were true, then *tm.k jw r.j* would be a fully exclusive writing of a substantival verb form.

It is generally accepted that *tm* never negates adverbial verb forms. I have contested this notion, however (Depuydt 1993b: 27–28). I believe that there is nothing inherently nominal (that is, substantival or adjectival) about *tm*. One clear example in which *tm.f* is not nominal is *nn tm.f jr bw-nfr* 'He will not fail to do good things' (Sinuhe B74–75). After the negation *nn*, one expects forms such as *jwt.f*, *jnt.f*, and *m3n.f*, that is, the so-called subjunctive, which I prefer to call the adverbial future *sḏm.f*. The assumption that *tm* can on occasion be adverbial would explain why the negated *tm.f sḏm* can denote a final clause or purpose clause, which is adverbial. Still, it is difficult to deny that *tm* typically serves as the negation of substantival and adjectival verb forms. What is needed is a satisfactory explanation of why *tm* is *not*

* Die Probe aufs Exempel müssten Fälle liefern, in denen beide *s.f*-Formen negiert sind: es wäre in beiden Gliedern *tm.f* zu erwarten.

common in adverbial verb forms. Plausible historical scenarios may provide an explanation.

I have proposed (Depuydt 1995c, 2002b) that adverbial verb forms came into existence in Old and Middle Egyptian by the simple omission of a characteristic sentence pause. Absence of a pause can also denote subordination in English, as in "I saw him he is sick," as opposed to "I saw him. He is sick." "He is sick" turns into a circumstance of a preceding main clause by omission of a pause in "I saw him he is sick." It is a fact that adverbial or circumstantial forms often appear identical to independent forms, as in *n sḏm.n.f*, which as an independent form means 'He does not hear' and as a dependent form 'while he does not hear'. Presumably, *n sḏm.n.f* is originally an independent form that became adverbial or dependent by omission of a feature marking the sentence boundary. That would explain why all the adverbial forms in question *follow* their main clause and do not *precede* it, even if some have defended the view that the forms in question can also precede.

Omission of a pause subordinates something that follows to something that precedes, and not the other way around. Indeed, in the process of omission between two independent sentences, the first independent sentence still begins as an independent sentence. Nothing has changed. Change occurs when one arrives at the place where the pause is omitted. The change therefore affects the status of the *second* sentence, which is turned from an independent sentence into a dependent clause. I am not alone in assuming that an adverbial verb form cannot *precede* its main clause. The scenario outlined here supports this assumption.

The affinity between independent and adverbial verb forms is undeniable in the Old and Middle Egyptian verbal system. The scenario proposed above would fully account for this affinity. It explains that, because independent forms are negated by *n* and *nn*, so are adverbial verbs. Unlike adverbial forms, substantival and adjectival forms do not exhibit any affinity with independent forms. It comes therefore as no surprise that substantival and adjectival forms are negated differently from independent and adverbial forms. The auxiliary verb *tm* typically performs the function of negating substantival and adjectival forms, but this does not mean that *tm* is somehow inherently nominal. *Tm* is potentially a fully declinable verb, like any Egyptian verb. It is just that, according to the scenario proposed above, the negation of independent and adverbial forms was already taken care of, as it were. There was no need for *tm* to perform this function. That does not mean that *tm* could not *sometimes* be neither substantival nor adjectival. The use of *tm* in final clauses is a case in point. This usage is well known, though examples are perhaps not as plentiful as we might like. There appear to be no certain examples in the earliest corpus, the Pyramid texts (Allen 1984: 225 §346). In final clauses, distinctive writings of the subjunctive are found, which I would call the adverbial future *sḏm.f*, such as *jwt.f* and *m3n.f*. These forms are not normally thought of as either substantival or adjectival. *Tm* would therefore be expected to exhibit the same nonnominal form.

In sum, *tm.k jw r.j* is perhaps not an exclusive writing of a substantival verb form. It owes its identification as such partly to *r.j*. *Tm.k jw r.f* could have been adjectival, meaning 'the one against whom you do not come'. But even *tm.k jw r.j* is not fully exclusive. It is a possible writing of a final clause, but it is difficult to see how *tm.k jw r.j* could be a final clause in the passage cited above.

8.2.2 CT 5.326g–h, Coffin B2L

jw.k r.j ḏd.j r.k
tmm.k (probably for *tm.k*) *jw r.j tm.j ḏd r.k*

This example is almost identical to that in §8.2.1. As a Coffin text, it has the advantage of being earlier than that one and therefore slightly more reliable; Middle Egyptian was no longer spoken when the Book of the Dead version was copied. The present example has the disadvantage, however, of requiring the emendation of *tmm.k* into *tm.k*. Then again, as a biliteral verb, *tm* does geminate in the past passive participle, as in *jr.n.k tmmt jrw* 'You have done that which has not been done' (CT 6.371f) (for other examples, see Depuydt 1996b: 10 with n. 12, and 22 top). The form *tmm* can also be used as an agentless indirect adjectival verb form, which takes its form from the passive participle, as in *tmmt bs Wr-m3w ḥr.s* '(one) into which no one has initiated the Great One of Seers' (CT 6.272a/273a). In short, the writing *tmm.k* is suspect. According to all else that we know about Old and Middle Egyptian grammar, the form is not really supposed to exist.

Discussion: Loprieno 1995: 216.

8.2.3 Pyr. §412b

mrr.f jrr.f

'When he wants (something), he does (it)' (name of a god)

The writings *mrr.f* and *jrr.f* are distinctive in that there are only two possible interpretations. The characteristic feature is gemination in third-weak verbs, which marks just two suffix-conjugation forms. Both are active in voice and aorist in tense. However, as regards function, they can be either substantival or adjectival. It can safely be concluded that the forms *mrr.f* and *jrr.f* are not adjectival, because in the immediately following parallel *msḏḏ.f n jr.n.f*, *n jr.n.f* is not. Moreover, if *mrr.f* and *jrr.f* were adjectival, translation as a nominal sentence, something like 'What he likes is what he does', would result. In that case, however, one would expect feminine adjectival forms, *mrrt.f* and *jrrt.f*, as in (along with the copula *pw*):

mrrt.ṯ pw jrrt.ṯ (CT 6.273b, Coffin G1T)

'It is (only) what you (f.sg.) like (to do) that you do (f.sg.)'.

In the present case, the nominal sentence also serves as a cleft sentence. The sentence derives its status as a cleft sentence from the fact of contrast with absence of emphasis. Indeed, the same thought might also have been rendered as 'You do what you like' rather than 'It is (only) what you like that you do'. Cleft sentences appear to derive from nominal sentence structures in all stages of Egyptian and sometimes they retain their structure. That is the case with cleft sentences with *pw* in earlier Egyptian and all cleft sentences in Demotic. In many other cases, the structure of cleft sentences has grown away from that of nominal sentences, as with cleft sentences introduced by *jn* in Old and Middle Egyptian and with cleft sentences generally in Coptic. A descriptive/historical study of these phenomena remains desirable.

Pyr. §412b may be the best-known and most often cited example of the balanced sentence. *Mrr.f jrr.f* is in fact the name of a god whose main distinctive characteristic seems to be that he does what he wants. Indeed, at Pyr. §412b, *mrr.f jrr.f* follows genitive *n* 'of'. *N mrr.f jrr.f* then means 'of *Mrr.f–jrr.f*'. The name also occurs in the Coffin texts. For example, at CT 3.386b (two versions on coffin S1C, in addition to versions on other coffins), the god himself states: *jnk mrr.f jrr.f jt nṯrw* 'I am *Mrr.f–jrr.f*, father of the gods'.

In Pyr. §412b, *mrr.f jrr.f* is followed by *msḏḏ.f n jr.n.f*, an adverbial sentence. Its literal translation is then 'That he hates (something) is while he does not do (it)'. I am aware that an alternative interpretation has enjoyed some following: an initial substantival verb form followed by a main sentence, with a literal translation something like 'As for the fact that he hates (it), he does not do (it)', but I am not sure that this construction even exists.

An additional point regarding *msḏḏ.f n jr.n.f* is worth noting. There seems to be a certain disharmony between the balanced sentence *mrr.f jrr.f* and the presumed adverbial sentence *msḏḏ.f n jr.n.f*. On the one hand, the balanced sentence seems reversible: his wanting equates to his doing and his doing therefore equates to his wanting. On the other hand, his hating is accompanied by the circumstance of his doing, but nothing is said about what his not doing is accompanied by. The adverbial sentence does not seem reversible in the way that the balanced sentence is. The logical implications of this state of affairs are examined in §5.3 above. It is concluded there that, logically speaking, the adverbial sentence in the second part of the statement can easily be interpreted as following naturally from the balanced sentence in the first part. Furthermore, there is no problem with an adverbial sentence being parallel to a balanced sentence. Consider the clause *nn ḏd.j r.k* in

> *jw.k r.j ḏd.j r.k tm.k jw r.j nn ḏd.j r.k* (CT 5.323h–i, Coffin B2L)
>
> 'You come against me, I say against you.
> (But) it is while I do not say against you
> that I do not come against you'.

I would interpret *nn ḏd.j r.k* as an adverbial or circumstantial clause that alternates with a substantival clause, *tm.j ḏd r.k*, in the same spell (CT 5.326h, Coffin B2L; and Book of the Dead variant in chap. 90 [Papyrus of Nu (Budge 1898: 192, lines 10–12)]). The difference is that the substantival clause involves reversibility whereas the adverbial clause does not. There is a logical difference, see §5.3.

> *Discussions*: See §§5.3, 9.3, Gardiner 1947: 99, 1957: 356 §442,4; Polotsky 1964: 282 (1971: 67); Vergote 1965: 353. This list is hardly exhaustive because Pyr. §412b is mentioned in almost every substantial discussion of the balanced sentence.

8.2.4 Pyr. §696d TN

> *jbb N jbb Nḫbt*
>
> 'When N is thirsty, Nekhbet is thirsty'.

The form *jbb* is distinctive to the same degree as *mrr.f* and *jrr.f* in §8.2.3. *Jbb* is no doubt active in voice and aorist in tense. The written feature marking the two forms as either substantival or adjectival (relative) is again gemination in third-weak verbs. For the reasons also adduced in §8.2.3, it can safely be concluded that *jbb* is not adjectival, even if the writing could be that of an adjectival verb form.

The present example differs from *mrr.f jrr.f* in that both clauses feature the same verb, namely *jb(y)* "become thirsty." The two substantival verb forms resemble one another much more closely than in the case of *mrr.f jrr.f*. Equality is a concept that naturally comes to mind when considering an example like this one. Such examples are therefore particularly supportive of the existence of the balanced sentence, that is, as two substantival clauses being equated with one another as part of a nominal sentence.

The present example also offers an illustration of the principle of parallelism first enunciated by Adolf Erman, according to which there is "only one way to make progress [in the study of the verbal system]. One must observe verbs with mutable stem such as second geminate verbs … and irregular verbs. When one of these verbs exhibits a distinct writing for a certain verb form, the existence of the same verb form may be extended to verbs that do not have a distinct written form" (for bibliography, see Depuydt 1993b: 11–12; see also Parker 1996: 19). Here follows a parallel to Pyr. §696d:

> *ḥḳr N ḥḳr Rwtj* (Pyr. §696c)
>
> 'When N is hungry, Ruti is hungry'.

Hieroglyphic writing provides no clues as to the parsing of the two verb forms *ḥḳr*. But *ḥḳr* is fully parallel to *jbb* in the next sentence, so it should be assumed that *jbb* and *ḥḳr* represent the same verb form.

From the fact that *mrr.f* and *jrr.f* in *mrr.f jrr.f* are not followed by adverbial phrases, Gardiner (1947) had concluded that geminating forms of third-weak verbs can function as the verb forms of independent sentences. Polotsky (1964: 282 [1971: 67]) turns this argument on its head, reasoning that, if it has not been proven that such a geminating form can function in this way, then it is more likely that *jbb* is also an abstract relative (i.e, substantival) form.

> As long as it has not been shown that *ỉbb*– can stand by itself, one must assume that the two instances of *ỉbb*– mutually condition one another by their grammatical form.*
>
> *Discussions*: Farina 1927: 153 §258 ("in the protasis and apodosis of two correlative conditional propositions"†; Polotsky 1964: 281 (1971: 66).

* Solange nicht nachgewiesen ist, dass es ein alleinstehendes *ỉbb*– geben kann, ist anzunehmen, dass die beiden *ỉbb*– sich auch in ihrer grammatischen Form gegenseitig bedingen.

† Dans la protase et dans l'apodose de deux propositions conditionnelles corrélatives.

8.2.5 CT 3.61f–k, Coffins B1C, B2L

prr.ṯn r pt m nrwt *prr.j ḥr tpt ḏnḥwy.ṯn* (B2L: *ḏnḥw.ṯn*)
prr.ṯn r pt m ḥfȝw *prr.j ḥr ḳȝbw.ṯn*
prr.ṯn r pt m jȝrwt *prr.j ḥr wpt.ṯn*

'Whenever you depart for heaven as vultures,
I depart on top of your wings.
Whenever you depart for heaven as snakes,
I depart on your coils.
Whenever you depart for heaven as uraei,
I depart on your forehead'.

This is a clear example which, for once, does not pose any philological problems whatsoever.

Discussions: See chap. 1 and §5.1.4 above; Gardiner 1957: 357 §444,1; Niccacci 1980: 199.

8.2.6 CT 6.302m–p, Coffin B1Bo

hȝȝ.sn m ḥfȝw *hȝȝ N pn ḥr ḳȝbw.sn*
ṯss.sn m bjkw *hȝȝ.f* (for *ṯss.f*) *hr dnhw.sn*

'Whenever they descend as snakes,
this N descend on their coils.
When they rise as falcons,
he rises on their wings'.

Like the preceding example, this one is straightforward. The version of Coffin B1Bo is reliable. But unlike the preceding example, this one is not entirely free of problems. The form *hȝȝ.f* is in all probability a mistake for *ṯss.f*. Still, the erroneous verb form *ṯss.f* still exhibits gemination.

Discussion: Polotsky 1984: 119.

8.2.7 CT 3.24a–25b, Coffins S1C, S2C

hȝȝ.sn m ḥfȝw *hȝjj.j m ḳȝbw.sn*
prr.sn r pt m bjkw *prr.j r ḏnḥw.sn*

"When they descend as snakes,
I (will?) descend on their coils.

When they ascend to heaven as falcons,
I ascend on their wings'.

A similar text appears at Pyr. §1056a P: [*h33.sn r t3*] *m ḫf3wy h3jj*(*.j*) *ḥr ḳ3*[*bw.sn*]. In the Coffin text example, only the second pair of substantival forms meets the criterion of examples gathered in §8.1, namely that both forms exhibit *unambiguous* writings of substantival verb forms. In the first pair, the identity of the second form, *h3jj.j*, is not fully clear. But still, the double reed leaf is a *possible* writing of a substantival form.

The authenticity of the writing *h3jj.j*, with double reed leaf, is guaranteed by the fact that the versions of several other coffins also exhibit the form. The double reed leaf appears regularly in two future forms of third-weak verbs, so it is clear that *h3jj.j* is an adverbial future *sḏm.f*. Schenkel (2000) has established that *h3jj.j* can also be a *substantival* future *sḏm.f*. There are no other forms in which the double reed leaf appears with regularity. For the adverbial aorist *sḏm.f*, one expects the writing *h3.f*, without double reed leaf, as in the compound *jw.f h3.f* 'he (generally) descends'. If *h3jj.j* is future substantival, the literal translation is something like 'That they descend as snakes means that I *will* descend on their coils'. If *h3jj.j* is adverbial, the literal translation is something like 'That they descend as snakes is while (in the circumstance that) I will descend on their coils'. The second pair of verb forms in this example is easily interpreted as a balanced sentence. However, versions on other coffins have *pr.j* and *prjj.j* for the second verb form. It may be emphasized again that the opposite does not occur, namely *absence* of gemination in the *first* verb form and *presence* of gemination in the *second* verb form. What has been said above about *h3jj.j* above also applies to *prjj.j*. As for the form *pr.j*, absence and presence of double reed leaf alternate as writings of the substantival and adverbial future *sḏm.f* of third-weak verbs.

Discussions: Niccacci 1980: 201; Junge 1989: 25 (E22).

8.2.8 CT 3.115e–h, Papyrus Gardiner II

h3⟨*.n*⟩*.sn r t3 m ḫf3w* *h3.n.j m ḳ3b.sn*
pr.n.sn r pt m bjkw *pr.n.j ḥr ḏnḥwy.sn*

'Whenever they have descended to earth as snakes,
I have descended in their coils.
Whenever they have ascended to heaven as falcons,
I have departed on their wings'.

Like the gemination in third-weak verbs cited in the preceding examples, the *sḏm.n.f* of verbs of motion unambiguously marks verb forms as substantival. The text exhibits a minor problem in that one of two *n*'s appears to have been omitted in the form *h3.n.sn*.

> *Discussions*: See chap. 1 above; Niccacci 1980: 199–200; Polotsky 1984: 118.

8.2.9 Temple at Deir el-Bahri (Naville 1901: plate CXIV, line 18 from the right)

ḫnt.k dd.tw n.k j3w
ḫdd.k dd.tw n.k s3-t3

'Whenever you sail south, one gives you praise.
Whenever you sail north, one pays respect to you'.

This example dating to Hatshepsut's reign is remarkably late—it may have been copied literally from an Old or Middle Kingdom text.

This example also affords an application of the principle of parallelism. *Ḫdd.k* must be substantival. Therefore parallel *ḫnt.k* is substantival as well, even if no empirical feature of its writing makes it identifiable as such.

> *Discussion*: Gardiner 1957: 357 §444,1.

8.2.10 A Special Case: Pyr. §193c Nt

s.k *sw Nt pn* (*sic*)
sw Nt pn *s.k*

'You will go (or: perish), this Neith will go.
This Neith will go, you will go'.

This example may belong in §8.1 if the two statements are taken together as one. In the first statement, the second verb form is marked as substantival. In the second statement, the first verb form is marked as substantival. If the two statements are taken as one, then *both* verb forms are marked as substantival. If the first statement is taken separately, it could be classified with the examples in §8.3. If the second statement is taken separately, it could be classified with the examples in §8.4.

In version W, *s.k s N pn s N pn s.k*, the writings of the verb forms are not distinctive. Edel (1955/64: 241 §517) cites the following variant in "Theban Tomb 319." I cite the text from his grammar.

sjj.k *sw N* *n sw N* *n sw.k*

'When you perish, N perishes. N will not perish.
You will not perish'.

The Middle Egyptian of the New Kingdom always requires a certain amount of caution. To what degree did New Kingdom scribes still master the earlier language? In the variant just quoted, the first half could easily be a balanced sentence, as in the Pyramid text version, but the second half is not. The second half does exhibit a certain symmetry and stylistic balance; but for this balance to be grammatical as well, it ought to include the negation verb *tm*, yielding something like **tm s*(*y*) *N tm.k s*(*y*). Note that the negatival complement *s*(*y*) precedes the actor expression *N* (see Gardiner 1957: 263 §343). There is no difficulty in interpreting *sjj.k* as a writing of the active future substantival form. The occasional occurrence of the affix *w* is perhaps its most typical characteristic, but it is now accepted that *jj* can be a variant of *w* in that form. Schenkel (2000) observes that this is typical before one-consonant suffixes, as in *sjj.k* here. The twofold appearance of *w* after the negation *n* is also not problematic. The *sḏmw.f* is normal after the negation *n*, but it seems peculiar that it is spelled both *sjj.k* and *sw.k* in the same text. I believe that the text transmits vestiges of *sḏmw.f* in contexts in which that form is expected, but the transmission is slightly garbled.

Discussions: See §5.1.1 above; Allen 1979a: 23; 1984: 21 §34, 131 §225, 178–79 §286.

8.2.11 *Urkunden* 4 305,8

wnn pt wnn.ṯ

'As long as heaven exists, you exist'.

This example is not as distinctive as the preceding ones. Its content is also slightly abstract and its meaning not fully clear. The main reason for its deficit in distinctiveness is that, with the irregular verb *wnn* 'be', the alternation between presence and absence of gemination is not well understood. Gemination is expected in the substantival *sḏm.f*, in both the aorist or general present and the future, but not in the adverbial future. Thus both verb forms appear to be substantival.

Discussion: de Cenival 1972: 44.

8.2.12 Book of the Dead, chapter 99, Papyrus of Iouiya (Naville 1908: plate XXIII, line 5 from the right)

rwjj.f *rwjj.j ṯs pẖr*

'He will leave, I will leave, and vice versa'.

It is now widely accepted that the writing *jj* (double reed leaf) is characteristic of the substantival future *sḏm.f* (the *sḏmw.f*; see Schenkel 2000, who specifies that *jj* appears before single-consonant suffixes). There is also no doubt that *jj* often appears in the adverbial future *sḏm.f* (or subjunctive) of third-weak verbs, leaving the interpretation of *rwjj.f* and *rwjj.j* here uncertain.

Discussion: Niccacci 1980: 210.

8.3 ONLY THE SECOND VERB FORM IS DISTINCTIVELY WRITTEN AS SUBSTANTIVAL

8.3.1 Pyr. §149a–b W (= Pyr. §§1276b, 1275b P)

dbḥ.k pr.k r pt prr.k …
dbḥ.k hꜣ.k jr njwt hꜣꜣ.k

'When you ask to ascend to heaven, you ascend. …
When you ask to go down to the under-heaven,
you go down'.

As Allen (1984: 138 §236) observes, version N (Pepi II) exhibits the writing *hꜣw.k*, with *w*, instead of *hꜣ.k*. *Hꜣw.k* is easily interpreted as a *sḏmw.f*. I believe, with Polotsky, that *sdmw.f* is always substantival, above all because I cannot see how a verb form can be simultaneously substantival and not substantival, or simultaneously adjectival (relative) and not adjectival. Being a substantival verb form, *sḏmw.f* is normal as a direct object of *dbḥ.k*.

Discussions: de Cenival 1972: 43; Edel 1955/64: 235–36 §506.

8.3.2 Berlin 1157 (*Aegyptische Inschriften Berlin* [1913]: 257–58, Sethe 1928: 83–84), the "Semneh Stela," line 12

ꜣd.tw r.f dd.f sꜣ.f

'When someone is angry against him, he shows his back'.

The distinctive characteristic in this example is the gemination in the third-weak verb *rd(y)* 'give'.

> *Discussions*: See §8.2.2; Gardiner 1957: 361; Polotsky 1964: 282 (1971: 67) n. 3; Vergote 1965: 353; Loprieno 1986: 272–73.

8.3.3 Louvre C30, lines 17–18 (Sethe 1928: 63–64)

m33 () *sw nṯrw dd.sn n.f j3w*

m33 () *sw 3ḫw ḳm3.sn n.f s3-t3*

'When the gods see him, they render homage unto him.
When the spirits see him, they generate reverence(?) for him'.

The distinctive characteristic is the gemination in *dd.sn*. Polotsky observes, presumably on the basis of his wide reading, that the hieroglyphic writing  can be a writing for both *m3* and *m33*. In light of the principle of parallelism, it may be assumed that *ḳm3.sn* is a substantival verb form because *dd.sn* is.

> *Discussion*: Polotsky 1964: 282 (1971: 67) n. 3.

8.3.4 CT 3.185c–d, Coffin S1C

ḥms.j r wnm.j t ḥmsw Rʿ r wnm t

'When(ever) I sit down to eat bread, Re will sit down to eat bread'.

The distinctive characteristic is the *w* in *sḏmw Rʿ*. The possibility cannot be entirely discounted that the example contains two emphatic sentences in which *r wnm t* is twice emphasized: 'It is to eat bread that I will sit down. It is to eat bread that Re will sit down'.

> *Discussions*: See chap. 1 above; Niccacci 1980: 210–11.

8.4 ONLY THE FIRST VERB FORM IS DISTINCTIVELY WRITTEN AS SUBSTANTIVAL

8.4.1 Pyr. §153c W

mrw.f mt.ṯn mt.ṯn mrw.f ʿnḫ.ṯn ʿnḫ.ṯn

'When(ever) he wants you to die, you will die. When(ever) he wants you to live, you will live'.

It may be assumed that *mt.ṯn* and *ʿnḫ.ṯn* are both twice substantival. The first instance of each is substantival because both verb forms function as direct object of *mrw.f.* The second is substantival because both forms function as second half of a balanced sentence. The second instance of both writings is in all probability future. It may be assumed that the first instance is future too.

As noted, cases in which the *second* verb is distinctively substantival are more securely identified as balanced sentences than those in which the *first* is. If the first form is substantival, the second might be adverbial. If the second form is substantival, the first form can only be interpreted as adverbial if the sentence pattern [adverbial form + substantival form] exists. I do not believe it does, but some do.

8.4.2 Pyr. §676b–c T

jṯṯ N ʿ.f jr.k mt.k jfḫ ṯw ʿ n N n ʿnḫ.k

'When N takes his arm to you, you will die. When the arm of N releases you, you will not live'.

The form *mt.k* cannot be identified with certainty. In the second sentence, *n ʿnḫ.k* is definitely not substantival, but compare the well-known expression *mrr.f jrr.f msḏḏ.f n jr.n.f* (§8.2.3), in which *jrr.f* in the first sentence is clearly substantival whereas *n jr.n.f* in the second sentence is not. It is therefore quite possible that Pyr. §676b–c "illustrates the principle noted in [§224B]: where affirmative and negative clauses are balanced, the second is adverbial" (Allen 1984: 221 §336).

Discussions: Farina 1927: 154 §259; Allen 1984: 221 §336.

8.4.3 Sinuhe B61

ṯ33.f jkm.f tjtj.f

'When he takes his shield, he tramples'.

Discussions: Gardiner 1957: 359 §446; Vergote 1965: 354.

8.5 NEITHER VERB FORM IS DISTINCTIVELY WRITTEN AS SUBSTANTIVAL (BUT OTHER FEATURES SUPPORT SUCH AN INTERPRETATION)

8.5.1 CT 3.347h, Coffin S1C[a], and similarly in three other coffins

ḫw.j sw ḫw.f wj ṯs pẖr

'I protect (will protect?) him, he protects (will protect?) me.
And vice versa'.

It seems clear that this is an abbreviated version of

**ḫw.j sw ḫw.f wj ḫw.f wj ḫw.j sw*

'Whenever I protect him, he protects me.
Whenever he protects me, I protect him'.

Discussion: See §5.1.1 above.

8.5.2 CT 4.246a, many coffins

wḏꜣ.j wḏꜣ.f ṯs pẖr

'I'm fine, he's fine, and vice versa'.

The full version is evidently as follows.

**wḏꜣ.j wḏꜣ.f wḏꜣ.f wḏꜣ.j*

'I'm fine, he's fine. He's fine, I'm fine'.

Discussion: See §5.1.1 above.

8.5.3 CT 6.338c–d, Coffin B2L

pr.f r pt pr.j ḏs.j ḥnꜥ.f r pt

'When he ascends to heaven
I myself will ascend to heaven with him'.

Niccacci (1980: 201) writes:

> The correlation between the two actions of the god and the deceased person is expressed here by two *sḏm.f* forms that are probably prospective.*

* La correlazione delle due azioni del dio e del defunto, è espressa qui con due *sḏm.f* probabilmente prospettivi.

The absence of gemination marks both an aorist adverbial and a future substantival. Because adverbials are hardly likely in the present case, the verb forms must be substantival.

Discussions: See §5.1.1 above; Niccacci 1980: 200–201.

8.5.4 CT 7.207k, Papyrus Gardiner II

ꜥnḫ.j ꜥnḫ Rꜥ ṯs pẖr

'I live, Re lives, and vice versa'.

The full version is evidently as follows.

**ꜥnḫ.j ꜥnḫ Rꜥ ꜥnḫ Rꜥ ꜥnḫ.j*

'I live, Re lives. Re lives, I live'.

Discussion: See §5.1.1 above.

8.5.5 CT 4.50f–g, Coffins B1C, B2L

jw pt m rswt ḥms.j m mḥwt ṯs pẖr

'Heaven comes as a south wind,
I will sit to the north of it, and vice versa'.

jw pt m jmnt ḥms.j m j3bt ṯs pẖr

'Heaven comes as a west wind,
I will sit to the east of it, and vice versa'.

This passage is discussed in detail in §5.1.1 above. The sentences are hybrids. Two sentences have been telescoped into one.

8.5.6 Book of the Dead, chapter 90 (Papyrus of Nu, Budge 1898: 192, lines 10–12) and parallels

jw.k r.j ḏd.j r.k
tm.k jw r.j tm.j ḏd r.k (Book of the Dead, Chapter 90)

'You come against me, I say against you.
You do not come against me, I do not say against you'.

(Literally: 'That you come against me means (the same as) that I say against you. That you do not come against me means (the same as) that I do not say against you'.) Future tense cannot be excluded: 'That you will come …'.

jw.k r.j *ḏd.j r.k*
tm.k jw r.j *nn ḏd.j r.k* (CT 5.323h–i, Coffin B2L)

'You come against me, I say against you.
You do not come against me, I will not say against you'.

Coffin B17C has affirmative *ḏd.j r.k*, perhaps in error. W. Schenkel (pers. comm.) suggests that the scribe forgot the negation at the change of lines.

jw.k r.j *ḏd.j r.k* (CT 5.326g–h, Coffin B2L
tmm.k (for *tm.k*?) *jw r.j* *tm.j ḏd r.k* [only this version preserved])

It is interesting that the same coffin exhibits both *nn ḏd.j r.k* and *tm.j ḏd r.k* in two closely related spells (453 and 454). The two forms *jw.k* and *ḏd.j* in CT 5.323h–i are not written distinctively. The parallelism with the negation *tm* in Book of the Dead, Chapter 90 and CT 5.326g–h, Coffin B2L suggests that they are substantival. It is difficult to ascertain whether the tense is aorist (general present) or future.

Discussions: See §5.2 above; Polotsky 1964: 282 (1971: 67).

8.5.7 CT 4.53b, Coffins B3L, B1L; 99g, Coffins S2P, S2C, S1C; and 178g, Coffin G1Be

wnn.j jm.k *ṯs pẖr* (CT 4.53b, Coffins B3L, B1L)

ḫpr.n.j m jrt Ḥr *ṯs pẖr* (CT 4.99g, Coffins S2P, S2C, S1C)

ḫpr.n.j m Rꜥ *ṯs pẖr* (CT 4.178g, Coffins G1Be)

The explicit versions are as follows.

**wnn.j jm.k wnn.k jm.j*

I am (*or:* will be?) you means you are (*or:* will be?) me'.

**ḫpr.n.j m jrt Ḥr ḫpr.n jrt Ḥr jm.j*

'I have come into existence from the Eye of Horus means that the Eye of Horus has come into existence from me'.

ḫpr.n.j m Rꜥ ḫpr.n Rꜥ jm.j

'I have come into existence from Re means that Re has come into existence from me'.

Two verbs are concerned, *wnn* 'be' and *ḫpr* 'become'. *Wnn* is a second-doubling verb but it does not seem to behave fully like others. The distribution of geminating *wnn* and non-geminating *wn* does not always appear to be the same, though this has not yet been studied. The uncertainty regarding *wnn* affects its occurrence here, making it less distinctive. As for *ḫpr* 'become', this verb is intransitive. The *sḏm.n.f* of intransitive verbs of motion is as a rule substantival, despite any rare or debatable exceptions. By contrast, the *sḏm.n.f* of other intransitive verbs does not always appear to be substantival. *Ḫpr.n.f*, for example, is attested after the particle *jw*, where the adverbial *sḏm.n.f* is expected.

8.6 DOUBTFUL EXAMPLES

8.6.1 Eloquent Peasant B1, 116–17 (old numbering: B1, 85–86)

This example is probably not an instance of the balanced sentence, but some have claimed that it is.

dd st mr pr wr Rnsy s3 Mrw dd.f st n ḫnms.f

I have used the recent edition by Parkinson (1991). There has been much discussion of this example. It initially appears to be a *Wechselsatz*. The literal translation would be 'Whenever the majordomo Rensy son of Meru gives it, he gives it to his friend' (literally: 'The fact that the majordomo Rensy son of Meru gives it is the same as the fact that he gives it to his friend'). By this interpretation, the substantival force of both *dd* and *dd.f* is exhausted by signaling that the forms function as substantival components of a nominal sentence (cf. §9.4 below). If *dd* and *dd.f* are substantival in order to allow *rd*(*y*) 'give' to enter into the substantival slots of what is basically a nominal sentence, then they cannot at the same time also be substantival in order to emphasize an adverbial element. The substantival function of substantival verb forms cannot, I believe, do "double duty."

An alternative interpretation of the same passage is as follows. The second form, *dd.f*, emphasizes *n ḫnms.f*: 'It is to his friend that he gives (it)'. The first form, *dd*, can then be interpreted as a conditional clause without *jr* 'if'. However, I am not sure that a conditional clause can occur without *jr*; the proposed examples can usually be explained otherwise. If it cannot, interpreting this example as a *Wechselsatz* becomes more attractive—but manuscript R may rule this out.

The version in R reads unproblematically *dd.tw n ḫnms*[*w.f*] *jn ḫnmsw.f dd n.f st* 'It is to his friends that it is given. It is his friends that give it to him' (R 18.6–7 [old numbering: 130–31]). When more than one version of a text is preserved, some might be pure and others corrupt. Because R seems problem-free, one wonders whether anything untoward happened to the transmission of the text in B1. Perhaps the scribe of B1 misunderstood what he was copying or committed some other error. All this renders the interpretation of B1 precarious. We cannot be certain that we have a correct text in front of us.

As regards earlier discussions of the text, Polotsky (1944: 81 [1971: 185]) was aware of the problem with B1 and prefers the version of R, but for B1 accepts Gardiner's translation (1923: 10), 'The high steward Rensi, the son of Meru, used to give it to a companion of his' and *ntf dd.f st* 'he used to give it to him', disregarding the first *dd.f st*, presumably as a corruption of the original text, which results in a regular emphatic sentence *dd st … Rnsy … n ḫms.f* 'It is to his friend that Rensy gives it'.

> *Discussions*: Polotsky 1944: 81 (1971: 185); de Cenival 1972: 43 (see §9.5 below); Vergote 1965: 354; Junge 1978: 115; Loprieno 1986: 273; Schenkel 1988: 284; Junge 1989: 26 (E24).

8.6.2 Pyr. §§1044a–b, 1045a–b N (Fragmentary in P)

h33 N m mw wṯs sw Wsjr … h33 N m t3 wṯs sw Gb …

'When N descends into water, Osiris raises him.
When N descends to earth, Geb raises him'.

The second verb form, *wṯs*, is not written distinctively. It could therefore be either substantival or adverbial.

> *Discussion*: de Cenival 1972: 43.

8.6.3 Additional Doubtful Examples

The following examples have been brought up in discussions of the balanced sentence, though some or all of them may not belong in that category. (1) The unclear example *jṯw N jnw.f* (Pyr. §1459c M) is discussed by Allen (1979a: 24). If it is a balanced sentence, it means something like 'Whenever N takes, he will also bring'. (2) Allen (2002: 91 n. 41) translates *gm.tw kfwt.s sbw.s r jm3ḫ* in Eloquent Peasant B1, 352–53 (old: 321–22) (= B2, 86–87) as 'When its revelation is made, it will lead to honor', which seems to reflect an analysis as a balanced sentence. If this is correct, then *sbw.s* would be a rare occurrence of the active future substantival *sḏm.f* (*sḏmw.f*) in the second member. (3) The example *d(y).f n.j d(y).j n.f* (CT 5.381f, Coffins B3L,

B1Bo), discussed by Niccacci (1980: 211), perhaps rather consists of two independent sentences, meaning something like 'May he give to me and may I give to him'. The writing *d(y)* is not normally associated with substantival function. As regards verb forms without affixes, gemination (as in 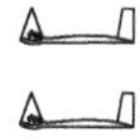 *dd*) and presence of *r* (as in  *rd*) can mark substantival function. (4) The example *rḫ.n.j ṯn rḫ.n.j rnw.ṯn* (CT 5.222b–c = 226j), discussed by Niccacci (1980: 211), which might be a balanced sentence, is a variant of *jw.j rḫ.kwj ṯn rḫ.kwj rnw.ṯn* 'I know you, I know your names' (CT 5.223d–e), which cannot be a balanced sentence. (5) The example *ḫdd.j ḫr jnw.s n nsw ṯnw rnpt prr.j jm m m3ʿ ḫrw* 'I come bearing its gifts for the king every year, I leave there as one justified of voice' (*Urkunden* 4 77.5–6 [Stela of *Ḥr-mnj*; Florence 2548; dated by Sethe to early Dynasty 18]), discussed by de Cenival (1972: 43–44), may well consist of two emphatic sentences with *ṯnw rnpt* and *m m3ʿ ḫrw* as the two emphasized adverbial phrases. Interpretation as a balanced sentence should at least be a factor in the following examples from the Pyramid texts (see Allen 1984: 126–32 §§224–25, 275–76 §409, 332 §493, 369 §542). (6) *Pr.n.ṯ jm.f pr.n.f jm.ṯ* (Pyr. §198a N) might well mean 'You (fem.) issued from him, he issued from you', a balanced sentence, but the proper translation might also be something like 'It is from him that you issued. It is from you that he issued', two separate independent sentences. (7) A similar passage is *pr.n.k m N pr.n N jm.k* (Pyr. §1875b N). (8) The following example, unclear in structure, reminds one of the contents of certain well-known balanced sentences.

pr.k h3.k h3w.k ḥnʿ Rʿ
pr.k h3w.k pr.k ḥnʿ Rʿ
pr.k h3.k h3w.k ḥnʿ Nbt-ḥwt
pr.k h3.k pr.k ḥnʿ 3st (Pyr. §209a–b, §210a–b W)

(9) Pyr. §§376a–379b WPM exhibits the kind of repetition of verbs that is characteristic of balanced sentences, as in *mr ṯn N nṯrw mr sw nṯrw* (Pyr. §378a). (10) The passage *pr N ḥr mnty 3st ḥfd N pn ḥr mnty Nbt-ḥwt* (Pyr. §379c WPM = Pyr. §996c) would be a balanced sentence if its translation is 'When N goes out on the thighs of Isis, this N climbs on the thighs of Nephthys', but alternative interpretations are 'It is on the thighs of Isis that N goes out …' or 'May N go out on the thighs of Isis …'. (11) If *j.p3.s* (N *p3.s*) *ḫnn.s* (Pyr. §494b WPMN) is a balanced sentence, then it means something like 'Whenever she flies up, she lands'. The related passage *j.p3 N ḫnjj*

N ḥr šwty jt.f Gb (Pyr. §1971 N) is probably not a balanced sentence. (12–13) Two more passages worthy of consideration are *dd.k ʿ.k jr jmnt dd.k ʿ.k jr N* and *dd.k ʿ.k jr jꜣbt dd.k ʿ.k jr N* (Pyr. §608a–b TP). If these are balanced sentences, then the meaning is 'Whenever you place your arm toward the west, you place your arm toward N. Whenever you place your arm toward the east, you place your arm toward N'. An alternative interpretation is that these are four independent sentences each emphasizing an adverbial element of its own; or perhaps the four substantival verb forms are all parallel, all emphasizing *mj nw* … 'like these things …', which follows in Pyr. §608c. (14) Is *j.nn.k N j.nn.f ṯw* (Pyr. §679c T) a balanced sentence, meaning something like 'When you turn away(?) N, he turns away(?) you'? (15) *J.mn N j.mn swt.f nfrt* (Pyr. §710b TMN) is a balanced sentence if it means 'When N endures, his good places endure'. (16) *J.šm.k j.šm Ḥr mdw.k mdw Stš* (Pyr. §798a PMN) is quite possibly a balanced sentence, meaning 'When you go, Horus goes. When you speak, Horus speaks'. (17) So is *ḳꜣ Mnṯw ḳꜣ.f ḥnʿ.f ḥjp Mnṯw ḥjp.f ḥnʿ.f* (Pyr. §1081a–b P), meaning 'When Montu is high, he is high with him. When Montu runs, he runs with him'. Future tense would also seem possible in this case. (18) Another probable instance of the balanced sentence is *mr.ṯ N mr ṯn N* (Pyr. §1230b PMN), which probably means 'When you (f.sg.) like N, N will like you'. (19) [*W*]*nn* […] *wnn.f ḏt* (Pyr. §1491c P) is fragmentary and therefore difficult to evaluate. (20) Interpretation as a balanced sentence comes easy in the case of *rs.k m ḥtp rs Rʿ m ḥtp rs.k m ḥtp rs Mdj m ḥtp* (Pyr. §1518b–c), which would then mean 'When you awaken in peace, Re will awaken in peace. When you awaken in peace, Medy will awaken in peace'. (21) The passage *ẖfʿ ṯw Tfnt ꜣmm ṯw Šw* (Pyr. §1739b M) perhaps means 'When Tefnut grasps you, Shu seizes you'. (22) The meaning of *nẖẖ.s ẖr.k wꜣꜣ.s ẖr.k* (Pyr. §1881a N) is uncertain. (23) The repetition of the verb *wsr* in *wsr.n.j m r.j wsr.n.j m fnḏ.j ḏs.j wsr.j m r.j wsr.j m fnḏ.j* (CT 4.390c–g, Coffin B2L[a] and other versions) may constitute a balanced sentence.

There are examples that exhibit balance without grammatically being balanced sentences. Allen (1984: 129–30) cites the following: (*ʿnẖ.f ʿnẖ N pn*) *n mt.f n mt N pn n sk.f n sk N pn* (Pyr. §167b–c W); *j*(*w*)*.ṯ* (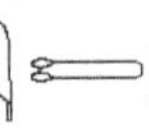) *jr pt jw* () *N jr pt* (Pyr. §534b TN); *nfr n N ḥnʿ.sn nfr n.sn* (Pyr. §909b P); *ḥkꜣ.k n.k ḥkꜣ n N n.f* (Pyr. 2030a N).

I conclude by citing a possible instance of the balanced sentence recently published by Edel (1995). The text in question, a crude rock inscription located on the west bank of the Nile near Hindallab, is as follows.

stjj.f r t3
sk3.f prt⟨.f⟩ stt.f
sʿnḫw nṯr.f k3.f

Edel cites the inscription mainly because it contains a clear instance of the active future substantival *sḏm.f*, or *sḏmw.f*, namely *sʿnḫw*. The form *stjj.f* corroborates the appearance of *sḏmw.f*. It is now widely accepted that *jj* can alternate with *w* in the future substantival *sḏm.f*, with *w* normal before nouns, as appears to be the case in *sʿnḫw nṯr.f*, but *jj* common before single-consonant suffix personal pronouns, as appears to be the case in *stjj.f* (see Schenkel 2000). No adverbial phrase follows *sʿnḫw* that could account for the use of the substantival form by being emphasized. One solution is to suppose that *stjj.f r t3* and *sʿnḫw nṯr.f k3.f* are members of a balanced sentence, so the meaning would be 'Whenever he pours (water) on earth, … his god will nourish his ka'. It then becomes difficult to interpret *sk3.f prt⟨.f⟩ stt.f* other than being in parallelism with *stjj.f r t3*. The translation is then as follows: 'Whenever he pours water on earth and cultivates his seed and *stt*-plant, his god will nourish his ka'. Such a translation seems plausible but not entirely certain.

8.7 RELATED EXAMPLES

Often, thoughts similar to those expressed by balanced sentences are expressed by other types of sentences, which should therefore be part of the larger discussion. In one type, discussed above, the first verb form is clearly substantival, as evidenced by presence of gemination, but the second is apparently not, as evidenced by absence of gemination. This type is therefore not a balanced sentence. An example is *prr Rʿ m j3bt gm.f N m 3ḫt* (Pyr. §919a PN), already cited by Farina (1927: 153 §257) alongside balanced sentences as an example of "two correlative conditional propositions";* Pyr. §919b–c contains similar statements. The association between the two types was therefore already made before the balanced sentence was even discovered. Pyr. §919a may be interpreted as an adverbial sentence. The literal translation is then 'That Re comes forth in the east is while he finds N in the horizon'.

The following example, cited by Allen (1984: 285 §422C), is similar in content, but the action takes place in the past.

pr.n N r pt
gm.n.f Rʿ ʿḥʿ ḫsf jm.f (Pyr. §812c–d PMN)

* Deux propositions conditionnelles corrélatives.

The form *pr.n*, featuring a verb of motion, is no doubt substantival. Pyr. §919a, cited above, also exhibits the combination of *pr*(*y*) 'go out' and *gm*(*y*) 'find', in that order. If Pyr. §812c–d is in any way parallel to Pyr. §919a, then *gm.n.f* is adverbial, but the literal translation is then 'It is after he had found Re standing in front of him that N went up to heaven'. The expected order of events would then be inverted. If *gm.n.f* is indeed adverbial, the only way of placing the event of finding *after* the event of going up to heaven is to assume the presence of contiguity (for the concept of contiguity, see Depuydt 1993a and the most detailed presentation of the mechanism in Depuydt 1999d: 693–729). The translation is then 'No sooner did N go up to heaven than he found Re …'. On the other hand, if *gm.n.f* is substantival, then the sentence is a balanced sentence and the translation is 'Whenever N has gone up up to heaven, he has found Re …'. Allen considers both the balanced and the adverbial interpretations possible. The following similar example also featuring *gm*(*y*) may also exhibit contiguity.

gm.n.tw Ḥr ḫrw.f m3ʿw rdw n.f j3t n jt.f (Hymn to Osiris, Louvre C286,18; see de Buck 1963: 112,4)

'No sooner had Horus been found justified of voice than his father's office was given to him'.

Gm.n.tw, an instance of the form *gm.n.tw.f*, is no doubt substantival (emphatic). It has normally been assumed that *ḫrw.f m3ʿ* is emphasized (so Polotsky 1957: 112–13 [1971: 46–47]), an alternative that cannot be excluded. Incidentally, it came to my attention recently that the very first sentence of Erman's venerable chrestomathy (Erman 1904) would also appear to be an instance of contiguity (for more new instances, see §9.10.3; cf. §9.10.2.6).

ḫpr.n tr n msjjt s3ḥ.n.j r dmj n [ng]3w d3.n.j m wsḫt (Sinuhe B11–13 = R36–38)

The verb form *ḫpr.n*, as a form without the particle *jw* and as a *sḏm.n.f* of an intransitive verb, is presumably substantival (emphatic). Accordingly, the literal translation of *ḫpr.n tr n msjjt* would be something like 'That the time of lunch happened is (while) …'. Consequently, what follows from *s3ḥ.n.j* onward must be a circumstantial or adverbial clause 'After I had arrived at the town of Negau …'. However, the order of events is then inverted. A translation exhibiting contiguity, such as 'No sooner did noontime happen than I arrived …', therefore imposes itself. Decades ago, grammarians would have interpreted *ḫpr.n* and *s3ḥ.n.j* as two juxtaposed main sentences, for lack of anything else to interpret it as. The arrival of substantival verb forms in

Egyptian grammar in the 1940s imposed a translation that inverts the order of the events. The concept of contiguity makes it possible to put the two events back in the right order.

The following example from the Coffin texts, cited by Niccacci (1980: 201), is similar in type to Pyr. §919a. The translation implies that the sentences are adverbial.

hꜣꜣ.sn r tꜣ m h̤fꜣw *hꜣjj.j m ḳꜣbw.sn* (CT 3.24a–25b,
prr.sn r pt m bjkw *pr.j ḥr ḏnḥw.sn* Coffin B2Bo)

‘No sooner do they descend to earth as snakes
than I descend in their coils.
No sooner do they go up to heaven as falcons
than I go up on their wings’.

It is a small step from this example to the following one, similar in content, which is cited and discussed by Junge (1989: 25 [E23]) and Niccacci (1980: 203–4). It is again assumed that the sentences are adverbial.

prr.sn r pt m bjkw *jw.j ḥr ḏnḥw.sn* (CT 3.100h–101b,
hꜣꜣ.sn r tꜣ m h̤fꜣw *jw[.j] ḥr ḳꜣbw.sn* Coffin S1C)

‘It is while I am on their wings
that they go up to heaven as falcons.
It is while I am on their coils
that they descend to earth as snakes’.

Jw plus suffix pronoun is quite often adverbial. It may in fact seem to function more often adverbially than independently. However, the proportion between the two functions would be difficult to quantify. It is again a small step from this example to the following one, which is discussed by Niccacci (1980: 204).

n-ntt pꜣ.n.sn js r pt m bjkw *jw N pn ḥr ḏnḥw.sn*
n-ntt hꜣ⟨.n⟩.sn js r tꜣ m h̤fꜣw *jw N pn ḥr ḳꜣbw.sn* (CT 6.283f–i, Coffin B1Bo)

In this case, the adverbial character of *jw* is supported by the fact that *pꜣ.n.sn* and *hꜣ⟨.n⟩.sn*, as *sḏm.n.f* forms of verbs of motion, are both substantival and require the presence of an emphasized adverbial element. The clause introduced by *jw* seems the most plausible candidate for being this adverbial element.

The presence of *js* goes hand in hand with the presence of *n-ntt*, but it is unlikely that *js* and *n-ntt* convey exactly the same meaning. The following logic applies. Three related patterns are well attested.

(1) *N js pw* 'while it is N'

(2) *n-ntt N pw* 'while it is N'

(3) *n-ntt N js pw* 'while it is N'

Comparing (1) and (2), it is obvious that *js* and *n-ntt* are similar in function. Both mark *N pw* as adverbial or circumstantial. In (3), *n-ntt* and *js* appear together. It is clear from the parallelism between (1) and (2) that *js* turns a sentence into an adverbial circumstantial clause, so in (3) *js* likely marks a feature *in addition to* adverbiality, namely, sentence anaphora (see Depuydt 1993c, 2008). Consequently, the meaning of (3) is something like 'and that while/because it is N'. The details regarding the concept of sentence anaphora exceed the scope of the present argument.

9 Past Research on the Balanced Sentence

9.1 Balanced Sentence Broadly and Narrowly Defined

9.1.1 General Remark

A distinction is necessary between broad and narrow definitions of the balanced sentence. In the broad definition, balanced sentences also occur in English, as in "I'm OK, you're OK." The sense of this example is clearly that the two propositions are being declared to mean the same thing. In that sense, the sentence may be called balanced. In the narrow definition, however, the term "balanced sentence" (*Wechselsatz*) denotes Old and Middle Egyptian nominal sentences whose two members both contain *substantival verb forms*. H. J. Polotsky put the narrow definition of the balanced sentence on the map of Egyptian grammar in the 1960s. Earlier, in the 1930s and 1940s, he had proposed the existence of the substantival verb forms themselves.

9.1.2 The narrow definition of the balanced sentence

The balanced sentence is a well-defined and well-delineated phenomenon. A substantival verb form appears in each of the two slots of a nominal sentence. The events denoted by these verb forms are therefore equated with one another. An example is *prr.ṯn r pt m nrwt prr.j ḥr tpt ḏnḥwy.ṯn* (CT 3.61f–g, Coffins B1C, B2L [B2L has *ḏnḥw.ṯn*]). The literal translation of this passage is something like 'The fact that you ascend to heaven as vultures equates to (*or:* means the same as) the fact that I ascend on top of your wings'. It seems difficult to confuse the balanced sentence with any other grammatical phenomenon. Its structure is easy to comprehend. The message it conveys is transparent: two events are said to mean exactly the same thing. I limit the term "balanced sentence" to instances in which two events *both denoted by substantival verb forms* are represented as identical; the presence of two substantival verb forms is essential to this definition.

9.1.3 The broad definition of the balanced sentence

It is common for grammarians to understand the balanced sentence in a wider sense, wherever two events are equated, whether by means of two substantival verb forms or not. Pascal Vernus, for example, on the whole subscribes to a narrow conception of the balanced sentence that is close to my own (1981: 76–88) but does allow for the possibility of an "extended" use of the term encompassing cases in which two events are equated but without the use of two substantival verb forms (1990: 18 n. 74). As an example of "an extension of the balanced sentence in which the two members are initial constructions," Vernus cites Eloquent Peasant B1, 332–33 (old: 301–2): *jw srw ḥr rd(y)t n.k jw.k ḥr jṯt* 'The magistrates are giving to you, you are taking'. I would not call this a balanced sentence. Schenkel (1988: 282) writes:

> [Vernus] would like to restrict the "*Wechselsatz*" to cases in which there is a relation of identity between two expressions.*

There is a slight contradiction in Schenkel's statement pertaining to the concept of "limiting" the definition of the balanced sentence to the "relation of identity." As I define the balanced sentence, the essential feature is the two substantival forms; the relation of identity takes second place to their presence. The concept of equation is *larger* and *more encompassing* than that of the two substantival forms. There are examples in which two events are equated with no substantival forms, but all examples with two substantival forms exhibit equation. The definition of the balanced sentence therefore needs to be *expanded* from the narrower concept of two substantival verbs to arrive at the broader concept of equation. However, Schenkel's statement speaks of limiting (*beschränken*) the definition to the relation of identity *(auf Identitäts-Beziehung*). I believe that this shift is better characterized as *expansion.*

9.2 THE BALANCED SENTENCE PRIOR TO THE FORMULATION OF ITS NARROW DEFINITION

Already before Polotsky, Old and Middle Egyptian balanced sentences were noticed and their general purport was understood, but as balanced sentences in the broad definition. Polotsky (1964: 281 [1971: 67]) recalled:

> In his lectures, Sethe used to cite Arabic *inte mabsut–ana mabsut* 'When you are happy, I am happy', which he had probably heard in Egypt.†

* [Vernus] möchte … den "Wechselsatz" auf Fälle beschränkt wissen, in denen zwischen zwei Ausdrücken eine "Identitäts"-Beziehung besteht.

† Sethe pflegte in seinen Vorlesungen arab. *inte mabsut–ana mabsut* "wenn du zufrieden bist, bin auch ich zufrieden" zu zitieren, das er wohl selber in Ägypten gehört hatte.

Polotsky also points to Farina's description (1927: §258) of a balanced sentence such as *jbb N jbb Nḫbt* 'N is thirsty, Nekhbet is thirsty' as "deux propositions conditionnelles corrélatives." Sander-Hansen (1941: §102) called it the "appositional identity sentence" ("appositioneller Identitätssatz") and Firchow (1953: 117) uses the terms "analogical parallelism" ("analogischer Parallelismus") and "turn sentence" ("Kehrsatz") (Polotsky 1969: 471 with nn. 2–3).

9.3 EMERGENCE OF THE BALANCED SENTENCE ON THE GRAMMATICAL SCENE

9.3.1 Gardiner's review (1947) of Polotsky's *Études* (1944)

Gardiner's often-cited review (1947) of Polotsky's *Études de syntaxe copte* mistakenly supposes that Polotsky's theory requires every geminating form of a third-weak verb, such as *mrr.f* and *ḥss.f*, to be followed by an emphasized adverbial phrase. True, the emphasizing function is prominently treated in the *Études* and is also typically the function of such forms when sentence-initial. But emphasizing adverbial phrases is just one of their functions. They are in fact rather substantival verb forms that assume many of the functions of substantives (nouns), only one of which remains after Middle Egyptian: emphasis of an adverbial phrase. It is easy to see how this fact regarding later Egyptian could have informed Gardiner's belief that the Old and Middle Egyptian forms were similarly restricted.

Gardiner believed that gemination in third-weak verbs marks repetitive or habitual action or the like, but Polotsky believed that the function of gemination is to mark the fact that verb forms are substantival. Polotsky calls the forms in question "abstract relative verb forms." Compare English "that" in "the man *that* is sick" versus "I see *that* he is sick." According to Polotsky, "that" is relative in both cases, but in the former, it is concrete, referring to a specific entity (the familiar meaning of "relative"). In the second example, "that" is also relative, but abstractly so. It does not refer to a concrete entity but to *the fact that* he is sick.

In his review, Gardiner felt the need to defend his own view that gemination marks repetitive action against Polotsky's theory that it does not. But under Gardiner's erroneous assumption that geminating forms of third-weak verbs must be followed by adverbial phrases, any exception would indicate that Polotsky's theory was in error. He concedes some ground by supposing that a form denoting repetitive or habitual action can inherently denote emphasis of an adverbial phrase. It is not clear to me, however, how such a combination of functions is possible. The verb form must have one or the other of the two functions—they are not compatible.

Gardiner (1947: 99) parades with some fanfare the passage *mrr.f jrr.f* 'He likes (something), he does (it)' (Pyr. §412b) (cf. §5.3 above). There is no adverbial phrase in this example, so Polotsky's theory seemed conclusively falsified.

9.3.2 Thacker's reaction (1954) to Gardiner's review (1947) and Vergote's reaction (1957–58) to Thacker and Gardiner

Gardiner's criticism found acceptance with Thacker and Vergote—note the use of the word "prove" in these quotations.

> Polotsky's attempt to see in all imperfective *śḏm·f* forms an imperfective relative form is to be rejected. As Gardiner has argued, the syntactical parallelism of the form *ḥss·f* to *n śḏm·n·f* (the latter being beyond all doubt a narrative verb-form) in such phrases as *mrr·f irr·f mśḏḏ·f n ir·n·f* 'whenever he likes he does; whenever he dislikes he does not', prove[s] that *ḥss·f* must be a narrative verb-form, on a par with *śḏm·n·f*. (Thacker 1954: 210–11)

> Further, examples occur where there is no adverbial phrase, such as 'if one is aggressive towards him, verily he turns his back *dd·f s3·f*' [Berlin 1157,12 (see §8.3.2 above)]. (ibid. 212)

> In his review of the *Études de syntaxe copte*, Sir Alan Gardiner has proved that the imperfective *śḏm·f* can function very well as a narrative form; he actually produces the following sentence in which parallelism with negated *n śḏm·n·f* guarantees that imperfective *śḏm·f* exhibits this characteristic: *mrr·f irr·f mśḏḏ·f n ir·n·f*, 'Whenever he wants (it) he does (it). Whenever he does not want (it), he does not do (it)'. Thacker rallies behind this point of view and cites a new example: 'When someone is aggressive against him, *dd·f s3·f*, he truly turns his back (to him).' Because the two clauses do not contain any adverbial complement, the imperfective cannot here be a substantival relative underscoring [the adverbial complement]. In the first citation, [the imperfective] has the meaning that has always been attributed to it: it is an 'aspect' that expresses repetition and continuity. In the second, it has emphatic value but [this value] pertains to the entire clause, not to a part [of it].* (Vergote 1957: 279–80)

* Dans son compte rendu des *Études de syntaxe copte* Sir Alan Gardiner a prouvé que le *śḏm·f* imperfectif peut bel et bien s'employer comme forme narrative; il produit en effet la phrase suivante dans laquelle le parallélisme avec la négation *n śḏm·n·f* lui guarantit ce caractère: *mrr·f irr·f mśḏḏ·f n ir·n·f*, chaque fois qu'il (le) veut il (le) fait, chaque fois qu'il n'aime pas il ne (le) fait pas. Thacker se rallie à ce point de vue

> In his review of the *Études*, Sir Alan Gardiner has proved that [the imperfective *śḏm.f*] does indeed have predicative or narrative function in the sentence *mrr.f irr.f mśḏḏ.f n ir.n.f.* He translates, 'Whenever he wants (it), he does (it). Whenever he does not want (it), he does not do (it),' and regards this as proof that the traditional view concerning the durative meaning of this form makes sense.* (Vergote 1957–58: 41)

As alternatives to Polotsky's theory, Thacker and Vergote propose highly idiosyncratic theories that have not found acceptance. Their most striking characteristic is an unsuccessful attempt to mix Gardiner's and Polotsky's views, which to me seem logically incompatible. The statements cited above show how sentences that would later be interpreted as balanced sentences formed the most serious obstacle to Polotsky's theory. The balanced sentence thus plays a prominent role in the history of Egyptian grammar.

9.3.3 Polotsky's postulation of the balanced sentence (1964)

To be sure, Polotsky's theory originally did not account for the contrary examples cited by Gardiner and Thacker. He proposed an explanation in 1964 (281–82 [1971: 66–67]) and in 1969 briefly returned to the balanced sentence in a review of Edel's grammar of Old Egyptian (1969: 471).

In Polotsky's explanation, in the disputed passages both verb forms are substantival, together forming a nominal sentence. To prove the substantival character of both verb forms, Polotsky adduces the use of the negation verb *tm* in an example such as *tm.k jw r.j tm.j ḏd r.k* (Book of the Dead, Chapter 90, §5.2 above).

9.3.4 Vergote's reaction (1965) to Polotsky's balanced sentence (1964)

Polotsky's 1964 article was a critical review of a 1960 work by Vergote on the vocalization of Egyptian verb forms. In his reply, Vergote (1965: 353–54)

et cite un nouvel exemple: "si quelqu'un est agressif envers lui, *dd·f š3·f*, en vérité il (lui) tourne le dos." Puisque les deux phrases ne comprennent pas de complément adverbial, l'imperfectif ne peut pas être ici une relative substantive soulignant celui-ci. Dans la première citation il a la signification qu'on lui a toujours attribuée: c'est un "aspect" exprimant la répétition et la continuité. Dans la seconde, il a une valeur emphatique mais celle-ci porte sur la proposition entière, non sur une partie.

* In zijn bespreking van de *Etudes* heeft Sir Alan Gardiner echter bewezen dat [de imperfectieve *śḏm.f*] wél een predicatieve of narratieve functie vervult in de zin *mrr.f irr.f mśḏḏ.f n ir.n.f.* Hij vertaalt "telkens als hij (het) wil doet hij (het), telkens als hij (het) niet lust doet hij (het) niet" en ziet hierin het bewijs dat de gebruikelijke opvatting betreffende de duratieve betekenis van deze vorm steek houdt.

persists in his denial of the existence of the balanced sentence (apparently the last as well as the first scholar to do so). Instead, he holds on to Gardiner's interpretation of the passages in question as *two* independent, juxtaposed sentences. If there is any balance between the two components, then this balance is stylistic, not grammatical.

Vergote's grounds for keeping rejecting the balanced sentence would appear to be the following. One of his and Thacker's pre-1964 objections is not addressed in Polotsky's article. Indeed, following Gardiner, Thacker and Vergote had voiced at least two distinct criticisms against Polotsky's theory. The first was addressed by Polotsky. It relates to sentence-initial geminating forms of third-weak verbs (see §9.3.1 above). The fact that such forms appear in pairs led Polotsky to propose that the two verb forms form a pair making up a balanced sentence, adequately addressing the first criticism. Still, Vergote tiptoes around Polotsky's proposal, neither accepting nor rejecting it, while portraying it with admirable accuracy (few understood the significance of Polotsky's discoveries so early and so clearly).

> Now Polotsky notes that one is dealing with "*Wechselsätze*," that is, correlative conditional clauses. Having found the sentence *iw.k r.i ḏd.i r.k tm.k iw r.i tm.i ḏd r.k* 'When you come against me, then I say against you (a saying)' in the Book of the Dead, he concludes that in this case the two clauses mutually condition one another by their grammatical form. The use of *tm* proves, according to him, that these are substantival clauses, which would have to be translated literally—if we understand him correctly—as 'the fact that you come against me is the fact that I say against you. The fact that you do not come against me is the fact that I do not say against you'.* (Vergote 1965: 353)

The second criticism is that, in an example such as the often-cited

mrr.f jrr.f msḏḏ.f n jr.n.f (Pyr. §412b)

'He likes (something), he does (it).
He hates (something), he does not do (it)',

* Or Polotsky fait remarquer qu'on a affaire ici à des "Wechselsätze," c.-à-d. à des propositions conditionnelles corrélatives. Ayant trouvé dans le Livre des Morts la phrase *iw.k r.i ḏd.i r.k tm.k iw r.i tm.i ḏd r.k*, "Wenn du gegen mich kommst, so sage ich gegen dich, wenn du nicht gegen mich kommst, so sage ich nicht gegen dich (ein Spruch)," il en conclut que dans ce cas les deux propositions se conditionnent mutuellement dans leur forme grammaticale. L'emploi de *tm* prouve, selon lui, qu'elles sont des propositions substantives, qu'on devrait traduire littéralement—si nous le comprenons bien—"le fait que tu viens contre moi est le fait que je dis contre toi; le fait que tu ne viens pas contre moi est le fait que je ne dis pas contre toi."

the verb form *jrr.f* is parallel to *n jr.n.f*. *Mrr.f jrr.f* as a whole is therefore parallel to *msḏḏ.f n jr.n.f*, which is definitely not a *Wechselsatz*. Then why would *mrr.f jrr.f* be? It is this second criticism that Polotsky did not address. A recent reflection of it is found in Schenkel 1990: 153.

> Polotsky successfully made [his theory] more complete by explaining the sentence *mrr.f jrr.f*, thrust into the debate by Gardiner, as an instance of a special sentence type. Then again, he had to grant the following [lit. 'add a drop of vermouth']: A negated counterpart of this *Wechselsatz*, which exhibits the form [*msḏḏ.f jr.n.f*] 'Whenever he does not want, he does not do (something)', cannot be explained as a *Wechselsatz*.*

N jr.n.f might be a main clause verb form that does not emphasize an adverbial phrase. Then why would parallel *jrr.f* not function in the same way? I am not alone in the belief that *msḏḏ.f n jr.n.f* may be an adverbial sentence consisting of the substantival verb form *msḏḏ.f* and the adverbial verb form *n jr.n.f*, the literal translation being 'That he hates (something) is while he does not do (it)'. I also believe that the logical compatibility of the two sentences, the balanced and the adverbial, can be explained (§5.3 above).

Vergote instead maintains Gardiner's conviction that examples adduced in favor of the balanced sentence rather demonstrate the opposite, that geminating verb forms are *not* always substantival. Vergote was one of the first to accept much of Polotsky's theory, but he maintains certain conflicting concepts and rejects some that are integral to Polotsky's theory (e.g. 1957–58, 1970: 544–49).

9.4 TWO ERRONEOUS VIEWS ON BALANCED SENTENCES

9.4.1 Emphasis on adverbial phrases

Few misconceptions about the balanced sentence have confused the discussion as much as that concerning instances in which the substantival verb forms appearing in balanced sentences are accompanied by adverbial phrases, as in the following example (cited by Polotsky 1984: 119):

> *hꜣꜣ.sn m h̤fꜣw hꜣꜣ N pn ḥr ḏnḥw.sn*
> *ṯss.sn m bjkw hꜣꜣ.f* (for *ṯss.f*) *ḥr ḏnḥw.sn* (CT 6.302m–p)

* Eine … wichtige Ergänzung [seiner Theorie] gelang Polotsky mit der Erklärung des von Gardiner in die Debatte geworfenen Satzes [*mrr.f jrr.f*] … als eines besonderen Satztyps. Allerdings mußte [Polotsky] … einen Wermutstropfen hinzugeben: Ein negiertes Pendant dieses Wechselsatzes der Form [*msḏḏ.f n jr.n.f*] "Will er nicht, so tut er nicht" läßt sich nicht als Wechselsatz erklären.

'They descend as snakes, this N descends on their wings.
They rise as falcons, he rises on their wings'.

In this example, substantival verb forms are distinctively written with gemination, and adverbial phrases follow substantival verb forms: *m ḥfꜣw* 'as snakes', *ḥr ḏnḥw.sn* 'on their wings', and *m bjkw* 'as falcons'. A main function of substantival verb forms is to emphasize adverbial phrases. The question arises: Are *m ḥfꜣw*, *ḥr ḏnḥw.sn*, and *m bjkw* emphasized ('It is as snakes that …', etc.)? Polotsky (1984: 120) thought so:

> Because the adverbial phrases ('as snakes', 'on their wings', and so on) probably bear the emphasis, the substantival [verb] forms will need to be understood as "emphatic."*

This assumption that adverbial phrases accompanying substantival verb forms in balanced sentences are emphasized is quite common, but I think it untenable, on the following grounds. When a verb appears in a certain form, the purpose is to denote a certain function. Verbs appear in substantival form in balanced sentences for a reason—that balanced sentences are substantival (nominal) sentences. Both members of such sentences are substantival (nominal). A verb appearing in either of those two members needs to adopt one of its substantival forms, and the substantivalness is fully exhausted by allowing the verb to enter a substantival position. The substantival verb form can therefore not in addition emphasize an adverbial phrase, a function that it exercises elsewhere. One linguistic signal cannot convey two messages. Assuming that substantival verb forms can do "double duty" makes little sense. If adverbial phrases in balanced sentences are emphasized nevertheless, then it can only be by some other sign, perhaps intonation, which cannot be observed.

A somewhat similar case concerns the Demotic form *jw.jr.f sḏm*, which has two well-known functions: (1) it appears in conditional clauses; (2) it marks emphasis on an adverbial phrase. However, I do not think that it can exercise both functions *at the same time*. Still, the contrary view that it can is not at all uncommon.

9.4.2 Gemination denotes repetitive action

The balanced sentence came onto the scene of Egyptian grammar in order to explain a phenomenon that seemed to threaten the foundation of Polotsky's theory (1964: 281 [1971: 66]):

* Da die adverbialen Bestimmungen ("als Schlangen," "auf ihren Flügeln" usw.) vermutlich den Ton tragen sollen, werden die substantivischen Formen als "emphatisch" zu verstehen sein.

> It happens, however, that the *mrr.f*-form is not accompanied by any adverbial phrase and neither depends from a preposition nor functions as subject or object of another verb.*

Gardiner (1947) pointed out this phenomenon, adducing it to support his view that the geminating form *mrr.f* is "imperfective," that is, denotes repeated or habitual action or the like (Gardiner 1957: 359 §446).

It is of course the repetition of the consonant, like that of *r* in forms such as *mrr.f* and in *jrr.f*, that iconically suggests that events denoted by verb forms exhibiting repetition are also repetitive. By contrast, in Polotsky's theory, the repetition of a consonant has nothing to do with repetition of the event.

A form such as *jrr.f* contrasts with other forms in more than one dimension, first time or tense.

jr.n.f	'that he has done'
jrr.f	'that he does'
jr.f	'that he will do'

The repetition of the consonant signals a specific tense contrasting with the *past* substantival *jr.n.f* '(the fact) that he has done' and the *future* substantival *jr.f* '(the fact) that he will do'. By considering the above paradigm in isolation, it is still possible to hold on to the notion that the repetition of the consonant in *jrr.f* denotes a repetitive event. However, a second dimension suggests otherwise:

jw.f jr.f	'he does'	INDEPENDENT
jrr.f	'that he does'	SUBSTANTIVAL
jrrt.f	'what he does'	ADJECTIVAL

It is obvious that *jrr.f* 'that he does' is nothing but the substantival conversion of *jw.f jr.f* 'he does'; everything else remains the same. Let us assume that the repetition of the consonant *r* in *jrr.f* denotes repetition of the event and that *jrr.f* means something like 'he repeatedly/habitually does'. The form *jw.f jr.f* expresses the same tense as *jrr.f*. If *jrr.f* denotes repetition of the event, then *jw.f jr.f* also ought to denote repetition. However, *jw.f jr.f* does not exhibit repetition of the consonant. How can repetition of the event be denoted by presence of gemination in *jrr.f* and at the same time by absence of gemination in *jw.f jr.f*? Thus the assumption appears to be false.

* Nun kommt es aber doch vor, dass die *mrr.f*–Form von keiner adverbialen Bestimmung begleitet ist und weder von einer Präposition abhängt noch auch als Subjekt oder Objekt eines andern Verbs fungiert.

Instead, it is much more plausible that the difference between the two forms is that the one with gemination is substantival whereas the one without gemination is not. The essence of the gemination is therefore substantivalness, as Polotsky proposes, and not some feature of tense, as Gardiner proposes.

The question remains: What kind of tense do *jw.f jr.f* and *jrr.f* jointly denote? The tense may best be described as a general present, also called aorist. The general present tense is expressed in English by the so-called simple present, as in "he eats," which denotes an action that one generally performs, at no definite time—"He eats at home" means that he does so yesterday, today, and tomorrow. The general present tense is opposed to the real present tense, which appears in "He is eating," an event that is happening right now in the present moment of time.

It cannot be excluded that a form such as *jrr.f* on occasion describes actions that are repetitive or habitual, but that does not mean that denoting repetition of the event is the function of the repetition of the consonant.

The notion is still very much alive in Egyptology that the repetition of the consonant in a geminating verb is somehow expressive of a repetition or higher intensity of the event expressed (e.g. Jansen-Winkeln 1997), but the evidence contradicts it. The following example makes it clear that repetition of a consonant has nothing to do with repetition or higher intensity of an event. Its significance was first noted by Polotsky (1944: 80–81 §28 [1971: 184–85], 1964: 281 [1971: 66], 1976: 17 §2.5.2).

jw.j sṯ.j *jw bꜣ.j sṯ.f* (CT 1.364/5b–
sṯṯ bꜣ.j m rmṯw jmyw Jw-nsrsr *sṯṯ.j ḏs.j m nṯrwt* 366/7b)

The verb forms *jw.f sḏm.f* (*jw.f jr.f*) and *sḏm.f* (*jrr.f*) are in parallelism. Polotsky (1964: 281 [1971: 66]) translated:

> Soleo concumbere soletque genius meus concumbere. Concumbit genius meus hominibus qui in Insula flammae sunt, concumbo ipse deabus.
>
> [I have the habit of having intercourse and my ba has the habit of having intercourse. It is with the people that are in the Island of the Flame that my ba has intercourse. It is with the goddesses that I myself have intercourse.]

—the first part differently from the second. The difference is the use of the verb *solere* 'have the habit of, be used to, be accustomed to' in the first line. It is the forms *without* gemination that are translated with this verb. The use of *solere*, however, slightly contradicts Polotsky's statement (ibid.) as to what the example is meant to demonstrate: "that the function of the *mrr.f*-form lies outside the categories tense-aspect-mood."* I would therefore rather translate the verb forms in the first line exactly like those of the second. The first line would then be *Concumbo concumbitque genius meus*, without *solere*, to avoid any semblance of contradiction.

The main point, however, is this. Sexual activity occurs in both parts of this passage, *sed dubitandum an vehementius in secunda parte.*

9.5 AN INDEPENDENT POSTULATION OF THE BALANCED SENTENCE: DE CENIVAL 1972

In an important characterization of geminating verb forms as substantival, de Cenival (1972: 43) classified the following example as a "clause associated in parallelism" ("phrase associée en parallèle") (cf. §8.6.1 above).

dd st mr pr wr Rnsy s3 Mrw dd.f st n ẖnms.f (Eloquent Peasant B1, 116–17 [old numbering: B1, 85–86])

At first sight, it would seem that de Cenival's "phrase associée en parallèle" refers to the balanced sentence. However, upon closer inspection, what seems to be meant is the balanced sentence in the broad definition (§9.1), since there is no reference to Polotsky's (1964, 1969) discussions of the balanced sentence in the narrow definition, only to his earlier discussion of the example (1944: 81 [1971: 185]), and since de Cenival adds to his term "clause associated in parallelism" the qualification "conditional clauses without *jr*" ("conditionnelles sans *jr*").

De Cenival does, however, formulate a definition of the balanced sentence in the narrow sense, independent of Polotsky. After discussing the Eloquent Peasant example, de Cenival cites well-known phrases from the Coffin texts of the type *prr.sn* … *prr.j* … 'They go out …, I go out …' in their many variations, noting also that there are instances in which the first verb form geminates and the second does not, of the type *prr.sn* … *pr.j* … 'When they go out …, I go out …'.

When both verb forms are geminating, de Cenival considers it "tempting" to treat them as "the juxtaposition of two substantives as subject and predicate," that is, as what he describes as a "nonverbal sentence with substantival predicate." He adduces Pyr. §1539a–b, which he interprets as a nominal sentence in which both verb forms are infinitives, which are obviously substantival verb forms.

* Dass die Funktion der *mrr.f*-Form ausserhalb der Kategorien Tempus-Aspekt-Modus liegt.

prt.k tn m pr.k Wsjr N pn prt Ḥr m sḫn.k Wsjr N pn (Pyr. §1539a–b P)

'This exit of yours from your house, this Osiris N, is (the same as) this exit of Horus to meet you, this Osiris N'.

De Cenival concludes with the "hope" that "an explanation based on the substantival character of *mrr.f* is also plausible" in examples of the type *prr.sn … prr.j…*, but he does not apply this definition to the Eloquent Peasant example. His confidence in his own definition is diminished (1972: 45) because he cannot account for instances in which the first verb form is geminating but the second is not, as in the type *prr.sn … pr.j …*:

> The syntactic link between the two members is not clear to me (nor is the one that unites a conditional clause beginning with *ỉr* and the "principal" [clause]).*

It is now widely assumed that the type *prr.j … pr.j …* is an adverbial sentence comprising a substantival and an adverbial verb form. De Cenival, perhaps more clearly than anyone before, including Polotsky himself, articulated substantivalness as the essence of the function of geminating verb forms of the type *mrr.f.* Emphasizing adverbial elements is just one usage of substantival *mrr.f,* but before de Cenival, this function had been so prominent that it was in danger of drowning out all the other roles that substantival *mrr.f* can play, let alone its essential substantivalness. Polotsky (1976) gave that substantival character full prominence.

9.6 EXPANDED DEFINITION: FRIEDRICH JUNGE

Friedrich Junge has studied the balanced sentence at some length in two related studies: his *Habilitationsschrift* (1978: 115–20), and a decade later a book on emphasis and the meaning of sentences (1989: 24–26, 54, 83, 85, 88–94); Junge 1986 is a preliminary version of the latter. Niccacci (1982) reviewed the earlier one, and I the later (Depuydt 1992a).

On the whole, Junge accepts the existence of the balanced sentence and its classical interpretation. One statement of the classical treatment that had appeared recently is Frandsen's (1975: 24–25 §1.3.8). Around the same time, Wente (1977: 310–11) wrote:

* Le lien syntactique entre les deux membres n'est pas clair pour moi (pas plus que celui qui unit une conditionnelle en *ỉr* et la "principale").

> Few grammarians of the new school would deny that the well-known and frequently paralleled *wnn pt wnn.ṯ ḫr.j*, "So long as heaven exists, you shall be with me" (*Urk*. IV, 348, line 9), involves two nominal forms of *wnn* in a "balanced" or "correlative" (p. 311) construction.

If Junge's treatment differs from the classical treatment in any way, it is in his classification of other types of sentences, including some adverbial sentences, as balanced sentences. He seems to call "balanced" any sentence structure in which two similar clauses exhibit some kind of parallelism.

But under the traditional definition, the balanced sentence consists of two juxtaposed substantival verb forms that are equated with each other. To subsume additional sentence patterns under the definition of the balanced sentence, Junge makes connections between sentence patterns on a theoretical level, many of them in "deep structure," a concept formerly used in generative grammar. Junge's account on occasion also deviates in other ways from standard views on the balanced sentence, such as concerning the category of number (cf. Depuydt 1992a: 332). Consider the example

> *h33.sn m ḥf3w h3y.j m ḳ3bw.sn*
> *prr.sn r pt m bjkw prr.j ḥr ḏnḥw.sn* (CT 3.24a–25b, Coffin S1C)
>
> 'When they descend as snakes, I descend in their coils.
> When they ascend to heaven as falcons,
> I ascend on their wings'.

Most would accept that the second line represents just one sentence, and the purpose of the two substantival forms *prr.sn* and *prr.j* is to allow the two verb forms to be used as the two members of one nominal sentence—a balanced sentence. The substantivalness of the forms cannot in addition denote emphasis on an adverbial phrase (§9.4). If *prr.sn* and *prr.j* did indeed mark emphasis, the sequence *prr.sn* … *prr.j* … ought to consist of two adverbial sentences, which is how Junge (1989: 25) treats the sequence: as two sentences, each exhibiting emphasis on an adverbial phrase. Viewing it as one sentence is the traditional interpretation.

In what follows, the traditional interpretation of the balanced sentence is accepted at the exclusion of any other. The traditional definition is so clear and unambiguous that calling any other phenomenon a balanced sentence as well can only cause confusion.

9.7 A COFFIN TEXT FORMULA: NICCACCI 1980

Among recent discussions of the balanced sentence, Niccacci's (1980) is perhaps the most detailed and the most original. Niccacci studies the variants of a Coffin text formula of which the following variant may be considered representative:

prr.ṯn r pt m nrwt *prr.j ḥr tpt ḏnḥwy.ṯn* (CT 3.61f–g)

'When you go up to heaven as falcons,
I go up on top of your wings'.

Several of the variants are balanced sentences. Among the many other variants, Niccacci recognizes several instances of the following construction.

SUBSTANTIVAL VERB FORM EXTRAPOSED ("AS FOR THE FACT THAT …") + MAIN CLAUSE

(A main clause is a construction that could stand by itself as an independent sentence.) As I have noted above, I am not sure that this construction exists; I would interpret many of Niccacci's examples as adverbial sentences. Two relevant variants are:

prr Rꜥ m j3bt *gm.f N m 3ẖt* (Pyr. §919a)

'When Re comes out in the east, he finds N in the horizon'.

prr.sn r pt m bjkw *jw.j ḥr ḏnḥw.sn* (CT 3.100h–i)

'When they go up to heaven as falcons, I am on their wings'.

Among Niccacci's other examples is the by now familiar

msḏḏ.f *n jr.n.f* (Pyr. §412b)

'(He likes something, he does it.)
He hates something, he does not do it'.

According to him, the three strings of words *gm.f N m 3ẖt*, *jw.j ḥr ḏnḥw.sn*, and *n jr.n.f* are main clauses, with literal translations

'As for the fact that Re comes out in the east,
he finds N in the horizon'.

'As for the fact that they go up to heaven as falcons,
I am on their wings'.

'As for the fact that he hates something, he does not do it'.

But I and Allen (1984: 127 §224) would prefer to treat them as adverbial clauses, with literal translations

> 'That Re comes out in the east
> is while finding (while he finds) N in the horizon'.
>
> 'That they go up to heaven as falcons
> is with me being (while I am) on their wings'.
>
> 'That he hates something
> is while not doing (while he does not do) it'.

9.8 THE 1980S

9.8.1 The *sḏmw.f* in the balanced sentence in the Pyramid texts: Allen 1979a

Allen (1979a: 23–24) discusses usages of the substantival future *sḏm.f* (the form sometimes exhibiting a characteristic *w*, hence called *sḏmw.f*) in an article anticipating results later published in his study of the verbal system of the Pyramid texts (1984). The balanced sentence is one of the usages.

9.8.2 Balanced sentences at Medinet Habu?: Piccione 1980

Piccione (1980: 107–8) interprets some passages in New Kingdom historical texts at Medinet Habu as possible balanced sentences. I do not believe that the balanced sentence survived after the Middle Kingdom, that is, into Late Egyptian, Demotic, and Coptic, except in copies of Old and Middle Egyptian texts.

9.8.3 Ways of combining clauses into sentences: Vernus 1981

Vernus (1981: 76–77) adduces the balanced sentence in a description of ways in which two verb forms can combine into a sentence, proposing the existence of a "Second scheme" (*second schème*), in which a circumstantial or adverbial verb form can stand at the beginning of the sentence. I do not believe that this Second scheme exists. In fact, in Late Egyptian and Demotic, circumstantial or adverbial *jw* never begins a sentence. It is only in Coptic that circumstantial **ⲉ** *e* (from *jw*) begins to appear at the beginning of sentences, so why would Old and Middle Egyptian predecessors such as adverbial *sḏm.n.f*, adverbial *sḏm.f*, or the stative as an adverbial verb form ever be initial? Adverbial *phrases* otherwise very rarely stand at the beginning of a sentence in Old and Middle Egyptian, unlike in many other languages, including English, in which initial adverbial phrases are common. So why would adverbial *clauses* begin sentences in Old and Middle Egyptian?

9.8.4 The future *sḏm.f* and the balanced sentence: Schenkel 1981

Schenkel (1981) examines Coffin text phrases of which certain variants are balanced sentences, but the principal aim of this study is to examine writings of the future *sḏm.f*, also called prospective *sḏm.f*.

9.8.5 The balanced sentence in the Pyramid texts: Allen 1984

In his study of the verbal system of the Pyramid texts, Allen (1984: 126–32 §§223–25 and elsewhere) catalogues and discusses unequivocal, possible, and doubtful instances of balanced sentences in those texts.

9.8.6 The balanced sentence as evidence for the substantival *sḏm.n.f*: Polotsky 1984

Polotsky (1984: 119–20) adduces the parallelism between the geminating *sḏm.f* forms at CT 6.302m–p and the *sḏm.n.f* of verbs of motion at CT 3.115e–h, both in balanced sentences, as an argument for the notion that there exists a substantival *sḏm.n.f*. But I disagree (cf. §9.4) with Polotsky when he states:

> Because the adverbial phrases probably bear stress, they are probably to be understood as emphatic.*

9.8.7 Surveys of the balanced sentence: Silverman 1985, Schenkel 1990

Silverman (1985: 278–80) and Schenkel (1990: 152–54, 167, 177–79) conveniently survey what is known about the balanced sentence.

9.8.8 A conspicuous absence of the balanced sentence: Schenkel 1985

Schenkel (1985) expands on his earlier analysis of the writings of future *sḏm.f* forms (§9.8.4). In a list of the three usages of the *sḏmw.f* (Schenkel 1985: 485), the balanced sentence is conspicuously absent:

(1) negated *n sḏmw.f* 'he will not hear';
(2) cleft sentence *jn X sḏmw.f* 'It is X who will hear'; and
(3) substantival *sḏmw.f* marking emphasis on an adverb or adverbial phrase.

* Da die adverbialen Bestimmungen … vermutlich den Ton tragen sollen, werden sie als "emphatisch" zu verstehen sein.

Of the three usages, only (3) is identified as substantival. Including the balanced sentence would have made it two of the four but, the other two are substantival as well. As regards (1), it may not be clear *why* the form following the negation *n* is substantival, but the negation *n* is otherwise followed by the passive *sḏm.n.tw.f* and the *sḏm.n.f* of verbs of motion, both forms unequivocally substantival. To be sure, substantival geminating verb forms, such as *mrr.f*, are not attested after the negation *n,* but it will be remembered that, by Gunn's rule, the equivalent of "He does not hear" is *n sḏm.n.f* (*n m.r.n.f*), with past tense affix *n*, and not *n sḏm.f.* The etymology of the negation *n*, which might have explained why a substantival verb form can follow it, is unknown. With regard to orthography, Schenkel (1985, 2000) formulates rules and discerns patterns regarding the appearance of *w* and *jj* (double reed leaf) while acknowledging that exceptions exist. There is much corruption in the transmitted texts, but if Schenkel's orthography standards are applied to syntax, then a few exceptions possibly owed to corruption do not invalidate a rule. Much evidence supports the notion that the *sḏmw.f* is always substantival (Depuydt 1993b: 24–27).

9.8.9 The balanced sentence and the emphatic nominal sentence: Depuydt 1986, 1988a

The balanced sentence is a nominal sentence in which two clauses A and B are equated with one another. It says that A means the same as B.

jbb N jbb Nḫbt (Pyr. §696d)

'When N is thirsty, Nekhbet is thirsty'.

It is stated that the fact that N is thirsty means exactly the same as the fact that Nekhbet is thirsty. The one means the same as the other.

In mathematics, no one would hesitate to rewrite A = B as B = A. However, when it comes to language, the following question arises: Does *jbb N jbb Nḫbt* 'When N is thirsty, Nekhbet is thirsty' mean exactly the same as *jbb Nḫbt jbb N* 'When Nekhbet is thirsty, N is thirsty'? It seems clear that, logically speaking, the latter automatically follows from the former and the former automatically follows from the latter. Examples of actual reversals in texts (see §5.1 above) support this logical reversibility.

Still, consider an example such as the following.

wḏ3.j wḏ3.f ṯs pḫr (CT 4.246a)

'I am fine, he is fine, and vice versa'.

If *t̠s ph̠r* 'vice versa' is resolved, the full version is:

**wd̠ꜣ.j wd̠ꜣ.f wd̠ꜣ.f wd̠ꜣ.j*

'I am fine, he is fine. He is fine, I am fine'.

Is *wd̠ꜣ.j wd̠ꜣ.f* 'I am fine, he is fine' exactly the same as *wd̠ꜣ.f wd̠ꜣ.j* 'He is fine, I am fine'? One difference is obvious: the word order. But what, if anything, does this mean? Is there a difference in the respective messages of the two sentences?

Few pairs of concepts are as popular in linguistics nowadays as subject and predicate, topic and comment, and theme and rheme. However, since 1988 (see Depuydt 1988a: 401–6), I have followed the likes of linguists such as Lucien Tesnière, Hugo Schuchardt, Louis Hjelmslev, and many others in rejecting the existence of these concepts. I have never seen any reason to go back on this decision. The concepts in question seem to me nothing but ghost concepts. I have stated this conviction on several occasions.

On the other hand, contrastive emphasis, or focus, or whatever one may wish to call it, is a *bona fide* empirical phenomenon with signals such as cleft sentences and so-called emphatic verb forms. I once suggested (1986: 366; cf 1988a: 404 n. 3) that the second clause of a balanced sentence is marked by contrastive emphasis. An example such as *jbb N jbb Nḫbt* 'When N is thirsty, Nekhbet is thirsty' would be the answer to an implied question "What does the fact that N is thirsty mean?" the answer being "It means that Nekhbet is (also) thirsty." That is what the first clause means, in contrast to all possible things that it could mean. Hence the contrastive emphasis.

Questions for specification ("who?" "what?" "which?" "where?" and so on) by definition denote contrastive emphasis (see Depuydt 2007). That explains why interrogative pronouns and adverbs appear in emphatic constructions in all stages of Egyptian. The purpose of questions for specification is to identify a person, place, time, and so on, as contrasted with other possible persons, places, or times. The answers to questions for specification must therefore also denote contrastive emphasis.

The suggestion that balanced sentences denote contrastive emphasis affecting the second clause must remain provisional. There is no positive evidence to prove the presence of emphasis—such would require an explicit signal of emphasis. This does not exist, so I propose that contrastive emphasis was expressed by a distinct intonation pattern. I have suggested elsewhere that certain instances of the structure A *pw* B denote contrastive emphasis affecting B (Depuydt 1999d: liii–lvii). A normal literal translation of the structure A *pw* B 'It is A, namely B'. It is probable that the ternary structure

A *pw* B is an expansion of the binary structure A *pw* 'It is A'. However, in some instances of A *pw* B, the translation of the part A *pw* is 'A is …' rather than 'It is A'. A *pw* B can then not (or no longer) be seen as an expansion of the binary structure. In those instances, A *pw* B is often found in parallelism with other, proven, emphatic constructions. It may be hypothesized that contrastive emphasis then affects B. Presumably a special pattern of intonation marked emphasis. Intonation is not represented in writing.

In sum, with regard to the example under discussion, there are at least three possibilities to be considered.

First, *wḏꜣ.j wḏꜣ.f* means exactly the same as *wḏꜣ.f wḏꜣ.j*. The difference in word order has no effect on the meaning. It is difficult to see what word order alone can do to the content of a message.

Second, the second clause bears contrastive emphasis. That would accord with the fact that, in certain common formulas, balanced sentences alternate with adverbial sentences in which an initial substantival verb form appears to mark emphasis on an adverbial verb form. Thus Coffin S1C bears both the balanced sentence

prr.sn r pt m bjkw prr.j r ḏnḥw.sn (CT 3.24c–25b),

of which the literal translation appears to be

> 'The fact that they ascend to heaven means the same as the fact that I ascend on their wings',

and

prr.sn r pt m bjkw jw.j ḥr ḏnḥw.sn (CT 3.100h–i),

which is presumably an adverbial sentence of which the literal translation is

> 'The fact that they ascend to heaven as falcons is while I am on their wings' (*or*: 'It is while I am on their wings that they ascend to heaven as falcons').

Third, some other nuance differentiates *wḏꜣ.j wḏꜣ.f* from *wḏꜣ.f wḏꜣ.j*, but it is difficult to see what that nuance would be. What would the nonwritten explicit signal marking that unknown nuance be?

I would exclude a fourth possibility, of associating a possible difference in nuance with concepts such as subject and predicate, topic and comment, or theme and rheme. I personally believe that these concepts are ghost concepts. The main reason for my believing this is that no theory assigns transmittable empirical signals to these concepts.

9.9 ENTANGLEMENT IN THE GRAMMATICAL DISCUSSION OF THE BALANCED SENTENCE WITH OTHER SENTENCE TYPES

9.9.1 Source of the entanglement: Difficulties in classifying sentence structures

When strictly defined as a nominal sentence containing two substantival verb forms, the balanced sentence fits in its own niche, hardly subject to confusion with other constructions in grammatical discourse. Historically, the balanced sentence did not evolve into another construction after Middle Egyptian. Its disappearance from the language underscores the well-delineated, even isolated place it occupies in Old and Middle Egyptian grammar.

Nonetheless, the balanced sentence has on occasion become confused with other sentence structures in recent discussions of Old and Middle Egyptian grammar, such as Schenkel 1988: 275–84. In later contributions, Schenkel (1997, 1998) has intertwined the balanced sentence with other sentence types in two ways: by classifying certain passages as balanced sentences that many others would not; and by classifying the balanced sentence as a subcategory of another sentence type. Thus, Schenkel has recently classified the balanced sentence as a member of a larger species of sentence construction that he calls the "rank-5-expansion" ("Rang-V-Erweiterung"). A survey of this "rank-5-expansion" appears in Schenkel's teaching grammar since the 1997 edition (1997: 258–71). The strict definition of the balanced sentence as two substantival verb forms relies on criteria of form, but any wider definition of the balanced sentence relies on criteria of meaning, mainly the concept of equation.

How has the balanced sentence become associated with structures that are clearly distinct from it? The cause seems to be the many persistent difficulties regarding classification of sentence structures in Old and Middle Egyptian. No one doubts that there are various ways for two clauses to combine into a sentence in Old and Middle Egyptian, but owing to the inadequacies of hieroglyphic writing, it is not fully clear exactly what these various ways are. One is the balanced sentence. Another is the adverbial sentence, as in *jj.n.j ḫr.k … jnt.k wj m3n.j nfrw.k* (Book of the Dead, chapter 25, Introduction; Budge 1898: 249,5–6). Here the substantival verb form *jj.n.j* appears in the substantival slot of the adverbial sentence and the verb forms *jnt.k* and *m3n.j*, as part of adverbial clauses, appear in its adverbial slot. The result is the following literal translation: 'That I have come to you is in order that you might bring me and that I might see your beauty'.

The question arises: In how many ways can two clauses containing verbs combine into sentences? Vernus (1981) did much to formulate the problem acutely and also presented certain new solutions, appropriately leaving the balanced sentence in its own niche.

9.9.2 Three elusive sentence structures

In recent discussions of how two clauses can combine into a sentence, three Old and Middle Egyptian sentence structures have received much attention, but I deny their existence.

9.9.2.1 The first elusive pattern: Vernus's "Second" scheme

Vernus (1981) has defended the existence of the first sentence structure, calling it the "Second" scheme (*second schème*) of combining two clauses containing verbs into sentences (the "First" scheme [*premier schème*] is the adverbial sentence, such as the just-cited Book of the Dead, chapter 25, Introduction).

In Vernus's Second scheme, an adverbial or circumstantial clause precedes the main clause. I, however, have never seen an unambiguous example of an adverbial or circumstantial clause preceding its main clause in Old and Middle Egyptian. Likewise, in Late Egyptian or Demotic, a clause introduced by circumstantial *jw* certainly never precedes its main clause. H. J. Polotsky and his student Mordechai Gilula (1987: 139) agree. The present writer belongs to the school of thought that believes that a circumstantial clause cannot stand at the beginning of a sentence unless it is preceded by a preposition (e.g. *ḥr m-ḫt*).

Schenkel (1990: 178) once expressed the same conviction:

> In this case Vernus interprets the first member of the sentence as an adverbial expression—which is hardly possible according to the rules of earlier Egyptian-Coptic."*

This is not the place to examine the entire problem in detail. One argument, anticipated above, concerns the Late Egyptian and Demotic equivalents of Old and Middle Egyptian adverbial or circumstantial clauses. They are marked by the circumstantial or adverbial converter *jw*, which never precedes its main clause in Late Egyptian or Demotic. Then why would its Old and Middle Egyptian equivalents, especially if so much effort seems necessary to locate unequivocal examples that everyone can agree on?

* Vernus sieht in diesem Falle im ersten Satzteil einen adverbialen Ausdruck—was nach den Regeln des älteren Äg[yptisch]-Kopt[isch] kaum möglich ist.

9.9.2.2 The second elusive structure: Niccacci's topicalized substantival clause

A second nonexistent sentence structure that has received much attention consists of a main clause preceded by an initial substantival verb form. The initial verb form is said to be placed in "anticipatory emphasis" or topicalized. In a literal translation, the sentence then begins with 'As for the fact that …' or the like. Niccacci (1980) has defended the existence of this pattern at length, but see §9.7. I believe that most of Niccacci's examples of a topicalized substantival verb form can easily be interpreted as adverbial sentences.

9.9.2.3 A third elusive structure: Noun + sḏm.f

The ambiguity of Old and Middle Egyptian writing, in which a single string of words can often be interpreted in different ways, may be illustrated by yet another elusive structure, namely Noun (or nominal phrase) + *sḏm.f*, as contrasted with simple *sḏm.f* (*sḏm* + Noun). This structure is widely accepted (e.g. Doret 1980, with bibliography), but, inspired by a private communication by Polotsky in the 1980s, I have shown that many purported instances of Noun + *sḏm.f* can be interpreted with hardly any difficulty as *jw* + Noun + *sḏm.f* minus *jw* (Depuydt 1991a: 50–52 §§25–28). Noun + *sḏm.f* is then not quite a construction in its own right, but rather the adverbial or circumstantial conversion of the pattern *jw* + Noun + *sḏm.f*. The two are therefore better classified together as just one pattern. Two arguments for this classification are: First, it has long been known (see especially Doret 1980) that the *sḏm.f* form is the same in both Noun + *sḏm.f* and *jw* + Noun + *sḏm.f*. There are therefore no morphological obstacles to associating them and classifying them together. Second, in the common pattern *jw.f sḏm.f* 'he hears', a noun or nominal phrase as a rule replaces the *first* suffix pronoun. The norm is *jw* + Noun + *sḏm.f*, not *jw sḏm* + Noun, let alone *jw.f sḏm* + Noun. By contrast, in the passive conversion, *jw sḏm.tw* + Noun appears to be normal. Thus grammarians might *a priori* have wondered what the adverbial or circumstantial conversion of *jw* + Noun + *sḏm.f* might be. Often in Old and Middle Egyptian, the absence of *jw* serves as the marker of the adverbial or circumstantial character of a clause. Noun + *sḏm.f* therefore naturally emerges as the expected circumstantial equivalent of *jw* + Noun + *sḏm.f*, although grammars do not normally classify Noun + *sḏm.f* in this manner. One exception is Allen (1984: 166–71 §§274–78), who considers it a possibility (§275).

> It seems reasonable to assume that subject plus *sḏm.f* can also occur without introductory particle, as an adverbial-predicate construction, with the same value as other such constructions.

I interpret this "adverbial-predicate construction" as "*jw* + Noun + *sḏm.f* minus *jw*." Noun + *sḏm.f* then has all the properties of *jw* + Noun + *sḏm.f*, except the independence marked by *jw*.

One possible example cited by Allen (ibid.) is as follows. In it, Noun + *sḏm.f* probably bears emphasis. The contrastive emphasis is marked by an initial substantival verb form (*pr.n*).

> *pr.n N m P ḫr b3w P …*
> *Wp-w3wt wp.f n.f w3t*
> *Šw f3.f sw*
> *b3w Jwnw sḳ.sn n.f rd …* (Pyr. §§1089a, 1090a–c P)
>
> 'It is … while Wepwawet opens the way for him,
> while Shu carries him, and
> while the ba's of Heliopolis set up a stairway for him …
> that N has come forth (from the city of) Pe near(?) the ba's
> of Pe…'.

It should be added that not every instance of Noun + *sḏm.f* in Old and Middle Egyptian is necessarily an instance of *jw* + Noun + *sḏm.f* minus *jw*. Three other possibilities are as follows.

(1) The initial noun or nominal phrase may be "topicalized." It can then be introduced by expressions such as "As for the fact that …" or "As regards …" in English. But in that case, Noun + *sḏm.f* is merely a subcategory of the much larger phenomenon of topicalization, which is not restricted to Noun followed by *sḏm.f*. The sentence proper begins in fact with *sḏm.f*. In some cases, topicalization seems to be the only possible interpretation, as in *t3.n pḥ.n sw* (Shipwrecked Sailor 11), which cannot be interpreted as **jw t3.n pḥ.n sw* minus *jw*, and therefore as an instance of *jw* + Noun + *sḏm.f* minus *jw*, because *t3.n* 'our land' is resumed by the direct object of *pḥ* 'reach', namely *sw* 'it'.

(2) The noun or nominal phrase in Noun + *sḏm.f* could be a vocative ('O …'). The sentence proper again begins with *sḏm.f*. There is then no significant absence of *jw*.

(3) Noun + *sḏm.f* is an incomplete sentence meaning 'N while he hears' (see Polotsky 1976: 31–32 §3.5.3). An example (cited by Polotsky) is a legend identifying a person in an inscription: *Mrw … d(y).f j3w n Wsjr* 'Meru as he gives praise to Osiris'.

9.9.3 Schenkel's inverted standard theory

9.9.3.1 Similarity of the inverted standard theory to Niccacci's topicalized substantival clause

Let us return, then, to Niccacci's structure, the topicalized substantival verb form followed by an initial clause, which has received much support from Schenkel, who interprets many sentences as instances of this construction in his more recent publications (1997, 1998, also to some extent earlier in 1990: 178). Along the same lines, Loprieno (1995: 197) writes about *h33.sn m ḥf3w h3jj.j ḥr k3bw.sn* (CT 3.24a–b):

> The topicalized V[erb] P[hrase] *h33=sn* is extraposed, and the predicate of the main sentence functioning as its comment is an independent verb form.

Schenkel's view is part of a wide-ranging theory unlike any ever proposed for Old and Middle Egyptian grammar, which he calls the Inverted Standard theory. One feature deserves attention here, a vast expansion of the domain of Niccacci's structure, in which an extraposed or topicalized substantival verb form precedes a main clause. Schenkel so interprets a large number of sentences that had widely been interpreted as adverbial sentences, such as:

> *jj.n.j ꜥ3 r njs r.k* (Papyrus Westcar 7,20)
>
> 'I have come here to summon you' (*or*: 'It is in order to summon you that I have come here').

If *jj.n.j ꜥ3 r njs r.k* contains an extraposed substantival clause, that clause ought to be *jj.n.j ꜥ3* 'that I have come here'. However, that leaves only *r njs r.k* 'in order to summon you' as the rest of the sentence. *R njs r.k* by itself cannot be a main clause, but to interpret the passage as an instance of Niccacci's construction, it would have to be one. To make it a main clause, Schenkel (1998: 147) makes the unusual theoretical move of postulating a vanished stative verb form *jj.kw* preceding *r njs r.k*. Accordingly, the original version of the sentence, presumably at some time in prehistory, would have been **jj.n.j ꜥ3 jj.kw r njs r.k* 'As for the fact that I have come here (*jj.n.j ꜥ3*), I have come (*jj.kw*) to summon you (*r njs r.k*)'. Schenkel (1998: 147) writes:

> Such an alternative explanation is indeed possible or at least: It is possible to say how this type of sentence has come into existence in the history of the language.*

* Eine solche Alternativerklärung ist in der Tat möglich oder zum mindesten: Es ist möglich zu sagen, wie der Satztyp sprachhistorisch zustande gekommen ist.

In my opinion, being able to state that something is possible seems a little sparse as an argumentation. The realm of that which is not impossible is very large. There is so much one can say is possible without being able to prove it true.

9.9.3.2 *Four arguments against the inverted standard theory*

Objections can be raised to Schenkel's structure. First, the pattern is never attested in the historical period; it is a pure postulate. Second, it is not clear why speakers would at some time in prehistory so often have repeated a verb (**jj.n.j ... jj.kw ...*). According to Schenkel, the repetition marks contrastive emphasis resting on the adverbial phrase. But it is not obvious how repetition produces such emphasis. Third, the question arises: If the marker of emphasis (namely repetition) has disappeared by historical times, what still signals emphasis in *jj.n.j ꜥ3 r njs r.k* in the historical period? True, the substantival verb form *jj.n.j* is retained, so that emphatic *jj.n.j ꜥ3 r njs r.k* 'It is in order to summon you that I have come here' is still differentiated from non-emphatic **jw.j jj.kw ꜥ3 r njs r.k* 'I have come here in order to summon you'. Still, by Schenkel's theory, the actual marker of emphasis would have vanished.

Fourth, Schenkel's theory dispenses completely with the concept that sentences with emphatic verb forms are *adverbial* sentences, a cornerstone of Polotsky's theory. In fact, by Polotsky's own report (see Depuydt 1995a), his observation of the close association between emphatic verb forms and adverbial phrases lay at the origin of his entire theory. Compare the following two passages from the Late Egyptian story of Horus and Seth.

EMPHATIC VERB FORM = ADVERBIAL *ḥr jḫ* 'why (because of what)?'

i.jr.k dyt d3jj.st ḥr jḫ (7,12–13)

'Why did you let her cross?'

OTHER VERB FORM = NOMINAL *jḫ* "what?"

jw.k ḏd n.s jḫ (7,8)

'What did you tell her?'

Polotsky observed that, clearly, there is a link between emphatic verb forms (*i.jr.k dyt*) and adverbial phrases (*ḥr jḫ*)—the two occur together. But when *jḫ* is direct object, then the verb form is not emphatic. When he first communicated this observation that lies at the root of his theory of the verbal system, Polotsky (1940: 244 [1971: 36]) described the absence of emphatic verb forms with interrogative pronouns that denote direct objects as "astonishing" (*étonnant*).

Additional illustrations of the contrast could be adduced. What matters here is that Schenkel's theory cannot account for the close association of emphatic verb forms with adverbial phrases. It offers no reason why emphatic verb forms should not emphasize nominal phrases, such as direct objects, just as they do adverbial phrases. Nothing in Schenkel's theory narrows the function of emphatic verb forms specifically to adverbial phrases. Yet, this narrow range of function is what appears in all stages of Egyptian. This narrow focus on adverbial phrases is a most striking characteristic of the data.

9.10 FOUR EXCURSUSES

9.10.1 Explanation of the absence of second tenses in certain questions for specification

In the same article in which he describes the absence of emphatic verb forms with interrogative pronouns that denote direct objects as "astonishing" (*étonnant*), Polotsky (1944) reports a rule pertaining to Sahidic Coptic that remains unexplained. It concerns instances in Coptic in which interrogative pronouns denote direct objects. It was noted above that in Late Egyptian emphatic verb forms are never used when interrogative pronouns denote direct objects. But in Sahidic Coptic, as Layton (2004: 355 §445) observes,

> Some standard phraseology occurs in focalizing conversion as a matter of course: e.g. **ⲉⲩ–ⲣ–ⲟⲩ** [*e=w-r u*] **ⲉⲩ–ⲛⲁ–ⲣ–ⲟⲩ** [*e=w-na-r u*] 'What are they doing/will they do?'

A present tense example is **ⲁⲛⲁⲩ ϫⲉ ⲉⲩⲣ̄ ⲟⲩ ϩⲛ̄ ⲛ̄ⲥⲁⲃⲃⲁⲧⲟⲛ** *anaw je e=w-r u hn n-sabbaton* 'Look what they are doing on the Sabbath' (Mark 2:24). Examples of other tenses are **ⲛ̄ⲧⲁϥⲣ̄ ⲟⲩ** *nta=f-r u* 'What has he done?' (Ps 10:3) and **ⲉϣⲁⲣⲉⲉⲣ ⲟⲩ ⲛⲁⲩ ⲛ̄ⲧⲟ** *ešare=Ø-er u na=w nto* 'What do you, for your part, do to them?' (Shenute ed. Amélineau I 108,9–10, cited from Layton 2004: 359 §449). Clearly, Coptic and Late Egyptian differ. But there is more. The statement about Coptic is incomplete without "an exception that I signal as a matter of curiosity without being able to provide an explanation" (Polotsky (1940: 245 [1971: 37]).* When **ⲣ̄ ⲟⲩ** *r u* appears in the question "How are you?" or the like, which asks after someone's well-being, first

* Une exception que je signale à titre de curiosité sans pouvoir en offrir une explication.

tenses and not second tenses are used. Polotsky cites the first tense in ⲦⲈⲦⲚ̄Ⲣ̄ ⲞⲨ *tetn=r u* 'How are you?' (Gen 43:27). The second tense ⲈⲦⲈⲦⲚⲢ̄ ⲞⲨ *e=tetn-r u* is not found here. Another example of the first tense is ⲤⲈⲢ̄ ⲞⲨ *se-r u* 'How are they?' (Acts 15:36). The second tense ⲈⲨⲢ̄ ⲞⲨ *e=w-r u* is not found here. Two examples cited by Polotsky from the Sayings of the Desert Fathers are ⲚⲈⲤⲚⲎⲨ Ⲣ̄ ⲞⲨ *nesnêw r u* 'How are the brothers?' and ⲚⲈⲔϨⲂⲎⲨ⟨Ⲉ⟩ Ⲣ̄ ⲞⲨ *ne=k-hbêw⟨e⟩ r u* 'How are your affairs?' A first tense also accompanies Ⲟ Ⲛ̄ⲀϢ Ⲛ̄ϨⲈ *o n-aš n-he* in the expression ⲠⲔⲞⲤⲘⲞⲤ Ⲟ Ⲛ̄ⲀϢ Ⲛ̄ϨⲈ *p-kosmos o n-aš n-he* 'How is the universe?', from the same source. It is tempting to interpret all these instances of the first tense as survivals of earlier Egyptian: in Late Egyptian, second tenses are never used when interrogative pronouns denote direct objects, but that explanation does not account for why first tenses are used in some instances and second tenses in others. The following alternative explanation is therefore proposed here to account for the presence of the first tenses. It is true that, *in form*, idioms such as "How are you?" are questions for specification. But, *in meaning*, isolating contrast has vanished from them. The function of "How are you?" is *not*— as it may seem at first sight—to ask in what way (how?) as contrasted with other ways one finds oneself. Indeed, the question often remains unanswered. The reply may in fact be a repetition of the question, namely "How are *you*?" In conclusion, isolating contrast is probably absent. And so therefore is the second tense. Likewise, in English, "How are you?" is not pronounced "*How* are you?", with isolating contrast, in the sense of "In what way as contrasted with other ways are you?" Indeed, "how" in "How are you doing?" does not mean "in what manner?"

9.10.2 The manifold interpretations of a single passage, Sinuhe B127, one of them as a balanced sentence

9.10.2.1 The text

Most instances of both Vernus's Second scheme and Niccacci's extraposed substantival verb form preceding a main clause can be interpreted as adverbial sentences, although the meaning and structure of many passages remains obscure, making the passages in question less than perfect evidence for grammatical argument. The following example illustrate how different observers can interpret the same string of words in different ways and how the balanced sentence can be drawn into the discussion.

sḏr.n.j ḳ3s.n.j pḏt.j (Sinuhe B127)

At first sight, it is difficult to see how so much controversy could exist about this short sentence. Two verbs are concerned, *sḏr* 'lie down, go to bed' and *ḳ3s* "string (a bow)'.

9.10.2.2 *Sḏr means 'lie (down)', not 'sleep'*

Sḏr never denotes sleep as a state of unconsciousness, which is rendered instead by *(n)ḳdd*, but rather 'lie (down)', that is, 'be in or come into a horizontal position'. In an interesting twist, *nḳdd* shifted in meaning from 'sleep' to 'lie (down)' by the time of Coptic. Coptic *nkotk* thus replaced *sḏr*, which became obsolete (Depuydt 2006). Two additional striking examples showing **ⲛ̄ⲕⲟⲧ̄ⲕ** *nkotk* in sharp contrast with verbs that mean 'sleep' have since come to my attention along with an observation by Polotsky about **ⲛ̄ⲕⲟⲧ̄ⲕ**:

ⲁⲓⲉⲛⲕⲁⲧⲕ ⲉⲙⲡⲓϩⲓⲛⲏⲃ (Bilabel & Grohmann
a=y-enkatk e-mp=i-hinêb 1934: no. 122,8–9)

'I lay down (**ⲉⲛⲕⲁⲧⲕ** *enkatk*) without being able to sleep (**ϩⲓⲛⲏⲃ** *hinêb*)'.

ⲁⲃϭⲓ ⲛⲟⲩⲃⲏⲛⲧ ⲉⲃⲛⲕⲁⲧ ⲉⲃⲁⲃϣ (ibid. no. 138,5–6)
a=b-ci n-u-bent e=b-nkat e=b-abš

'He took a worm that was lying there (**ⲛⲕⲁⲧ** *nkat*) sleeping (**ⲁⲃϣ** *abš*)'.

(standard Sahidic: **ⲁϥϭⲓ ⲛ̄ⲟⲩϥ̄ⲛ̄ⲧ ⲉϥⲛ̄ⲕⲟⲧ̄ⲕ ⲉϥⲟⲃ̄ϣ**
a=f-ci n-u-fnt e=f-nkotk e=f-obš)

Both examples, from magical texts, are cited by Polotsky (1935: 422 [1971: 281]). Polotsky rightly and uniquely notes about the first example that **ⲛⲕⲁⲧ** cannot mean both "sleep" and "lie down to sleep,"* but he does not address said historical shift.

9.10.2.3 *Source of the multiplication of interpretations: Sḏr.n.f is substantival*

Both verb forms in Sinuhe B127 are instances of the *sḏm.n.f* form, which denotes past tense. If the two pasts are simply translated in sequence, the result is 'I went to bed, I strung my bow'. That has generally been assumed to mean that Sinuhe, the story's hero, strung his bow around bedtime in preparation for the next morning. In fact, the next sentence refers to the following dawn. The first verb form, *sḏr.n.j*, is often treated as a kind of auxiliary with a free translation something like 'in the evening' or 'at night'.

* **ⲛⲕⲁⲧ** [bedeutet] nicht sowohl 'schlafen' als 'sich zum Schlafen hinlegen'.

This passage became controversial as a result of a proposal made by H. J. Polotsky in 1944 that the initial *sḏm.n.f* of verbs of motion in main clauses is emphatic (1944: 84–85 §29 [1971: 188–89]).

> [Gardiner ([3]1957 [[1]1927]: 331 §414,4)] has observed that, in the case in which the purpose for which one has come is specified, the verb 'come' is used in the *sḏm·n·f*, and not in the pseudo-participle, the latter being used when one only points out that one is there (classic example: [*mk wj jj.kw*] *Westcar* 8, 12). According to him, this usage is part of the general rule that the *sḏm·n·f* of verbs of motion is used when one envisages more the action than the completed state. This explanation does not appear satisfying to us …. We believe … that this is a cleft sentence and that *ii·n·i* is an "emphatic" *sḏm·n·f*.*

Gardiner had already noted that the *sḏm.n.f* of verbs of motion is typically found when the aim of the motion is mentioned and the stative when it is not. Accordingly, the *sḏm.n.f* of verbs of motion must always be either substantival (which includes the emphatic use) or adjectival (that is, relative). This principle imposes rather strict limitations on the interpretation of instances of the *sḏm.n.f* of verbs of motion in texts. Polotsky later consolidated this principle (1957 [1971: 43–51]).

Polotsky's observation was later expanded to encompass all intransitive verbs, not just verbs of motion. However, it is now widely accepted that, whereas verbs of motion seem to be *always* either substantival or adjectival, exceptions occur with other types of intransitive verbs—but the three principal verbs of posture, *ꜥḥꜥ* 'stand, rise', *ḥms*(*y*) 'sit (down)', and *sḏr* 'lie (down)', belong with verbs of motion by the present criteria. In other words, the verb forms *ꜥḥꜥ.n.f*, *ḥms.n.f*, and *sḏr.n.f* are always either substantival or adjectival.

Verbs of posture do denote motion, but movement occurs while the body stays in place, such as a change from a sitting position to a standing position, from a standing position to a lying position, and so on. By contrast, other verbs of motion denote displacement of the body.

* [Gardiner (1957: 331 §414,4)] a constaté que, dans le cas où l'on déclare le but pour lequel on est venu, on emploie le verbe "venir" au *sḏm·n·f*, et non pas au pseudo-participe, ce dernier s'employant quand on n'insiste que sur le fait qu'on est là (exemple classique: [*mk wj jj.kw*] *Westcar* 8, 12). Selon lui, cet emploi rentrerait dans la règle générale que le *sḏm·n·f* des verbes de mouvement s'emploie quand on envisage plutôt l'action que l'état achevé. Cette explication ne nous paraît pas satisfaisante …. Nous croyons … qu'il s'agit d'une phrase coupée et que *ii·n·i* est un *sḏm·n·f* "emphatique."

9.10.2.4 Interpretation of Sinuhe B127 by Vernus's First and Second schemes

Turning to recent interpretations of the grammatical structure of our passage, the first two are according to Vernus's First and Second schemes (1981: 79; cf. Schenkel 1988: 279). Vernus considers both interpretations possible. The Second scheme is more characteristic of Vernus's views on Old and Middle Egyptian sentence structure. The literal translations according to the two schemes are as follows.

> (First scheme) 'It is after having strung my bow that I spent the night'. (I strung my bow *in the evening*, before going to bed.)
>
> (Second scheme) 'After having spent the night, I strung my bow'. (I strung my bow *in the morning*.)

The interpretation by the First scheme represented nothing new when Vernus presented it once again in 1981. *Sḏr.n.j* is a substantival verb form and *ḳ3s.n.j* is an adverbial verb form and the two together form an adverbial sentence. The substantival and adverbial *sḏm.n.f* are written alike. The interpretations of *sḏr.n.j* as substantival and *ḳ3s.n.j* as adverbial are therefore a matter of inference.

The interpretation according to the Second scheme is characteristic of Vernus's approach to Old and Middle Egyptian sentence structure. According to the Second scheme, the first *sḏm.n.f* (here *sḏr.n.j*), "whether substantival or not, … assumes circumstantial function" (1981: 79). This verb form is followed by "an independent verb form (*forme verbale non subordonnée*)" (76) in the apodosis (here *ḳ3s.n.j*). Vernus (79) writes as follows.

> [The second scheme] must be understood syntactically as comprising in the protasis a verb form, substantival or not, which assumes circumstantial function, plus a nonsubordinate verb form in the apodosis.*

By this interpretation, the verb form *ḳ3s.n.j* is "independent." It denotes a main clause. It is not clear whether Vernus interprets *sḏr.n.j* as substantival. Perhaps he takes it as a substantival verb form with adverbial function. If so, then our passage becomes more or less an instance of Niccacci's construction, in which an initial substantival verb form precedes an independent verb form. However, Vernus's Second scheme is a wider phenomenon than Niccacci's construction.

* Un tel tour doit être compris syntaxiquement comme comportant dans la protase une forme verbale, substantive ou non, assumant la fonction circonstancielle + une forme verbale non-subordonnée dans l'apodose.

9.10.2.5 Interpretation of Sinuhe B127 as a balanced sentence

It was Schenkel (1988: 279) who introduced the balanced sentence into the discussion of Sinuhe B127. According to this interpretation, both *sḏr.n.j* and *ḳ3s.n.j* are substantival verb forms and the two are equated. His literal translation is '(The fact) that I spent the night, was: (the fact) that I tied my bow' ('Daß ich die Nacht verbrachte, war: daß ich meinen Bogen band').

9.10.2.6 Interpretation of Sinuhe B127 as an adverbial sentence denoting contiguity

I would eliminate two of these three interpretations on the following grounds. (a) Vernus's Second scheme does not exist (§9.8.3).

(b) Interpreting Sinuhe B127 as a balanced sentence is not suitable with regard to meaning. If the translation is modeled on the other balanced sentences, then it ought to mean 'Whenever I have gone to bed, I have strung my bow'. The general sense is then that my going to bed has always meant the same as my stringing my bow. Two events are equated with one another. One has never happened without the other. That does not, however, seem to be the meaning of Sinuhe B127. The reference is to a single narrative event in a chain of events. There seems to be no reason why the narrative would suddenly be disrupted to announce that two events, going to bed and stringing a bow, have always coincided.

That leaves only the third option, the adverbial sentence. The literal translation is 'It is after having strung my bow that I went to bed'. But this is not my preferred interpretation, either. The following problem remains. Why would it be stated with emphasis that it was in this or that circumstance that I went to bed or that evening came? This seems like an awkward mode of expression. The statement would clumsily answer the implied question as to which of all possible circumstances prevailed when I went to bed. That seems to turn the message on its head.

All this has led me to adopt another interpretation of Sinuhe B127. I would retain the adverbial sentence structure, as in Vernus's First scheme—it is unnecessary to postulate a new sentence pattern.

I propose that the adverbial sentence is used here to convey contiguity (Depuydt 1999d: 693–729). In summary, contiguity has no signal of its own; it is expressed by a combination of signals:

CONTIGUITY = (1) INVERSION + (2) EMPHASIS

Contiguity may be applied as follows to B127. The verb *sḏr* commonly means 'lie down' or 'go to bed'. But in the present example, it is better interpreted as denoting the transition from daytime to nighttime, when one normally reclines for sleep. The meaning of *sḏr* is then something like 'turn into evening' or 'become at night'. What is unusual from the point of view of modern European languages is that the event of turning into evening can be said about a person. In the present case, *sḏr.n.j* signifies 'that I turned into evening' or 'that I became at night'.

The literal translation of *sḏr.n.j ḳ3s.n.j pḏt.j* is then 'That I entered the night is while having (already) strung my bow'. The peculiar nuance of contiguity is : So soon did the stringing of the bow follow the arrival of evening that it seemed as if it was *after* stringing the bow that evening arrived. In other words, the sequence of the events is *inverted* and this inverted order is then marked by contrastive emphasis, as if it had been unexpected. The effect is that the second event very soon follows the first. The two events are contiguous. In English, the expression "No sooner … than …" denotes contiguity, as in "No sooner did it become evening than I strung my bow." A freer translation of Sinuhe B127 would be: 'In the evening (that is, after the activities of the day and before going to sleep) I strung my bow'. By this interpretation, Sinuhe would prepare his bow in the evening before going to sleep for his fight the next day.

9.10.2.7 An Arabic parallel to sḏr.n.j *in Sinuhe B127*

Arabic exhibits an interesting expression that would appear to be parallel to Sinuhe B127. An example is cited in Brockelmann's Arabic grammar (1982: 181 §146 n. b): *amsa ahluha ḥtamalu*. The literal translation of this example is 'Their people entered into evening, (and) they left'. The author notes how the absence of *wa* 'and' between the two verb forms denotes that the second action is "an action that immediately follows (the first)" (*eine unmittelbar anschließende Handlung*). In other words, the effect is that of contiguity. One interesting fact that I did not sufficiently stress elsewhere (Depuydt 1993a: 165 n. 64) is that *amsa* 'entered into evening' applies to people, namely *ahluha* 'their people', apparently just as *sḏr* 'enter into evening' does to *j* 'I' in *sḏr.n.j* 'that I entered into evening'.

9.10.3 Additional potential instances of contiguity

9.10.3.0 General remark

Arguments concerning "contiguity" and many examples have been presented elsewhere (for bibliographical references, see Depuydt 1999d: lxviii; for the most detailed explanation, see ibid. 693–729). Additional examples are found in several references cited there and also in Depuydt 1999e: 312, 2000, 2001b: 83, 2002b: 602, in press: §6. There have been some reactions, but not many. It remains a fact that, owing to the defective nature of hieroglyphic writing, there will always be a gap between establishing contiguity as a grammatical category and verifying its presence in a specific instance. Examples adduced will therefore often be merely potential examples. Further potential examples are presented here.

All this does not preclude that one might change one's mind about the interpretation of individual passages. Thus I once thought that *njs.n.tw n wʿ jm jst wj ʿḥʿ.kw* … at Sinuhe R24 (B1) might mean something like 'No sooner had one of them been called out than I got up …' (see Depuydt 1993a: 181). I am now convinced that *n wʿ jm* is emphasized by the *sḏm.n.tw.f* form: 'It is to one of them that the call went out, and that while I was standing …'. The reason is that I am now convinced that a clause introduced by the particle *jst* is *never* marked as emphasized by a preceding substantival verb form. I hope to treat this in detail elsewhere.

Two examples cited in Depuydt in press may be repeated here in anticipation of their publication (cf. Kruchten 1999: 11 n. 49). They seem especially convincing.

ẖpr dw3 n 7 hrw jw jn sn p3 ẖpš n mn-m3ʿt rʿ (*KRI* 1.103, 10,12)

ẖpr.n ʿḥʿ hrw *jw.w ẖr n b3w ḥm.f* (*KRI* 1.12,13–14)

The past *sḏm.f* of intransitive verbs is emphatic in Late Egyptian. The first of these passages therefore literally means 'It is after *Mn-m3ʿt-rʿ*'s strong arm had brought them that morning arrived seven days later'. *Ḫpr dw3* is emphatic. *Jw jn (sn) p3 ẖpš* is past circumstantial. Likewise, *ẖpr.n ʿḥʿ hrw jw.w ẖr n b3w ḥm.f* literally means 'It is while they had (already) fallen for his majesty's power that day-time arrived'. These literal translations present the sequence of events in reverse order. Contiguity allows interpreting them in their proper order.

9.10.3.1 *Teachings of Ani B 22,2–3 and B 22,5–6*

Most recently, in a review of my Middle Egyptian grammar (Depuydt 1999d), Quack (2003: 264) has expressed a lack of confidence in the concept.

> The argumentation that an event could follow so closely *after* another [event] that it takes place *before* it in the consciousness of the speaker does not really persuade [this reviewer]. Depuydt may well be alone in his opinion at this time.*

* [Den Rezensenten] überzeugt … die Argumentation an sich nicht recht ein Ereignis könne so dicht auf ein anderes folgen, daß es im Bewußtsein des Sprechers bereits vor ihm stattfinde. Hier dürfte Depuydt mit seiner Meinung derzeit ziemlich allein stehen.

First of all, I do not think that I am entirely alone. In terms of antecedents, the concept of contiguity is rooted in certain observations by Polotsky, who as early as 1957 explicitly encouraged the construction of a full-fledged and comprehensive theory that would account for certain facts of Old and Middle Egyptian (cf. Depuydt 1999d: 728–29). There is in the scholarly community some degree of tolerance towards theoretical trial balloons. Dwelling in the vast paradise of the non-impossible or not-entirely-impossible is often indulged. Consequently, many a theory has died only with the death of its author. Recantations are rare, especially if one considers that not every theory ever proposed can be right. As regards contiguity, looking again and again at the arguments, I see no need for a recantation a decade after having proposed the concept. I would rather raise the stakes by affirming my strong belief in the absolute veracity of the concept. Then again, truth is based on consensus. What ultimately happens to a theory is to a certain extent beyond the control of its author. Ideally, rehearsing the main arguments in favor of contiguity would seem like the most effective defense, but I do not want to take up the space here. Suffice it to return to just one argument. If the phenomenon of contiguity seems complex and twisted, it should be remembered that many or most languages exhibit unusual expressions that denote contiguity and in which the order of events is inverted, such as "No sooner … than …" in English. If in English, then why not also in Egyptian?

In fact, I might add a striking piece of evidence in favor of contiguity from a text that no one knows better than Quack himself, namely the Teachings of Ani (see Quack 1994). On the one hand, I have proposed that the well-known pattern *wnn.f* (*ḥr*) *sḏm jw.f* (*ḥr*) *sḏm* denotes contiguity. *Wnn.f* (*ḥr*) *sḏm* is substantival (emphatic) and *jw.f* (*ḥr*) *sḏm* is adverbial, as in the common expression *wnn t3jj.j šˁt ḥr spr r.j jw.k ḥr* … 'No sooner will my letter reach you than you will …'. It seems clear the second action follows soon upon the first. The literal translation is something like 'That my letter will reach you is while you already…'. On the other hand, I have also proposed that the pattern *i.jr.f sḏm jw sḏm.f* can likewise denote contiguity, as in *i.jr.f jjt jw jr.f* … 'No sooner had he arrived than he did…'. *'I.jr.f sḏm* is substantival (emphatic) and *jw sḏm.f* introduces circumstantial clauses.

It would be desirable to possess an instance in which a similar sequence of events is expressed in the same text in close vicinity once by the pattern *wnn.f* (*ḥr*) *sḏm jw.f* (*ḥr*) *sḏm* and once by the pattern *i.jr.f sḏm jw* …. This in fact happens in the Teachings of Ani, with both patterns referring to a similar sequence of events.

wnn.f | [3] *dyt ẖnwjj m pr.f*
jw ḥ3tjj tjmḥw (Ani B 22,2–3)

i.jr p3 s 3bb dyt ẖnwjj m pr.f
jw bw gm.f p3 | [6]*jj.f ḥ3wttj m rʿw* (Ani B 22,5–6)

The overall message of these passages is not fully clear to me. The central concept seems to be the absence or presence of quarrel at home (*ẖnwjj m pr.f*). In the first example, the clause introduced by *jw* denotes an event that follows soon upon the event expressed by the first clause. Owing to the parallelism between the two passages, it may be assumed that the same is the case for *jw* in the second example, which is of the type *i.jr.f sḏm jw* ….

Most diagnostic of the presence of contiguity is the unusual combination of and contrast between two features: a substantival verb form emphasizes an adverbial element—but no overt contrastive emphasis affects the adverbial element. Instead, the contrastive emphasis serves to signal contiguity according to a sophisticated mechanism (see Depuydt 1999a: 693–729; cf. Depuydt 1993a: 119). Thus it is obvious that *jw.f ḥr sḏm* is emphasized in the pattern *wnn.f* (*ḥr*) *sḏm jw.f* (*ḥr*) *sḏm*. But the effect is contiguity, not contrast. By virtue of the parallelism of the two examples in the Teachings of Ani, the same should also apply to the example of the pattern *i.jr.f sḏm jw* …. Contiguity therefore seems to be the only way to explain both the parallelism between the sequences *wnn.f* … *jw* and *i.jr* … *jw* … and the absence in them of the overt contrastive emphasis that is normally expected when substantival verb forms combine with an adverbial element.

9.10.3.2 Papyrus Harris I, section 75

Another possible example of contiguity is found in section 75 of the "great" Papyrus Harris (Erichsen 1933). Chaos reigns in Egypt. Rival chiefs rule concurrently. Then the following happens.

kjj h3w ḫpr ḥr-s3.f m rnpwt šwjj
jw i.jr sw ʿ3mw wʿ ḫ3rw m-dy.w m wr jw dy.f p3 t3 ḏr.f m ḫrpw r-ḥ3t.f

'After that came another period of empty years when (*jw*) no sooner had an Asiatic, some Syrian, made himself among them into ruler than he placed the entire land as controller in front of him'. (lit. 'It is *after* he placed the entire land … that he made himself…').

Few Egyptian texts are free of problems of interpretation. In the present case, the meaning of *ʿ3m* in *ʿ3m wʿ ḫ3r* and of the idiom *rd(y) p3 t3 m ḫrpw* is not fully clear to me. Nor is it clear whether *r-ḥ3t.f* 'before him/it' is spatial or temporal. There has also been some discussion as to whether *i̓.jr sw* (*ʿ3mw*) is a verb form or a personal name (Spiegelberg 1899, Kruchten 1981). In his new edition and commentary, Grandet (1994–99, 2: 222) prefers interpreting *i̓.jr sw* as a personal name because "the use of an emphatic form would render the proposition absurd."* However, the concept of contiguity absorbs the effect of emphasis meaningfully, as follows.

First, the Syrian mentioned in the text probably first had to become ruler (*jr sw m-dj.w m wr*) before controlling Egypt (*ḫrp p3 t3 ḏr.f*). If that is the case, the literal translation presents the events in reverse order. Second, the isolating contrast marked by *i̓.jr*, if it is indeed an emphatic verb, seems to affect no specific adverbial phrase. These two symptoms as a rule appear together whenever emphasis is used in the service of contiguity.

9.10.3.3 A Syropalestinian battle scene dating to Ramses II

This potential instance of contiguity occurs in a text accompanying a depiction of Ramses II engaged in a battle in Syropalestine near a Hittite stronghold called  *Twnpt*. The text and the scene are self-contained and isolated, not part of the topically related Qadesh Battle inscriptions. The text survives in two fragmentary versions, found at two of Thebes' most prominent temples, the Ramesseum on the west bank (version R) and (better preserved) the Luxor temple on the east bank (version L). Sethe (1907) in publishing L made the connection with the previously published R and proposed an interpretation different from prior ones, and it in turn will be challenged below.

Neither the Ramesseum nor the Luxor temple has been fully published by modern epigraphical standards, but both versions R and L are conveniently accessible in K*RI* 2.175,3–12; for an English translation, see Kitchen 1996, 2:47.

Version R is found on what remains of the southeast wall of the Ramesseum's hypostyle hall (see Porter and Moss 1972: 438) and is better accessible to detailed scrutiny than version L owing to its publication as Youssef, Leblanc, and Maher 1977, no. 9a, in photograph (plate X) and facsimile (plate XI).

* L'emploi … d'une forme emphatique donnerait à ladite proposition un sens absurde.

Version L is found on the western outside wall, to the north of the side entrance, in the Luxor temple's first colonnaded court (see Porter and Moss 1972: 333 no. 202, I: "three scenes at north end" with plans XXX–XXXII; for a drawing that clarifies the precise location of the inscription, see K*RI* 2.170 no. 19). To verify version L, I have relied on the photographs in Wreszinski 1923–35, 2: plates 77–78.

What follows is the text in L. The columns of the original have been converted into lines—although such a procedure is always in danger of falsifying the intended order of the signs. For example, the column arrangement  of the original could be rearranged either into the line arrangement 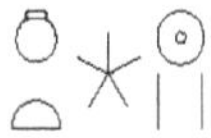 (Sethe 1907: 38) or kept as the line arrangement 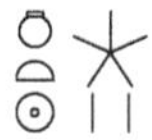 (K*RI*, 2.175). The former rearrangement is preferable.

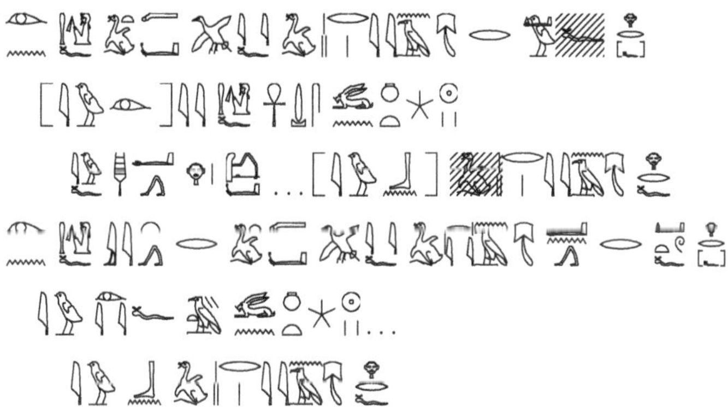

jr.n ḥm.f ṯ3t p3jj.f t3-r-jj-n3 r dyw.f (*sic*) *ḥr*[*.f*]
[*jw jr*]*jj ḥm.f ꜥnḫ wḏ3 snb wnwt 2*
jw.f ꜥḥꜥ ḥr ꜥḥ3 … [*jw bn*] *ṯ3-r-jj-n3 ḥr.f*
jr.n ḥm.f jjt r ṯ3t p3jj.f ṯ3-rjj-n3 ꜥn r dyt.f ḥr[*.f*]
jw jrjj.f t3y wnwt 2 …
jw bn ṯ3-r-jj-n3 ḥr.f

The text is either abraded or slightly fragmentary in places. For some of the readings, support can be derived from version R (in seven columns), which is overall more fragmentary. Only one matter pertaining to the reading of the text requires comment here, but it is essential to the ensuing grammatical analysis, and that concerns the reading 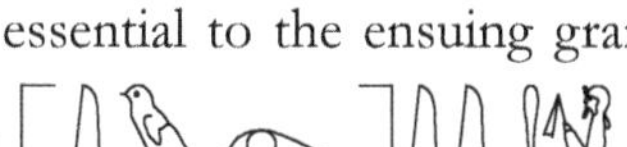 [*jw jr*]*jj ḥm.f* in the second line. Version R is entirely fragmentary in this spot. Both

Sethe (1907: 38) and Kitchen (*KRI*, 2.175) leave a blank between the two square brackets. But it is clear from an interpretative footnote that Sethe at least meant to read *jrjj.f*, without *jw* (39 n. 2). Polotsky (1944: 75 [1971: 170]) follows Sethe in this respect. In the same footnote, Sethe assumes that *sḏm.f* can be circumstantial like *sḏm.f* in Late Egyptian. But this is not normally the case.

An inspection of the photograph in Wreszinski's atlas leaves no doubt that there is room for an additional element in front of *jrjj ḥm.f*, and measuring the photograph shows that the available space would accommodate *jw*. It is nearly certain, therefore, that the text is to be restored as [*jw jr*]*jj ḥm.f*. This reading would harmonize with *jw jrjj.f* in the parallel statement in the second half of the text.

There is no doubt that Late Egyptian *jw sḏm.f* is both past tense and circumstantial, literally meaning 'he having heard'. Sethe therefore translates the first half as 'His Majesty seized his armor to put it on (only) after his Majesty (life, health, prosperity) had spent two hours standing (there) fighting without wearing armor'. A key element of this translation is that the king first fights two hours without wearing armor and only then seizes his armor to put it on.

Polotsky (1944: 75 [1971: 170]) supports this order of events but adds contrastive emphasis as a nuance. He had in the same work proposed that the past tense *jr.n ḥm.f ṯ3t*, with the auxiliary *jr*(*y*), is emphatic. In Late Egyptian, *jr.n.f sḏm* is a hybrid writing combining features of Middle Egyptian *sḏm.n.f* and Late Egyptian *i.jr.f sḏm*. It is equivalent in function to both. Polotsky's proposal has been generally accepted.

Sethe made much of the fact that the king shown in the relief does not wear any armor, taking it as striking confirmation of his interpretation of the text. However, Wreszinski (1923–35: plate 78 n. 5) observes that representations of the king in armor are rare and suggests that armor was painted rather than sculpted in reliefs ("die Wiedergabe im Panzer ist selten … vielleicht war dieser sonst immer nur aufgemalt").

There is a problem with the order of events proposed by Sethe and accepted by Polotsky. The first part of the text states, by Sethe's interpretation, that the king put on his armor *after* fighting for two hours without armor. But then, the second part says that the king comes to seize his armor 'again' (*ʿn*) in order to put it on. The natural inference is that the king had *not yet* put his armor on at the end of the first part. This impression that the king is still not wearing armor is confirmed by the subordinate clause of the second part of the text, if one translates, as Sethe does, 'after he had spent two hours … without wearing armor'. Clearly, the second part contradicts the first part if one believes, with Sethe, that the king had put on his armor by the end of the first part. Sethe (1907: 39) solves this problem by suggesting that the second half of the text is merely a repetition of the same events as the first half. The purpose of this repetition is emphasis. Yet, at face value, the text communicates two similar but distinct events: the king makes two attempts to put on armor. The adverb *ʿn* 'again' in the second half supports this interpretation; it is not known to be used for emphasizing a previous event by means of repetition. Rather, it denotes a *new* event that is a repetition of a previous event.

Is it possible, then, that the king came to seize his armor but did *not* put it on for some reason not mentioned? In fact, the text only states that the king *seized* his armor. It does not say that he actually *put it on.*

The literal translation of the text must be:

> *jr.n ḥm.f ṯ3t p3jj.f ṯ3-r-jj-n3 r dyw.f* (*sic*) *ḥr*[*.f*]
> That his Majesty took his armor to put it on himself (is)
> [*jw jr*]*jj ḥm.f ʿ.w.s. wnwt 2*
> after his Majesty l.p.h. had spent two hours
> *jw.f ʿḥʿ ḥr ʿḥ3* … [*jw bn*] *ṯ3-r-jj-n3 ḥr.f*
> standing while fighting …
> while there was no armor on him.
> *jr.n ḥm.f jjt r ṯ3t p3jj.f ṯ3-rjj-n3 ʿn r dyt.f ḥr*[*.f*]
> That his Majesty came to take his armor again
> to put it on himself (is)
> *jw jrjj.f t3y wnwt 2* …
> after he had spent these two hours
> *jw bn ṯ3-r-jj-n3 ḥr.f*
> while there was no armor on him.

Staying close to this literal translation, the first half of the text emphasizes by contrast that it is (only) after fighting two hours without armor that the king seized his armor. But apparently, he did not put it on; in the second half of the text, he is still without armor. That the king was courageous enough to fight without armor seems worth noting and emphasizing, but to state that it is specifically in the state of not wearing armor that he seized his armor and came to seize it again may seem slightly awkward and too obvious. Indeed, why would the king come to seize his armor if he *was* wearing it?

On the other hand, if contrastive emphasis serves to express contiguity, by the precise mechanism analyzed elsewhere (most fully in Depuydt 1999d: 639–729), then the free translation is:

As soon his Majesty seized his armor in order to put it on,
his Majesty l.p.h. (instead)
spent two hours standing (there) fighting …
without wearing armor.
As soon as his Majesty came to seize his armor again
in order to put it on,
he (instead) spent (also) these two hours …
without wearing armor.

Or:

No sooner had his Majesty seized his armor
in order to put it on
than his Majesty l.p.h. spent two hours standing
while fighting …
without wearing armor.
No sooner did his Majesty come to take his armor again
in order to put it on
than he spent (also) these two hours …
without wearing armor.

It remains unclear why the king did not put on his armor or did not succeed in putting it on after seizing it. But at least, supposing that he did not put it on makes it possible to understand why the king *returned* to seize his armor *again* in order to try and put it on.

9.10.3.4 The Stela of Kay

Another possible instance of contiguity is found in the Middle Kingdom stela of Kay (*K3jj*), Berlin 22820 (Anthes 1930; color photograph, Willeitner 2003: 17, plate 19). The text is as follows.

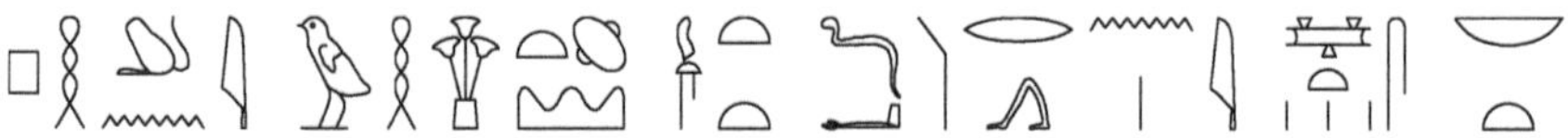

pḥ.n.j wḥ3t jmntt ḏʿr.n.j w3wt.s nbt

There is no doubt that a new sentence begins with *pḥ.n.j*. *Pḥ* means 'reach'. Verbs meaning 'reach' are especially common in examples proposed as instances of contiguity (see Depuydt 1993a: 167–74). It must be determined whether *pḥ.n.j* is substantival or not. The absence of the particle *jw* in a Middle Kingdom text favors substantival function. If *pḥ.n.j* is initial and substantival, then an adverbial element must be emphasized. *Pḥ.n.j* itself is not accompanied by an adverbial phrase. The emphasized element must therefore be *dʿr.n.j*, which is then an adverbial verb form. The resulting literal meaning is ‘It is after I had explored all its roads that I reached the western oasis’, with the events in the wrong order. Taking it as contiguity instead yields the translation ‘No sooner did I reach the western oasis than I explored all its roads’. It does take a certain leap of faith to assume that the two *sḏm.n.f* forms, *pḥ.n.j* and *ḏʿr.n.j*, are of different kinds even though they are written the same.

9.10.3.5 Four more examples with an initial verb of motion

(1) Siut III 10;
(2) Herdsman’s Tale (Papyrus Berlin 3024), 24–25;
(3) Louvre C30 and Cairo 20498;
(4) Tomb of Qereri (Tomb Q15 at El-Hawawish), East Wall, South of the Entrance, Lines 2–3 (see already §8.0.2.3.e)

Plausible examples of contiguity often begin with a verb of motion. Movement often takes place with the intention of beginning another action immediately upon arrival, with a characteristic closeness between the two actions. But when the second action is presented as a *concomitant or prior circumstance* of the movement, the sequence is slightly misrepresented, as it were, in order to emphasize the closeness, or contiguity, of the two events. Four examples may be mentioned here that I have not discussed elsewhere.

> *jjw* () *ẖ3w sḏr ḥr mṯn ḥr rd(y)t njw wnn.f mj s m pr.f* (Siut III 10)
>
> ‘When night falls, he who sleeps on the road is giving *njw*. He is like a man in his home’.

The verb form *jjw* may be interpreted as substantival; the absence of the particle *jw* makes *sḏr ḥr mṯn ḥr rd(y)t njw* most probably adverbial or circumstantial.

In the second example, there is no doubt that *jj.n.s* is substantival:

> *jj.n.s ḥ3.s m ḥbsw.s ṯḥṯḥ.s šnw.s* (Herdsman’s Tale [Papyrus Berlin 3024], 24–25; cited by Westendorf 1953: 73)

Since *jj.n.s* is not accompanied by an adverbial phrase, the verb forms *ḥ3.s* and *ṯḥṯḥ.s* must be adverbial and emphasized. As *sḏm.f* forms, these two verb forms in principle denote simultaneous events. However, simultaneity cannot be but a slight misrepresentation. Indeed, the two events in question apparently take place as the movement expressed by *jj.n.s* is coming to an end. The false representation of the order of events serves to denote the closeness between the two events. The resulting free translation is 'When she came, she undressed and shook loose her hair'. The literal translation is rather: 'That she came is while she undressed and shook loose her hair'.

In the third example, *jj.n.j* is also no doubt substantival:

(Horus speaking) *jj.n(.j) ḥw.n(.j) n.k ḥww* ⟨*tw* (*or:* *ṯw*)?⟩

This statement is found in both Louvre C30 and Cairo 20498 (see Sethe 1928: 64,19). The suggestion to reconstruct *tw* 'you' is Sethe's (1927: 95). The Cairo version (Lange and Schäfer 1908: 90), which is better preserved, has 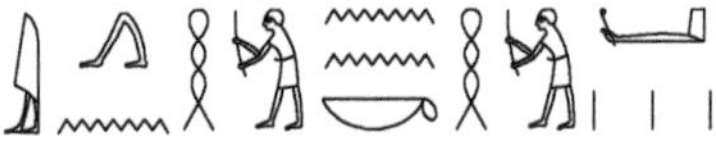. Like *jj.n.s* in the previous example, *jj.n(.j)* is not accompanied by an adverbial phrase. The verb form *ḥw.n(.j)* ought therefore to be emphasized and adverbial, meaning something like 'I having slain (the slayers)'. Accordingly, the literal translation is 'That I came is while I had already slain for you the slayers'. It seems more likely that the action denoted by *ḥw.n(.j)* takes place *after* or *at the end of* the movement expressed by *jj.n(.j)*. The inversion of events in the literal translation then misrepresents the sequence. It is assumed here that this misrepresentation serves to denote the closeness between the two events, a bit as in English "No sooner did I come than I slew for you the slayers." An alternative translation would simply be as follows: 'When I came, I slew for you the slayers'.

It is not immediately obvious that a verb form like *jj.n(.j)* forms a single sentence with a verb form that follows. The main reason for supposing that it does is that the *sḏm.n.f* of a verb of motion must be substantival and therefore cannot by itself form an independent sentence; compare an expression occurring in many variants in New Kingdom texts in so-called traditional Egyptian (Depuydt 1993a: 197) in which it is not immediately apparent that the initial verb forms a single sentence with a verb that follows. A typical instance is *wbn.k sḥḏ.n.k tꜣwy*. If the initial intransitive verb forms a sentence together with what follows, then it could be translated as 'No sooner do you rise than you illuminate the two lands' or 'No sooner have you risen than you have illuminated the two lands'. The tense is not certain, but there is a variant that proves that *wbn.k* can form a single sentence with a verb form that follows: *jrwy.sn wbn.k ptr.sn jm.k* (Sandman 1938: 12,8, 10, 12). *Wbn.k* and *ptr.sn* belong to the same sentence because *sn* in *ptr.sn* refers to *jrwy.sn*. *Jrwy.sn wbn.k* is incomplete without *ptr.sn*. Consequently, the translation must be something like 'When you rise, their eyes see by means of you', or 'As soon as you rise, their eyes see by means of you' (lit. 'It is while they see you that you rise'). In a similar way, Polotsky (1976: 26–27) concluded that *m33.sn* must be adverbial in an expression like *ḥ⁽⁽w nṯrw m33.sn sw* '(the one) at the sight of whom the gods rejoice' (CT 4.61o). The reason is that *ḥ⁽⁽w* is not complete without *sw*. The intervening verb *m33.sn* must therefore be subordinated to *ḥ⁽⁽w*.

In the fourth example, consider line 2 of a passage inscribed on the east wall, south of the entrance, in the tomb of Qereri, tomb Q15 in the cemetery of Akhmim at El-Hawawish (Vandier 1936: 36–37, inscription B; Kanawati 1986: 49, 51, plates 3b, 8c, figure 20c [reproduced at the end, on p. 373]). The tomb dates to Dynasty 6. Pharaoh Meryre (Pepi I) is mentioned.

jj.n(.j) ḥm r ẖrt-nṯr

It seems unlikely that the meaning of this string is 'It is truly(?) to the necropolis that I have come', with emphasis on the adverbial phrase *r ẖrt-nṯr* 'to the necropolis'. Why would the tomb-owner emphasize that it is to the necropolis that he has come? The following reasoning then applies. There is still need for an emphasized adverbial element. That emphasized element must therefore be *š3d.n(.j)* |[3] *š n mḥ 100 ḥr m3⁽.f nb nhwt 10 jm.f* in lines 2–3, *r ẖrt-nṯr* not being suitable as the emphasized element. The literal translation is accordingly 'It is after I had dug a pool of 100 cubits on each side with 10 sycamores in it that I truly(?) came to the necropolis'. After *jm.f*, a new sentence begins with *jr s3(.j) nb* … 'As for every one of my sons…'. As regards the order of events, it would seem that the coming preceded the digging, as is also supposed by Vandier (1936: 36 [for 'je suis allé' read 'je suis venu']), Kanawati (1986: 49), and Kloth (2002: 61 ['Ich bin wirklich zur Nekropole gekommen. Ich grub einen Teich …']). But the literal translation presents the sequence of events in inverted order. The explanation that I would propose is that (1) isolating contrast and (2) inverted order of events together conspire, as it were, to denote contiguity, which does not have a sign of its own (Depuydt 1999d: 723–27). The resulting translation is 'When I came …, I dug …' (*or*: 'As soon as I came, I dug …').

The inscription is of inferior quality but carefully traced. The proposed grammatical interpretation is supported by a statement in lines 1–2 of an inscription on the same wall north of the entrance (Vandier 1936: 34–35, inscription A; Kanawati 1986: 48, 51, plate 8b, figure 20b [see book's end]), which contains two transitive verbs in the *sḏm.n.f*, just like *š3d* 'dig' above:

jw jr.n(.j) ḥmt bnnt jw rd(y).n.sn |[2] *⁽3 200*

'I have made (that is, acquired [?]) female donkeys *bnnt*. They have given (birth to) 200 donkeys'.

A new sentence then begins with *jr s nb* … 'as for every man…'. The absence of *jw* before *šꜣd.n.j* after the previous example contrasts with the presence of *jw* in *jw jr.n(.j)* and *jw rd(y).n.sn* here. The contrast seems deliberate and must therefore be meaningful in some way. Neither *jw jr.n(.j)* nor *jw rd(y).n.sn* is followed by an adverbial phrase. Neither is therefore emphatic (substantival). It would appear to be part of the function of *jw* to mark that they are not emphatic. What then is the significance of the absence of *jw* in front of (nonrelative) *šꜣd.n(.j)*? According to traditional Polotskyan theory, *šꜣd.n(.j)* could be either substantival or adverbial. The adverbial phrase *jm.f* hardly seems like a suitable candidate for emphasis. The remaining solution to give meaning to the significant absence of *jw* is that *šꜣd.n(.j)* is adverbial or circumstantial, meaning 'I having dug'. Accordingly, *šꜣd.n.j* must be anterior in time to *jj.n.j* in a literal translation. The concept of contiguity allows an interpretation according to which coming is first and digging is second.

9.10.3.6 Papyrus Bankes I 4–7

> *ỉ.jr.k wḏ dy m njwt*
> *jw swḏ(.j) n.k ḥmt T. ỉrm ḥm G. pꜣy.s šrj*
> |[5] *jw d(y).k sn n wḥꜥ P.* |[6] *šmsw Ḥ.*
> *jw ḏd.w n.k jw.j ꜥḥꜥ.tw:*
> *"ỉ.jr.tw ṯꜣw tꜣy* |[7] *rmṯ m ṯꜣw" ỉ.n.w n.k*
> *jw.k (ḥr) ḏd: "ꜥḏꜣ …"*

There is no doubt that *ỉ.jr.k wḏ* is emphatic, and it is hardly likely that *m njwt* 'from Thebes' is emphasized. Therefore, *jw* … must be. Accordingly, past interpreters (Edwards 1982; Navailles and Neveu 1988) have rendered the passage along the following lines: 'It is after I delivered T. and G. to you and after you gave them to P. and H. and after you were told, "This woman has been stolen," that you departed here from Thebes'. This makes little sense. What is the point of specifying by contrast when exactly the person addressed ("you") departed from Thebes? Applying the concept of contiguity, I therefore propose 'When you departed here from Thebes, I delivered to you … and I gave to you … and you were told…'. Another argument is the continuative *jw.k (ḥr) ḏd: 'ꜥḏꜣ …'*, which is probably not emphasized—the resulting sequence of events would pose a problem: 'It is after you were told … that you left … and then you replied, "That is false"'. By this translation, the person addressed ("you") could not have replied because he would have left Thebes. Another is that *jw.j ꜥḥꜥ.tw* might mean 'while I was not there' (for *ꜥḥꜥ* 'be absent', see Valbelle 1977: 131–35).

9.10.3.7 Other potentially relevant examples

The following examples from Late Egyptian have in common that an emphatic verb form apparently serves to emphasize a *jw sḏm.f* or *jw.f* + stative (for a list of examples of this kind, see Cassonnet 2000: 280–82). For most of the following examples, the purport is not fully clear. The partial translations assume contiguity: *ỉ.jr.j* |[12] *šm jw ṯꜣy.j tw* (*text*: *tꜣy.tw.j*) *jrm.j* (*KRI* 6.776,11–12 [Papyrus BM 10052 r. 4,10]) 'when I go (*or*: went?) …, I took …'; *jw ỉ.jr.k wḏ dy m pꜣ thm pꜣ mšꜥ jw ḏd.k* |[6] *n.j*: "…" (*KRI* 3.255,5–6 [Papyrus Cairo 58056,5–6]) '… while, when you departed …, you said to me "…"'; *ỉ.jr.k jj jw rẖ.k p ꜥ n nꜣ mšš* |[9] *m jb.w* (*KRI* 6.523,8–9 [Papyrus Louvre 3169,6–7]) 'when you came, you learned about …'; *js bn jr.n* (equivalent of *ỉ.jr*) *rmṯ sꜥꜣ.f m njwt.f jw.f jw jrjj.f ḳn m-bꜣḥ nb.f* (*KRI* 2.79,1–4 [Qadesh Poem 255–56]) 'When a man has honored himself …, has he (then) not come… .?'.

No one would contest the literal translation of *ỉ.dy.j* [*ꜥḳ*] *wpt-ḥb jw dy.j pꜣy šꜣtj* (= *šbd*) [*n*] *p*[*ꜣy.f nb*] 'It is after I have given the staff to its owner that I will allow the feast to happen' (Ostracon Deir el Medina 59,3–6); for similar examples involving a deadline, see Depuydt 1997: 35–36 and Cassonnet 2000: 206–7). This passage has usually been interpreted as conveying that the staff needs to be returned *before* the feast (e.g., ibid.), although return *after* the feast has also been suggested (Menu 1973: 80). However, the concept of contiguity would allow the translation 'I will make it happen that, as soon as the feast arrives, I will give the staff to its owner', that is, 'by the feast'.

Additional passages are relevant. Is the sequence of events in *spr.n.j r jj dy m-dy.k* (*m*) *3 hrw jw A. wḏ m nꜣy.f wpt r tꜣy rjt n P.* adequately rendered by the translation 'When I succeeded in arriving here at your place in three days, A. departed on his mission to the vicinity of P.' (*KRI* 3.251,14–252,1 [Papyrus Cairo 58059,2–3]; Cassonnet 2000: 149, ex. 193)? Might *ỉ.jr.j n.f nn r-ḏr.w jw ḫꜣꜥ.f ḥꜣ.f r jꜣwt* (*LEM* 37,5–6 [Papyrus Anastasi IV 3,2]) mean 'No sooner have I done all these things for him than he neglects his office'? Could the passage *ỉ.dy.tn šbb ḥr mw jw pḥ.j jw.j wḏꜣ.k* (Papyrus Anastasi I 26,2–3) possibly mean 'No sooner will you give me food and water than I will end up being OK'?

Of three passages in the Story of Wenamun in which *jw sḏm.f* closely follows *ỉ.jr.f sḏm*, the use of *ẖr-ḥꜣt* 'earlier' in one suggests that contiguity does *not* play a role: *ỉ.jr Jmn grg* {*n*} *nꜣ tꜣwy ḏr.w ỉ.jr.f grg.w jw grg.f pꜣ tꜣ n Kmt pꜣ jw.k n-jm ẖr-ḥꜣt* (Wenamun 2,19–20 [*LES* 69,1–2]) 'Amon founded all other lands—he founded them only after he had earlier (*ẖr-ḥꜣt*) founded the land of Egypt, the one from which you have come'. *Ỉ.jr.f grg.w* is here taken as a repetition of *ỉ.jr Jmn grg* {*n*} *nꜣ tꜣwy ḏr.w*. Perhaps *ẖr-ḥꜣt* serves as an explicit marker of the *absence* of contiguity. Since the same construction serves for the expression of both absence and presence of contiguity, some unwritten feature of pronunciation must have marked the difference between the two. [But see now Depuydt in press: §6.]

In the second of the three Wenamun passages, contiguity can be excluded on the basis of context. Two events are concerned: Pharaoh receives goods from the prince of Byblos; and Pharaoh pays for the goods. As regards the order of these two events, it is clear that the prince wants to be paid first before delivering goods. Consequently, the sentence *jjꜣ ỉ.jr nꜣjj.w jr(t) pꜣy sḥn jw dy pr-ꜥꜣ ꜥ.w.s. jn.tw 6 br … jw.w šw r nꜣjj.w wḏꜣ* (Wenamun 2,6–7 [*LES* 67,14–16]) means 'Verily, my people have done their job only after Pharaoh l.p.h. sent six boats … and they were emptied into the storehouses'. [But see now Depuydt in press: §6.]

The third passages is not clear to me. If the statement *ỉ.jr Jmn ḫrw m tꜣ pt jw dy.f Swtḫ m rk.f* (Wenamun 2,19 [*LES* 68,16]) expresses contiguity, then it would mean something like 'When Amon makes thunder, he places Seth in his reign'. [But see now Depuydt in press: §6.]

9.10.4 Reply to an objection to one proposed instance of contiguity

Schenkel (1997: 265) objects to my proposed interpretation of Eloquent Peasant R 9.3–5 (old numbering: R 59–61) as exhibiting contiguity (Depuydt 1993a: 121–22, 168–70). The discussion revolves around what needs to be restored in a lacuna in the text.

> *pḥ.n.f rf mdt* [*tn*] *mḥ.n wꜥ m nꜣ n ꜥꜣ r.f m bꜣt nt jt*
>
> 'No sooner had he finished [this] speech than one of the donkeys filled his mouth with a wisp of barley'.

 *tn jw* has traditionally been restored in the lacuna. Parkinson (1991: 11a) considers the restoration of *tn jw* "practically certain." I doubt that such certainty as to what belongs in the lacuna is possible on the basis of the traces of the text. Only a small speck of ink survives at the end of the lacuna, before *mḥ.n.* This speck could perhaps be read as part of *w*, which might be part of the particle [*j*]*w*.

As I have noted (1993a: 168–70), *mḥ.n* denotes an event in the narrative chain of a story; but the compound *jw sḏm.n.f* never denotes events in a narrative chain—as *jw mḥ.n* would if *jw* were restored. The variant in manuscript B1 exhibits *ꜥḥꜥ.n mḥ.n*. The element *ꜥḥꜥ.n* is the narrative auxiliary *par excellence* and is also the polar opposite of *jw*. If *jw* is restored in R, then this passage would present the only known case in which *jw* alternates with *ꜥḥꜥ.n*. For that reason alone, restoring *jw* seems difficult to accept, so I restore a large writing of *tn* 'this', written as in line 103 of manuscript B1 (old numbering: B1, 72), where it is in a column, not a line as in R.

An additional argument (ibid.) was that B1 and R are written by the same scribe, but this was based on misreading a remark in Gardiner's and Vogelsang's edition (1908), which concerns B1 and B2 rather than B1 and R as I had thought. This error was pointed out by Joachim Friedrich Quack (apud Schenkel 1997: 265). Schenkel had earlier (1994: 238) also preferred *tn* over *tn jw* as a restoration. But the loss of this point does not justify the reading *tn jw* instead of *tn*, as Schenkel seems to imply.

> Leo Depuydt's reading without *iw* needs to be again replaced by that with *iw* that was common before, because the paleographic argument at least … is not valid owing to a confusion of manuscript R with manuscript B2 (information from J. F. Quack).*

* Die … Lesung Leo Depuydts ohne *iw* ist wieder durch die zuvor übliche mit *iw* ersetzt worden, da zumindestens das paläographische Argument … infolge einer Verwechslung der Hs. R mit der Hs. B 2 nicht stichhaltig ist (Hinweis [J.F.] Quack).

10 *Jr sḏm.f sḏm.f* as a Development from *sḏm.f sḏm.f*

10.1 On Growth in Logical Complexity

The present chapter demonstrates a strong affinity between conditional sentences of the types *jr sḏm.f sḏm.f* and *sḏm.f sḏm.f* (the balanced sentence), which might account for the ability of the substantival future *sḏm.f* (the *sḏmw.f*) to appear as the second verb form in *jr sḏm.f sḏm.f* *without* emphasizing an adverbial element. The hypothesis is that this type developed from the balanced sentence by adding *jr*. In §10.7, a particularly suggestive set of examples is presented in support.

The type "When you're fine, I'm fine," with *jr*, would thus have evolved from the type "You're fine, I'm fine"—the balanced sentence. This process may be described as an increase in logical sophistication in the history of thought and language. The type "You're fine, I'm fine" is a simple equation. "You're fine" is equated with "I'm fine." The two are considered the same. Because of the equation, the inversion, namely "I'm fine, you're fine," naturally follows from and means the same as "You're fine, I'm fine." By contrast, the type "When you're fine, I'm fine" is not an equation. There is an unbalance or asymmetry. "When I'm fine, you're fine" does not naturally follow from "When you're fine, I'm fine." Indeed, likewise, from "When it rains, I stay inside," it does not follow by inversion that "When I stay inside, it rains." It may or may not rain when I stay inside. The asymmetry may be interpreted as a higher degree of logical complexity, a more sophisticated nuance. By contrast, the balanced sentence reminds one, though somewhat distantly, of the fictional, and presumably primitive, "Me Tarzan, you Jane." The balanced sentence became obsolete after Middle Egyptian, as one might perhaps expect of a more rudimentary construction. The other conditional type remained in existence. The development postulated here would involve an increase in logical complexity and precision in the evolution of human civilization.

10.2 THE IDENTITY OF THE FOUR *sḏm.f* FORMS IN *jr sḏm.f sḏm.f* AND *sḏm.f sḏm.f*

A scenario in which *jr sḏm.f sḏm.f* and the balanced sentence are related is difficult to support if the two types exhibit no strong affinities. How similar do they appear in the texts? The focus is on identifying the two verb forms of both types. What are the facts?

The *sḏm.f* is a verbal *formation*. It is not a verb *form*. The *sḏm.f* formation encompasses *many* verb forms. Until 1884, only one *sḏm.f* verb form was known. In that year, Erman postulated that the *sḏm.f* form following *rd*(*y*) 'let, cause' is a verb form in its own right. Earlier, Erman (1881) had come close to postulating different forms within the *sḏm.f* formation when he discovered the relative forms. This discovery came when Erman noted the absence of *t* in *js pn jr.n.j* 'this tomb that I made' (Leiden stela V 4), in which *jr.n.j* follows the masculine noun *js* 'tomb', vs. the presence of *t* in *mꜥḥꜥt tn jrt.n.j* 'this tomb that I made' (Louvre stela C170), in which *jrt.n.j* follows the feminine noun *mꜥḥꜥt* 'tomb'. Some of Erman's examples of relative verb forms belong to the *sḏm.f* formation, but he is vague about what this means for distinctions within it. At that time, the notion that the *sḏm.f* might notate more than one verb form had not yet taken hold.

Six decades after the decipherment of hieroglyphic writing, the monolithic *sḏm.f* had split into two, and in the following decades, new *sḏm.f* verb forms kept being added, most recently the *sḏmw.f* (Edel 1955/64: 238–48 §§511–31; 1984; 1995). This form clearly denotes future tense, and I know of no usage that cannot be explained as substantival (Depuydt 1993b: 24–27); sometimes its substantivalness is abundantly obvious. Surely, a verb form could hardly serve one function in many instances and something completely different in other instances, so *sḏmw.f* must *always* be substantival. The only alternative would be that there are *two sḏmw.f* forms, one substantival and one not. Some have hinted at such a distinction but without pursuing its logical conclusions or potential contradictions.

It comes as no surprise that authors who have assumed that *sḏmw.f* can be both substantival and not substantival have struggled. Thus, Vernus's (1990: 19) position seems particularly tormented.

> There is ... no reason not to consider prospective *sḏm.f* [which is Vernus's name for the *sḏmw.f*] as a unity, all the more since the variety of its functions can stem from a single basic meaning.

When it comes to determining that "single basic meaning," Vernus addresses the "vexed question" of the *sḏmw.f* functioning as a "that"-form or finite substantival form (Polotsky). Vernus acknowledges that the *sḏmw.f* "shares some functions of a 'that'-form," but he does not accept the "that"-form function as the "single basic meaning." Instead, he opts for "a value of non-actuality and dependence." The concept of "non-actuality" is not entirely clear to me. Actuality sounds unusually abstract as a concept. *Absence* of actuality ("non-actuality") seems even more so. Even if there is such a thing as Vernus's non-actuality, it would seem difficult to prove the existence of such a subtle nuance. Moreover, how can a form be a "that"-form and not a "that"-form at the same time, given all that we know about "that"-forms in all the stages of Egyptian? Such a contradictory assumption can only remain alive, I believe, by hiding behind the many obscurities afflicting the interpretation of Egyptian texts.

The master list of generic *sḏm.f* forms results from many decades of study of Old and Middle Egyptian, and each addition to it has had to be justified individually; I believe it is now complete, and no new generic *sḏm.f* forms are likely to be discovered. Each one is represented by specific attestations. The task of the philologist is to associate each *sḏm.f* form in a text with one item in the master list. However, most of the attested *sḏm.f* forms will never be classified with certainty. But this disappointing fact regarding individual cases does not affect the reliability of the master list of generic *sḏm.f* forms.

There is no doubt that all four verb forms in both conditional types under discussion can be the substantival future ("prospective") *sḏm.f* (or *sḏmw.f*).

10.3 THE *sḏm.f* FORMS IN THE BALANCED SENTENCE

The identity of the two verb forms of the balanced sentence is now securely established. Both are always substantival. As regards tense, there are three possibilities: past, present, and future. In the future tense, both forms are the substantival future *sḏmw.f*.

10.4 THE *sḏm.f* FORM IN *jr sḏm.f*

Until the early 1980s, it was not clear whether the *sḏm.f* in *jr sḏm.f* is substantival or adverbial. But then, consensus began to widen regarding the following essential point. The verb form in *jr sḏm.f* is as a rule the substantival future *sḏm.f* (or *sḏmw.f*; e.g., Allen 1979a: 25; Malaise 1985). It is unsurprising that a form following *jr* 'as for (the fact that)' would be substantival; it is normal to find nouns after *jr*. Nor is a future tense unexpected in a conditional clause, which denotes a hypothetical event, one that may or may not occur. Parker (1996: 116) concluded:

The *sḏm.f* as protasis of the conditional sentence (*ỉr* + *sḏm.f*) is the Prospective-Nominal *sḏm.f*.

Much evidence supporting this proposition has come to light recently, and contrary evidence has been refuted. Any remaining exceptions do not confute the statistical predominance of *sḏmw.f* appearing after *jr*; for the history of research, see Depuydt 1993b.

10.5 THE SECOND *sḏm.f* IN *jr sḏm.f sḏm.f*

10.5.1 Four propositions

The argument in this section is summarized as follows.

(1) §§10.5.2, 10.5.3. The substantival *sḏm.f* (the *sḏmw.f*) can follow *jr sḏmw.f* in the apodosis.
(2) §10.6. When *sḏmw.f* follows *jr sḏmw.f* in the apodosis, it—perhaps unexpectedly—does not necessarily emphasize an adverbial phrase.
(3) If (1) and (2) are true, then the sole difference between *jr sḏmw.f sḏmw.f* and the future balanced sentence *sḏmw.f sḏmw.f* is the particle *jr*.
(4) §10.7. The strong similarity described in (3) favors a historical hypothesis by which the balanced sentence as a more primitive type of conditional sentence evolved into *jr sḏmw.f sḏmw.f* as a more sophisticated type. This hypothesis only works if *sḏmw.f* does not necessarily emphasize an adverbial phrase in the apodosis. Its substantivalness then readily points to a nominal sentence origin.

10.5.2 The *sḏmw.f* as second verb form in *jr sḏm.f sḏm.f*

10.5.2.1 Distinctive writings of sḏm.f *in* jr sḏm.f sḏm.f

The possibility that the second *sḏm.f* in *jr sḏm.f sḏm.f* might be the substantival future *sḏm.f* (the *sḏmw.f*) was first raised by H. J. Polotsky in 1974–75 lectures, but it is not mentioned in their publication (1976). Niccacci (1980: 206–7 n. 25; cf. Vernus 1990: 32 n. 21) reports:

> It is not clear whether such a *sḏm.f* [that is, the second *sḏm.f* in *jr sḏm.f sḏm.f*] is identical to Edel's *sḏmw.f* … (H. J. Polotsky orally in 1975).*

* Non è chiaro se tale *sḏm.f* [that is, the second *sḏm.f* in *jr sḏm.f sḏm.f*] sia identico al *sḏmw.f* di Edel … (H.J. Polotsky a voce nel 1975).

In this connection, Niccacci erroneously lists Ptahhotep 577 as an instance of the stem *rd(y)* (with *r*). The example rather features the stem *mꜣ* (without gemination) (see Polotsky 1976: 46).

Polotsky (ibid.) remains agnostic regarding the identity of this verb form but does in fact discuss it, to observe that this *sḏm.f* exhibits a characteristic combination of the following two features: (1) the stem *rd(y)* (with *r*!); (2) the stem *mꜣ* (without gemination!). The same combination is found in two other contexts: (1) after the negation *n* in the past tense (*n rd.f*, *n mꜣ.f*); (2) in the Old Egyptian narrative *sḏm.f*. Earlier, Polotsky (1964: 273 [1971: 58] with n. 1) had provisionally entertained the notion that the *sḏm.f* found in the apodosis following a conditional clause introduced by *jr* 'if' is the so-called "perfective" *sḏm.f*, the one that presumably also appears in the past negated verb form *n stp.f* 'he has not heard'. Polotsky had first written about verb forms in main clauses following subordinate *jr*-clauses as early as 1939 (p. 159 [1971: 30]), in discussing a text found in Siut tomb "IV," lines 24–29 (Griffith 1889).

In CT 2.205a, *rd N* follows *jr* + noun, not *jr* + verb form (Polotsky 1964: 273 [1971: 58] n. 1; see also Parker 1996: 116, ex. 124). Polotsky (1976: 46) appears to classify *jr* + noun and *jr* + verb form as two members of a single species when he calls the clause that follows either of them an "apodosis": "the apodosis of a clause introduced by *jr*, whether *jr* governs a substantive … or a *sḏm.f* …."* I would limit the term "apodosis" to instances in which *jr* is followed by a verb form because *jr* + noun 'as for N' is not a protasis.

The evidence for the identification of the second *sḏm.f* in *jr sḏm.f sḏm.f* is sparse. We begin with two examples of the stem *rd*.

10.5.2.2 Rd.j in Shipwrecked Sailor 70–72

A first possible instance of *rd* in the apodosis is found in the Tale of the Shipwrecked Sailor (Blackman 1932). The form *rd(y).j* () appears in the apodosis at 70–72 (see also Junge 1978: 100).

jr wdf.k m ḏd n.j jn tw r jw pn rd.j rẖ.k tw jw.k m ss

'If you hesitate to tell me who has brought you to this island,
I will cause you to know yourself as being ash'.

* L'apodose d'une phrase introduite par *jr*, soit que *jr* amène un substantif (*CT* II 205a *rdj-*, *Ptahh[otep]* 577 *mꜣ.f*) ou un *sḏm.f* (*Naufragé* 72 *rdj.j*, *ibid.* 134 *mꜣ.k*).

Rd.j clearly refers to the future. It is widely accepted that in the master list of *sḏm.f* forms there are only two futures besides the adjectival (relative) future *sḏm(t).f*: (1) the substantival future *sḏm.f* (*sḏmw.f*); (2) the adverbial future *sḏm.f* (often called subjunctive). Since form (2) would be written  *d(y).j* or the like, and since there is now much evidence that  *r* is as a rule written out in the case of (1), the reasonable conclusion is that *rd.j* is substantival and future in our passage.

However, the reason for the appearance of the substantival *rd.j* form may be that *jw.k m ss* '(I) being ash' is emphasized. *Rd.j* would then not be substantival because of its position in *jr sḏm.f sḏm.f*, reducing its value as evidence for the identity of the second *sḏm.f* in that construction.

10.5.2.3 Rd.k in Papyrus Purches 2–4

A second possible instance of *rd* in the apodosis appears in a Middle Kingdom letter, Papyrus Purches, lines 2–4.

> *jr sꜥb.k ḥmt tꜣ ḥmt 2 bꜣkt* |[3] *ḏdt.n.j n.k ḥr.s rd.k jn.t(w) mj bw nb nfr jrrw sš.k mk swt ḏd n.j wnn.k* |[4] *ḥr spr* …

> 'Now, if you will collect the copper of those two female slave-laborers I spoke to you about, you will be so kind as to have (it) brought (lit. you will have it brought like all the goodness your scribe does)—though I have been told you (yourself) will be arriving [where I am]'.

The translation is that of Allen (2002: 21).

As so often in ancient letters, the general meaning of a sentence is not fully clear and the sentence is therefore not suitable as grammatical evidence. An alternative interpretation is that *rd.k jn.t(w)* is a quote, yielding something like 'about which I said to you, "You will have it brought …" …'. Or, *rd(y).k jn.t(w)* might be a future substantival *sḏm.f* (*sḏmw.f*) that functions as the direct object of *ḏdt.n.j n.k*, yielding something like 'about which you said you would have it brought'. In fact, the next sentence contains a verb form subordinated to *ḏd* 'say' in the expression *ḏd n.j wnn.k* |[4] *ḥr spr* 'It has been said to me that you will be arriving'. *Mk swt ḏd n.j wnn.k ḥr spr* … would then need to be the main clause, but this may not be possible. The use of *ḏd* 'say' in this example brings up the distinction between *ḏd* and *mdw*.

10.5.2.4 Excursus: The contrast between ḏd 'say' and mdw 'speak'

It needs to be stressed that *ḏd* means 'say' and not 'speak' (Depuydt 1993d: 305). The existence of such a distinction has recently been doubted. Saying presupposes speaking. Saying is speaking while at the same time transmitting a message. In this respect, the difference in Egyptian between *ḏd* (Coptic **ϫⲱ** *jô*) 'say' and *mdw* (Coptic **ϣⲁϫⲉ** *šaje*) 'speak' is much sharper than the difference in English between 'say' and 'speak'. One cannot, therefore, use the meaning ranges of "say" and "speak" in English to evaluate the meaning of *ḏd* and *mdw* in Egyptian. One striking illustration of the difference between Egyptian and English is that English "speak" can sometimes be transitive and be followed by a direct object, as in the expression "speak words." In that sense, there is a grey area or twilight zone, as it were, between "say" and "speak" in English. In this grey area, the verb "speak" sometimes seems to behave like the verb "say" does. By sharp contrast, Egyptian *mdw* 'speak' and Coptic **ϣⲁϫⲉ** *šaje* 'speak' are never transitive (never take a direct object).

Contradicting the existence of such sharp contrast between *ḏd* 'say' and *mdw* 'speak', Parkinson (1996) has recently suggested that the verb *ḏd* sometimes means 'speak' and therefore behaves as *mdw* otherwise does. That would be the opposite of what happens to 'speak' in English, where 'speak' sometimes behaves like 'say'. In Egyptian, the equivalent of 'say' would according to Parkinson sometimes behave like 'speak'.

Parkinson adduces four examples in which he proposes that *ḏd* means 'speak', which I do not find convincing. None of them is fully clear, so they do not persuasively establish 'speak' as a meaning of *ḏd*. It must be admitted that *ḏd* does not appear to have a direct object in any of them, which at first makes *ḏd* look like *mdw* 'speak'. However, in the first two examples, the presumed *ḏd* 'speak' appears alongside a clear instance of *ḏd* 'say'. How could *ḏd* function in such two distinct ways in a single passage?

The first example can be translated as:

> *n ḏd ḏd ḏd ḏd.tj.fj gm kjj ḏd.tj.f* (Words of Khakheperreseneb [Gardiner 1909], r. 5)
>
> 'He who has said (things?) has not said what he who will say (things) says. May someone else find out what he will say'.

The meaning of the passage is not entirely clear to me (Parkinson [p. 390] renders it 'No speaker has spoken yet—may one who will speak now speak and another find what he will speak good'), but obviously, *ḏd* is transitive in the future relative *ḏd.tj.f* 'what he will say'. I would therefore prefer to treat the instances of *ḏd* without direct object as cases in which the direct object is not expressed. An undeniable example of omission of the direct object is:

> *jw.k r.j* *ḏd.j r.k*
> *tm.k jw r.j* *tm.j ḏd r.k* (Book of the Dead, chapter 90, Papyrus of Nu [Budge 1898: 192, lines 10–12])
>
> 'You come against me, I say against you.
> You do not come against me, I do not say against you'.

As regards non-expressed direct objects, one is reminded as well of the well-known *mrr.f jrr.f* 'Whenever he wants (something), he does (something)' (Pyr. §412b). Parkinson's own edition of the Eloquent Peasant (1991) includes an example of omission of the direct object: *jr jn.k jmj n sn.k* 'When you bring (it), give (it) to your brother' (B1, 283 [old numbering: B1, 252] = B2, 5–6). A similar example from Papyrus Purches (Allen 2002), line 5, is *ntf jn.f n.j* 'It is he who will bring (it) to me'.

In Parkinson's second example of a presumed *ḏd* 'speak', *ḏd* appears twice, in the first occurrence clearly meaning 'say' as it normally does and as Parker too accepts. Then why would *ḏd* not mean "say" the second time?

> *rḫ ḥr* (*ḏd*): *ḏd.j n m* (Papyrus Ramesseum II [BM 10755; Barns 1956], v. ii 4)
>
> 'The wise man is (saying): "To whom will I say (it?)?"'

The related statement *ḏd n m mjn* 'To whom will I say (it, namely the following) today?' serves fifteen times as a refrain in the well-known Man Tired of Life in Papyrus Berlin 3024 (Erman 1896; Goedicke 1970). In this text, a disillusioned man to his soul (*ba*), lamenting the miseries of the world and apparently contemplating suicide. I have proposed (Depuydt 1993d) that *ḏd* means 'say' as it always does and that the refrain *ḏd.j n m mjn* therefore introduces a quote that follows. The refrain can then be translated as: 'To whom can I say today: "[Quote follows here.]"?' However, the sentence has always been understood as "To whom can I speak today?", in the sense 'There is no one I can talk to today'. Parkinson points out that *ḏd.j n m* in his example above is followed by a verse-point. He believes that a verse-point cannot separate the verb 'say' from a quote that it introduces; *ḏd.j n m* must therefore mean 'To whom will I speak?' I would contest this. *Ḏd.j n m* is followed by a lacuna in the text, so we can only guess as to whether a quote did actually follow it originally.

In Parkinson's third example, Shipwrecked Sailor 20–21, *ḏd n.k* is perhaps not 'speak to you', but rather *ḏd.n.k* 'that which you have said'. The passage, with a provisional translation, is as follows: *srwḏ pw ḏd.n.k* 'What you have said is something tiring'. *Srwḏ* may be a participle. Shipwrecked Sailor 149 reads *nn ḏd.n.j* 'that which I had said'. A similar expression is *n3 ḏd.n.j n.k* 'that which I have said to you' (Heqanakhte Papers [Allen 2002], 1. 17).

As regards Parkinson's fourth example, it is not clear to me what exactly *ḏd.j sḏm.k* means at Eloquent Peasant B1, 99 (old numbering: B1, 68) and R 16.3 (old numbering: R 112), possibly 'I say, so that you might hear, *jr m3ʿt* "Do justice".'

10.5.2.5 An apparent example of sḏmw.f *in CT 5.3c–4c, Coffin B4C*

None of the examples above provides an identification of the second *sḏm.f* in *jr sḏm.f sḏm.f*. A form with *w*, however, occurs in one variant of a text discussed earlier (see also Parker 1996: 105, ex. 112).

> *jr jw pt m mḥt ḥmsw N tn m rs.s*
> *jr jw pt m rsw ḥmsw N [tn …]* (CT 5.3c–4c, Coffin B4C)
>
> 'If heaven comes as a north wind,
> this N will sit to the south of it.
> If heaven comes as a south wind, this N will sit …'.

But note that, at CT 5.5b, in the same Coffin B4C, *w* is absent in *ḥms N tn m j3bt.s* 'This N will sit to the east of it'. One variant with *w* is not much, but in several others, *ḥms*, without *w*, is followed by a suffixed personal pronoun. It has long been known that *w* is especially prone to appear before a noun or a noun phrase (see also Schenkel 2000). Compare *ḥms.j*, without *w*, and *ḥmsw Rʿ*, with *w*, in *ḥms.j r wnm.j t ḥmsw Rʿ r wnm t* 'When(ever) I will sit down to eat bread, Re will sit down to eat bread' (CT 3.185c–d, Coffin S1C). All this enhances the significance of the appearance of *w* in Coffin B4C of CT 5.3c–4c.

In the next two examples, *sḏm.f* forms with the affix *w* appear in the apodosis.

> *jr gs.tw mnt nbt snbw ḥr ʿwy* (Papyrus Berlin [Medical] 3038, 9,1)
>
> 'If any sore spot is anointed (with it), (it) becomes instantly healthy'.

> *jr ḫ3.k* (written *wḫ3.k*) *wbnw pf jnrw r.f* (Papyrus Smith 10,5)
>
> 'If you examine this wound, (it) trembles'.

It is not certain, however, that *snbw* and *jnrw* are instances of *sḏmw.f*. Westendorf (1962: 140 §201) classified both verb forms as prospective, that is, future. He accepted the existence of only one future *sḏm.f* other than the adjectival (relative) future.

The examples exhibit three peculiar characteristics that limit their relevance for this point. First, the suffix pronoun is omitted in both verb forms. Second, *w* in *snbw* is spelled with the plural strokes. Third, *jnrw* is spelled alternatively as *jnrjj* in the same expression at Papyrus Smith 8,1.

The first characteristic is slightly less peculiar considering that the suffix pronoun is also absent in the related passage *snb.ḫr ḥr ʿwy* 'Then (he) becomes healthy' (Papyrus Ebers 75,13–14). The second characteristic is somewhat less peculiar given that *w* is spelled normally in the related passage *swr jn s snbw.f ḥr ʿwy* 'Drinking by a man. He instantly becomes healthy' (3038, 21,11). The third characteristic is not so peculiar given that *jj* can alternate with *w* in the *sḏmw.f*.

One problem remains, however. In the passage just cited, the affix *w* in *snbw.f* appears even though there is no *jr*-clause. In this case, at least, the appearance of *w* is unrelated to an *jr*-clause. Why, therefore, would the appearance of *w* in *snbw* in *jr gs.tw mnt nbt snbw ḥr ʿwy* bear any relation to the *jr*-clause?

10.5.3 Other instances of *sḏm.f* as second verb form in *jr sḏm.f sḏm.f*

There are more instances of *sḏm.f* as the second verb form in *jr sḏm.f sḏm.f* than those cited in §10.5.2. All those that can be securely identified are substantival and general present (aorist), since they are third-weak verbs with gemination, as in the type *mrr.f*. Parker (1996: 55–57) cites CT 5.350d–351a, 6.310i, 6.333i–j, and 7.449c–450c. In all the known instances of this type, there is an adverbial phrase. Assuming contrastive emphasis affecting that adverbial phrase is not difficult. Emphasis would then explain the use of a substantival verb form in these instances.

Still further instances of *sḏm.f* in the apodosis following *jr sḏm.f*, besides those exhibiting gemination, are (cf. Vernus 1990: 32–33):

jr jḳr.k grg.k pr.k (Wisdom of Ptahhotep 325)

'If you excel, you will establish your house(hold)'.

jr ḳ3s.s ms.s (Papyrus Berlin [Medical] 3038, v. 1,5–6)

'If she vomits, she shall give birth'.

jr w3ḏ.f w3d t3 r-ḏr.f (from the Tebtunis Onomasticon)

'If he flourishes, the entire land flourishes'.

jr kꜥ.f mwt.f (Vernus 1990: 32 n. 21 from Sauneron 1989: 17 [1,25])

'If he vomits, he dies'.

The example from the Tebtunis Onomasticon appears in Fragment V, col. 4, line 10 of Papyrus I (pCarlsberg 180 + pBerlin 10465 and 14475 + PSI I 76) (for the text, see Osing 1998, vol. 1: 150–51; for a photograph and a facsimile, see *ibid.*, vol. 2: plates 12 and 12a). The Tebtunis Onomasticon dates to the Late period, but the construction seems pure Middle Egyptian. Perhaps the sentence was copied from a genuine Middle Egyptian text. The grammar of the sentence has been discussed by Osing (1987; 1998, 1:63) and myself (Depuydt 1999c: 38–40 §1.1.3.1).

Three of the verb forms, *grg.k*, *w3ḏ* + Noun, and *mwt.f*, exhibit no distinctive written characteristics, so there is no basis for identifying their forms. The fourth, *ms.s*, lacks gemination (as in *mss.s*), which eliminates some possibilities.

Let us turn to *ms.s* first. Which *sḏm.f* is it likely to be? Only two *sḏm.f* forms are attested in the apodosis following *jr sḏm.f*: (1) the substantival general present (of the type *mrr.f*) and (2) the substantival future *sḏm.f* (*sḏmw.f*). Parker (1996: 116–17) writes:

> Previous examples have shown that the apodosis of the conditional construction *ir* + *sḏm.f* + *sḏm.f* can be either a Nominal or Prospective-Nominal *sdm.f*.

Clearly, *ms.f* cannot be an instance of the type *mrr.f*. Furthermore, there are only two futures in the master list of *sḏm.f* forms. Vernus (1990: 16–17) has found no examples of one of the two, the adverbial future (also called subjunctive), otherwise a conceivable candidate—no distinctive writings such as *jwt.f*, *jnt.f*, *m3n.f*, or *d(y).f* (without *r*) are attested in the apodosis following *jr sḏm.f*. If *ms.f* is of neither type *mrr.f* or *jwt.f*, then it must be *sḏmw.f*. The other three verb forms may thus be defective writings (without *w*) of *sḏmw.f*.

A crucial problem arises, however. *Ms.s* is not followed by an adverbial phrase. How can it be substantival yet not emphasize an adverbial phrase? The explanation could then be applied to the forms listed in §10.5.2 above as sharing the same characteristics. Yet Schenkel (1992: 379) proposed that *sḏmw.f* appears in the apodosis only when an adverbial phrase is emphasized.

> It is a fact that an adverbial phrase follows the verb form in every instance. The special case of the main clause following *jr* [as a phenomenon in its own right] would therefore need to be removed from [Egyptian] grammar.*

It is not clear how Schenkel would interpret the four verb forms *grg.k*, *w3ḏ* + Noun, *mwt.f*, and *ms.s* cited above. Is an explanation possible?

10.6 COMBINATION OF SUBSTANTIVAL CHARACTER AND LACK OF EMPHASIS IN THE APODOSIS

It has already been noted in §10.2 that every use of the *sḏmw.f* is substantival and that substantival character is clear for certain uses of *sḏmw.f*. It is difficult to accept that the same verb form could be both substantival and not substantival at the same time. This consideration invites an explanation of all uses of *sḏmw.f* as substantival.

The substantivalness seems reasonably acceptable, but that does not mean that the *deeper motive* for the substantivalness is always clear. A case in point is the negated *n sḏmw.f* (see Schenkel 2000). Even if it is reasonable *that sḏmw.f* is substantival in *n sḏmw.f*, it is not fully certain *why*. That would require an understanding of the negation word *n*, but its historical origin seems impenetrable. It is easy to understand *that sḏmw.f* is substantival in *n sḏmw.f*. Indeed, the negation word *n* is often followed by such clearly substantival forms as the passive *sḏm.n.tw.f* and the *sḏm.n.f* of verbs of motion. Then why not also *sḏmw.f*? **N jrr.f* is not attested. By Gunn's rule, *n jr.n.f* is used instead.

A second relevant and well-attested use of *sḏmw.f* is in the future tense cleft sentence, as in *ntf sḏmw.f* 'it is he who will hear' and *jn* Noun *sḏmw.f* 'It is [Noun] who will hear' (Schenkel 2000). Here too, it seems easy to understand how *sḏmw.f* can be substantival, even if the historical origin of the pattern is not clear. The structure of cleft sentences is to a degree opaque in many languages (see Depuydt 1993b: 25–26 with nn. 45–46). Suffice it to note that Gardiner (1934) associated the structure *ntf sḏmw.f* 'it is he who will hear' with the structure *ntf nbw* 'his is the gold'. This interpretation is cited favorably by Polotsky (1969: 474), though I consider it incorrect (see Depuydt 1993b: 25 n. 45). Gardiner's explanation contradicts what is known about *sḏm.f* in *ntf sḏm.f*. Gardiner derives the future tense from the possessive character of the construction, adducing French *il entendra* 'he will hear' as

* Tatsache ist … daß in allen … Fällen ein adverbialer Ausdruck folgt. Die Besonderheit des Hauptsatzes nach *ỉr*-Einleitung wäre somit aus der Grammatik wieder zu streichen.

a parallel, which derives from Latin *intendere habet* 'he has hearing (to do)'. The Latin *habet* is present tense; the French future derives from the possessive construction. However, it is now certain that *sḏm.f* in the construction *ntf sḏm.f* is itself a future *sḏm.f*, and that is the source of its futurity, not some supposed possessiveness.

The corresponding Late Egyptian cleft sentence structure is *ntf i.jr.f sḏm* 'it is he who will hear', which is always future. Incidentally, the existence of this distinctly future structure may be viewed as a strong argument in favor of the following two propositions. First, there exists a separate future emphatic *i.jr.f sḏm* in Late Egyptian, and therefore also a distinct present emphatic *i.jr.f sḏm* and a distinct past emphatic *i.jr.f sḏm*. Second, the three verb forms are written alike. Some, including myself, accept this otherwise invisible distinction. Most, it would seem at the time of writing, do not.

To return to *jr sḏm.f sḏm.f*. If the *sḏm.f* in the apodosis is the substantival *sḏmw.f*, and if no adverbial phrase is emphasized, how can its substantivalness be accounted for? My own solution is to suppose that the two *sḏmw.f* forms of the structure *jr sḏmw.f sḏmw.f* relate, at least vestigially, as the two substantival members of a nominal or substantival sentence, X = Y. Two scholars have adopted this position.

First, Vernus (1990: 32) writes of *jr w3ḏ.f w3ḏ t3 r-ḏr.f* (Tebtynis Onomasticon, see §10.5.3), "One could also understand the apodosis [*w3ḏ t3 r-ḏr.f*] as [the second half of] a balanced sentence." The balanced sentence *wḏ3.j wḏ3.f ṯs pẖr* 'I'm fine, he's fine, and vice versa' (CT 4.246, see §5.1.1.2), may be adduced to clarify what Vernus seems to be hinting at.

Second, Parker (1996: 121) associates the construction *jr sḏm.f sḏm.f* with the pattern *jr* Noun + Noun. He admits, though, that this nominal sentence type, which is common in Late Egyptian, is rare in Middle Egyptian. Parker interprets *jr sḏm.f sḏm.f* as some kind of nominal sentence in order to explain why the second verb form is substantival while emphasis is absent.

If *jr sḏm.f sḏm.f* can contain two substantival verb forms neither of which denotes emphasis, then the association with the balanced sentence becomes attractive, albeit hypothetical.

There are two further possible examples. In *jr sḏm.k nn ḏd.n.j n.k wnn sẖr.k nb r-ḥ3t* 'When you listen to these things that I have said to you, …' (Ptahhotep 507–8), the clause *wnn sẖr.k nb r-ḥ3t* seems to be the substantival equivalent of **jw sẖr.k nb r-ḥ3t*. Yet the adverbial phrase *r-ḥ3t* does not seem to be emphasized. Unfortunately, the meaning of the apodosis is unclear. In *jr ḫpr sp nfr m wnn m ḥr-tp wnn.f mnḫ n nḥḥ jw s33.f nb r ḏt* 'If a good act happens through(?) him who is chief, he will be *mnḫ* forever, and his every wisdom will be forever' (Ptahhotep 520–23), the clause *wnn.f mnḫ* (with *mnḫ* as a stative!) seems to be the substantival equivalent of **jw.f mnḫ*. The adverbial phrase *n nḥḥ* 'forever' is probably not emphasized, since the parallel *r ḏt* 'forever' is not emphasized in the next sentence. If it were, one would have expected *wnn s33.f nb r ḏt* instead of *jw s33.f nb r ḏt*.

The balanced sentence exhibits full equation between the two members. The equation is reversible from $x = y$ to $y = x$. By contrast, the presence of *jr* makes the conditional sentence unbalanced ("When it rains, I stay inside" is unbalanced because the reverse, "When I stay inside, it rains" is not valid). The reversibility appears to have been lost owing to the addition of *jr*. Let us assume, then, that the conditional sentence with *jr* developed from the simple construction of a balanced sentence containing two substantival verb forms. If this is the case, then the conditional sentence with *jr* would have grown even farther away from the balanced sentence when other verb forms began appearing in the apodosis, including *jw.f r sḏm*, *sḏm.ḫr.f*, and *sḏm.k3.f*. These verb forms are all well attested in the apodosis.

Proof remains desirable that *sḏmw.f* does not emphasize an adverbial phrase in the apodosis. The absence of emphasis requires positive verification. Ideal proof would be unambiguous instances of *sḏmw.f* in the apodosis without any adverbial phrases. No such instances have come to my attention. *Ms.s* comes close because there is no clear evidence for the adverbial future *sḏm.f* (subjunctive) in the apodosis and the absence of gemination in *ms.s* eliminates the type *mrr.f*. That leaves substantival future *sḏm.f* as the most likely identification of *ms.s*.

10.7 AN ITEM OF EVIDENCE THAT *SḎMW.F* CAN BE SUBSTANTIVAL WITHOUT EMPHASIZING AN ADVERBIAL ELEMENT IN THE APODOSIS OF *JR SḎM.F SḎM.F*

The hypothesis outlined in §§10.4–5 implies the existence of occurrences of *sḏmw.f* in the apodosis in which *sḏmw.f* does not emphasize an adverbial phrase. Some evidence does point in that direction.

Compare the following two statements, a balanced sentence and a conditional sentence introduced by *jr*.

(1) BALANCED SENTENCE

jw pt m rswt ḥms.j m mḥwt (CT 4.50f, Coffins B1C, B2L)

'Heaven will come as a south wind,
I will sit to the north of it'.

"To the north of it" means 'to the north (of the house), out of the wind'. But *ḥr mḥt.s* is expected as an equivalent. The construction is hybrid (see §5.1.1.3).

(2) CONDITIONAL SENTENCE WITH *jr*

jr jw pt m rsw ḥmsw N [*tn* …] (CT 5.4b–c, Coffin B4C)

'If heaven comes as a south wind, this N will sit …'

Two assumptions may be made. First, the substantivalness of *ḥms* in (1) allows *ḥms* to function as a substantival member of a nominal sentence. That means that the substantivalness cannot in addition exercise a second function of emphasizing an adverbial phrase (on the fallacy of "double duty," see §9.4). Therefore, if the adverbial phrase *m mḥwt* (*ḥr mḥt.s* in this hybrid formulation [see §5.1.1.3]) is emphasized, it is only by some means not accessible to observation, such as intonation.

Second, if (1) and (2) basically express the same thought, then because of parallelism it is difficult to accept that the adverbial phrase following *ḥms* would be emphasized in (2) but not in (1). It has been established that the phrase is not emphasized in (1) (see first assumption). It is therefore logical to conclude that the phrase is not emphasized in (2) either. The form *ḥmsw* in (2) is an instance of substantival *sḏmw.f*—the affix *w* allows no doubt. But in this case, the substantival *sḏmw.f* does not emphasize the adverbial phrase that accompanies it.

The significance of the following facts (included for completeness) is not entirely clear. Clear instances of the balanced sentence *prr.sn* … *prr.j* … (CT 3.24c–25b, Coffins S1C, S2C) alternate with the construction of the type *prr.sn* … *pr.j* … (ibid., Coffin B2Bo), the latter presumably an adverbial sentence. The balanced sentences further alternate with *r prt.sn* … *pr.j* … (B17C), *r pr.sn* … *pr.j* … (B1C), *r prw.sn* … *pr.j* … (B2L), and *r prt.sn* … *prjj.j* … (B1Be). If *r* is associated with *jr*, which is by no means certain, then CT 3.24c–25b might reveal a certain affinity between the balanced sentence and the construction with *jr*.

11 Language Evidencing Growth in the Expression of Logical Thought

11.1 Are We Smarter than the Ancient Egyptians?

Are we, people of the modern age, smarter than the ancient Egyptians? Or perhaps not smarter, but more sophisticated or more complex? What do the terms "smart," "complex," or "sophisticated" mean? The topic at hand is of large proportions: the evolution of civilization. Where do we begin? The task is to describe the possible growth in sophistication in empirically unequivocal terms, and to explain it. In Cartesian fashion, an observation that is difficult if not impossible to deny may serve as point of departure. The observation is that civilization has changed. The task is to describe change in unequivocal terms: how the state of affairs was earlier, how it was later, and how the difference is best characterized.

It is also desirable not to focus on everything at once. The present focus is on a more abstract facet of civilization: thought itself. Thought is expressed in language. Thought and language are distinct, yet they are also inextricably intertwined. The nature of the relationship between thought and language is not fully understood.

If one reads Egyptian texts, for example, the Story of Sinuhe, the impression is difficult to escape that discourse does not proceed as a modern prose text would. Putting these intuitively but clearly sensed differences into words is another matter. By contrast with Egyptian prose texts, Greek writers such as Herodotus or Thucydides already write more or less as we do today, and they were among the first in history to do so. There is something distinctly modern to their way of writing. That does not mean that pre-Greek thought is without interest. In fact, the difference between Egyptian thought, on the one hand, and Greek and modern thought, on the other, should be all the more reason for recommending the study of ancient Egypt more widely.

In choosing metaphors to describe the evolution of civilization, one hesitates to call early civilizations more primitive. Intimations of superiority or inferiority are better avoided. A recommended metaphor is growth. All stages of a growing process are indispensable—and therefore of equal importance. Like life, civilization is an organism. Organisms grow and inevitably need to pass through certain stages. The time span of human history, which does not exceed five millennia or so, seems too short for postulating any fundamental biological changes to the brain. Any growth in complexity described below probably rather applies to the structure of thought and language as it is handed down from generation to generation, and not to the chemical composition of the brain.

The present focus is on thought and language, but further narrowing of the focus is necessary, to one facet of thought only: logical thought. Other facets of thought are poetic, religious, and literary.

11.2 THE INEVITABILITY OF LOGICAL INFERENCES

Logic can be defined as the ability to draw unavoidable inferences from given statements. In other words, logic is deductive, not inductive. How better to illustrate the inevitability of logical inferences than by means of an example found in the foundational work of modern logic, George Boole's *Investigation into the Laws of Thought*? The example (1854: 84ff.) concerns the Jewish precept (Lev 11:3, Deut 14:6) "Clean beasts are those that both divide the hoof and chew the cud." By the laws of logic, certain inferences follow from this proposition.

Clean beasts are those that both divide the hoof and chew the cud.

x = clean beasts;
y = beasts dividing the hoof;
z = beasts chewing the cud.

The above proposition may be represented by the equation

$$x = yz.$$

(i) $xy\,(1 - z) = 0$, which is equivalent to:

Clean beasts (x) dividing the hoof (y)
but not chewing the cud ($1 - z$) do not exist ($= 0$).

$xz\,(1 - y) = 0$, which is equivalent to:

Clean beasts (x) chewing the cud (z)
but not dividing the hoof ($1 - y$) do not exist ($= 0$).

$x\,(1 - y)\,(1 - z) = 0$, which is equivalent to:

Clean beasts (x) neither dividing the hoof $(1 - y)$
nor chewing the cud $(1 - z)$ do not exist $(= 0)$.

$(1 - x)\, yz = 0,$ which is equivalent to:

Unclean beasts $(1 - x)$ dividing the hoof (y)
and chewing the cud (z) do not exist $(= 0)$.

(ii) $1 = xyz + (1 - x)\, y\, (1 - z) + (1 - x)(1 - y)\, z + (1 - x)(1 - y)(1 - z),$
which is equivalent to:

All existing things or the entire universe (1)
consists of (=) either

clean beasts (x) dividing the hoof (y)
and chewing the cud (z), or

unclean beasts $(1 - x)$ dividing the hoof (y)
but not chewing the cud $(1 - z)$, or

unclean beasts $(1 - x)$ not dividing the hoof $(1 - y)$
but chewing the cud (z), or

things that are neither clean beasts $(1 - x)$ nor dividers of the
hoof $(1 - y)$ nor chewers of the cud $(1 - z)$.

(iii) $z = xy + v\, (1 - x)(1 - y),$
which is equivalent to:

Chewers of the cud (z) either are (=)

clean beasts (x) dividing the hoof (y), or

belong to an indefinite remainder (v) of objects that are
neither clean beasts $(1 - x)$ nor dividers of the
hoof $(1 - y)$.

(iv) $x\, (1 - y) = 0,$
which is equivalent to:

Clean beasts (x) not dividing the hoof $(1 - y)$
do not exist $(= 0)$.

In Boolean algebra, these inferences are obtained simply by executing certain algebraic operations. The results of these operations are always true. Clearly, there is something invariable about the workings of the mind.

11.3 THE EGYPTIAN LANGUAGE IN RELATION TO THE EVOLUTION OF LANGUAGE AND THOUGHT

Language originated tens of thousands of years ago, long before the beginning of Egyptian history. Presumably, the first manifestations of language were less sophisticated than what came later. The alternative is the unattractive assumption that language came about in all its complexity all at once. Direct evidence for the origin and early evolution of language is, of course, wanting. Then again, Egyptian has the longest attested history of any language. It is also one of the earliest attested. The question arises: Can one discern an increase in sophistication in Egyptian as the centuries progress? The reigning assumption is that the Egyptian of the Pyramid texts, the earliest attested stage of the language, is as complex as Coptic, the latest stage.

Language is one thing. What about logic? The crucial question is: Is there positive evidence of growth in logical thought in the surviving texts? Did the Egyptians get smarter, as it were? Language remains relevant in as far as logic is expressed in language.

The aim of this investigation is to point out some evidence of such growth.

11.4 THE THREE MAIN TYPES OF LOGICAL PROPOSITION

There are three main types of logical proposition, which can be represented by the following diagrams.

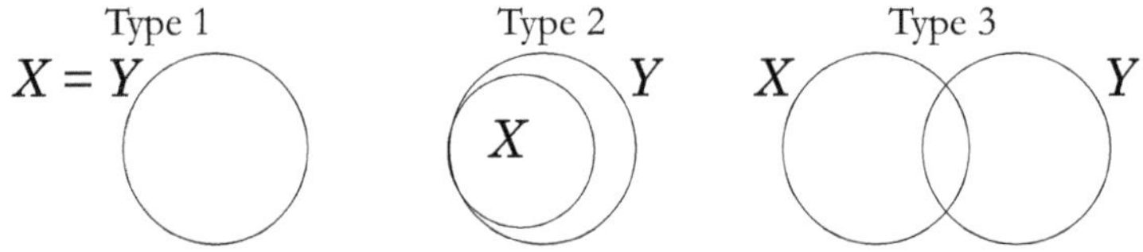

These three types are fundamental to the way we think. They look simple enough, yet they pervade logical thought. An increase in complexity may be observed from Type 1 to Type 2 to Type 3. Type 1 has only two classes: things that are both *X* and *Y* and things that are neither *X* nor *Y*. Type 2 has three classes: things that are both *X* and *Y*, things that are *Y* but not *X*, and things that are neither *X* nor *Y*. Type 3 has four classes: things that are both *X* and *Y*, things that are *Y* but not *X*, things that are *X* but not *Y*, and things that are neither *X* nor *Y*.

Type 2 exhibits an interesting unbalance or lopsidedness. There is *X* that is not *Y*, but there is no *Y* that is not *X*—the little circle is entirely encompassed by the big circle. Thought proceeds mainly on two levels: there are primary propositions and secondary propositions. Primary propositions describe relations between things. Secondary propositions describe relations between events.

LOGICAL PROPOSITIONS OF TYPE 2 (**denial of existence**)
PRIMARY proposition (relation between THINGS):
"(All) swallows are birds."
Existence denied: *swallows that are not birds*
SECONDARY proposition (relation between EVENTS):
"When it rains, I stay inside."
Existence denied: *occasions when it rains and I am not inside*

The primary proposition "All birds are swallows," which would be a nominal sentence in Egyptian, and the secondary proposition "When it rains, I stay inside," which is a conditional sentence, exhibit the same logical structure and therefore in some sense belong together in grammars. Both sentences exhibit the lopsidedness of the little circle inside the big circle: All birds are swallows, but not all swallows are birds. Likewise, every time it rains I stay inside, but not every time I stay inside does it rain. It may or may not rain when I stay inside.

In symbolic logic, one class divides the universe of thought into two parts, which are themselves classes. Two parts divide the universe into four classes. Three classes divide divide the universe into eight parts. And so on.

11.5 TYPE 3 AND THE AFFIRMATION (AS OPPOSED TO THE NEGATION) OF EXISTENCE

Type 3 is more complex than Type 2. Venn discusses the place of Type 3 ("particular" propositions) in logical discourse in penetrating fashion in chapters 6 and 7 of his *Symbolic Logic* (1894). Type 3 has four classes and Type 2 has only three classes, but Type 3 exhibits a fundamental difference from both Types 1 and 2 in another respect. There, the existence of certain classes is absolutely denied—in Type 1, the existence of *two* classes is absolutely denied (there is no *Y* that is not *X* and there is no *X* that is not *Y*, because *X* and *Y* are the same). In Type 2, the existence of just *one* class is absolutely denied (there is no *Y* that is not *X*). Venn emphasized the importance of these absolute denials and the sharpness that they lend to the logical content of a proposition. By contrast, in logical propositions of Type 3, there are no absolute denials of existence. Then what is the purpose of Type 3?

LOGICAL PROPOSITIONS OF TYPE 3 (**affirmation of existence**)
PRIMARY proposition (relation between THINGS):
"Some doctors are French."
Affirmation of existence: *There are French doctors.*
SECONDARY proposition (relation between EVENTS):
"Sometimes when it rains, I stay inside."
Affirmation of existence: *It happens that it rains and I stay inside.*

In pre-modern logic, "Some doctors are French" would be called a *particular* statement. Examples of *universal* statements would be "All doctors are French" and "No doctors are French." Universal statements belong to Type 2, denying the existence of a class; in that sense they are categorical and precise. The statement "All doctors are French" implies that there is no such thing as non-French doctors. By contrast, statements of Type 3 are much vaguer. In "Some doctors are French," the meaning of "some" is imprecise. "Some" could be just "two"; or it could be quite a few. The possibility is not excluded that all doctors are French. Venn (1894: 147) writes as follows.

> An assertion confined to "some" of a class generally rests upon observation or testimony rather than on reasoning or imagination, and therefore almost necessarily postulates existent data, though the nature of this observation and consequent existence is, as already remarked, a perfectly open question.

All the statement says is that at least some doctors are French—there is such a thing as French doctors. The class of French doctors is said to exist.

On the whole, Type 3 seems rather inconclusive and even philosophical, as in statements such as "Accidents happen," in the sense of "There are accidents." Venn (1894: 180) says:

> [Type 3] can extinguish no class and establish no class, and has therefore no categorical information to give.

but also:

> [Propositions of Type 3] have, so to say, a tone or reality and of sober fact about them which cannot always be claimed for universals.

11.6 THE FUNDAMENTAL DIFFERENCE BETWEEN AFFIRMING EXISTENCE AND NEGATING EXISTENCE

At first sight, negating the existence of a class seems like the counterpart or opposite of affirming the existence of a class. Compare *jw wn* 'there is' and *nn wn* or *n wnt* 'there is not' in Old and Middle Egyptian. The two seem tantalizingly close, like the two sides of a coin. In English, "there is not" is obtained by simply adding "not" to "there is." But concepts such as negation and existence belong to the domain of logic, where affirming existence is entirely different from negating existence—negation of existence characterizes Type 2, in which one circle encompasses another; affirmation of existence characterizes Type 3, in which two circles intersect. Negating existence appears to be more categorical and precise than affirming existence. Is this difference between negation and affirmation reflected in the facts of languages, including Egyptian? It is striking that expressions of non-existence are far more frequent than expressions of existence in Old and Middle Egyptian. Affirmative *jw wn* is remarkably rare, the only two examples in a main clause known to me being *jw wn nḏs Ḏdj rn.f* 'There was a commoner whose name was Djedi' (Papyrus Westcar 6,62–7,1) and *jw wn ḥf3w ḥr wpt ḏw pf* 'There is a snake on top of that mountain' (CT 2.377c). By contrast, *nn wn* and *n wnt* are extremely common (cf. Gardiner 1957: 82–83 §§107–8). Gunn (1924: 122–24, 164–68) presents a large collection of examples of *nn wn*. He describes *nn wn* as "common" (122) and *n wnt* as "quite common" (164).

By the time of Coptic, however, this huge gap in frequency of occurrence between *jw wn* 'there is' and *nn wn* or *n wnt* 'there is not' has not only closed, but in the Sahidic Coptic New Testament, there are about half as many more instances of **ⲟⲩⲛ̄** *wn* 'there is' as of **ⲙ̄ⲛ** *mn* (also **ⲙ̄ⲙⲛ̄** *mmn*) 'there is not' (see Wilmet 1957: indexes s.vv.).

Two questions arise: (1) Why does *jw wn* behave differently from *nn wn* or *n wnt* in Old and Middle Egyptian? (2) Why is Coptic **ⲟⲩⲛ̄** *wn* much more frequent than its Middle Egyptian ancestor?

As regards question (1), it does not surprise that the behavior of *jw wn* initially differed from that of *nn wn*. Negation of existence exhibits an entirely different logical structure from affirmation of existence. The two are expressed by different types of logical proposition, negation by Type 2 and affirmation by Type 3.

A distinction is worth noting at this point. Types 2 and 3 concern the relation between *two* classes, as between "French" and "doctors" in "All doctors are French" (Type 2) and "Some doctors are French" (Type 3). However, in expressions of existence or non-existence, there may be just *one* class, as in *nn m3ʿtyw* 'There are no righteous ones' (Man Tired of Life [Papyrus Berlin 3024], 122). Presumably negating the existence of *one* class is related to the negation of existence found in Type 2, in which there are *two* classes, and affirming the existence of *one* class is related to the affirmation of existence found in Type 3, in which there are *two* classes.

As regards question (2), the increase in affirmations of existence may be interpreted as an increase of Type 3 (see §11.8.2).

11.7 A NEW GRAMMATICAL RULE CONCERNING AFFIRMATION AND NEGATION OF EXISTENCE, AND ITS EXPLANATION

11.7.0 The three components of the rule

It is well known that **ОΥN̄** *wn* 'there is' and **M̄N** *mn* 'there is not' need to be followed by indefinite expressions (see Layton 2004: 248–50 §§322–24, 381–90 §§476–85), but what seems not to have been noticed is a contrast in their behavior regarding the indefinite article. This can be seen gradually taking shape in Late Egyptian and Demotic. This gradual process is of the statistical kind and would need to be documented in detail. The process is parallel to the gradual rise in frequency of the indefinite article in general. By Coptic, the rule is firmly in place. The following description focuses on Coptic.

There are three components to the rule. The *indefinite* article *cannot* occur after **M̄N** *mn* 'there is not' (§§11.7.1–11.7.2). The *indefinite* article, or an equivalent expression, occurs *very often* after **ОΥN̄** *wn* 'there is' (§§11.7.3–11.7.4). The *indefinite* article *cannot* occur after **ОΥN̄** *wn* when only a noun follows (§§11.7.5–11.7.6).

The contrast between **ОΥN̄** *wn* 'there is' and **M̄N** *mn* 'there is not' described in these three rules is somewhat more pronounced than the quantifier "often" in the following description by Layton (2000 [2004]: 248 §322) might lead one to believe.

> Subjects in the negative [following **M̄N** *mn*] often have a zero article, whereas the corresponding affirmatives [following **ОΥN̄** *wn*] often have indef[inite] **ОΥ–** [*u*-] or **ϩЄN–** [*hen*-].

11.7.1 First component

This first component pertains to **M̄N** *mn* 'there is not' and consists of two complementary parts. On the one hand, the indefinite article (that is, singular **ОΥ-** *u*- or plural **ϩЄN-** *hen*-) *cannot* follow **M̄N** *mn*. "[In Coptic] as in Demotic," writes Shisha-Halevy (1989: 42), "the indefinite article is conspicuous by its absence [after **M̄N** *mn*] (for reasons of logico-semantic incompatibility?)" (on existential statements in Coptic, see also Shisha-Halevy 1984). Indeed, with certain well-defined exceptions, *absence* of an article or "zero"-article is the norm after **M̄N** *mn*.

> **M̄N ϩM̄ϩAΛ ЄϤΧОСЄ ЄПЄϤΧОЄІС** (Matthew 10:24)
> *mn hmhal e=f-jose e-pe=f-joys*

'There is no servant who is more elevated than his master' (*or*: 'No servant is more elevated than his master').

A sequence *ⲙ̄ⲛ ⲟⲩϩ̄ⲙϩⲁⲗ *mn u-hmhal* … 'there is not a servant …' does not exist. In the very rare occurrences in which the singular indefinite article **ⲟⲩ-** *u-* does seem to follow **ⲙ̄ⲛ** *mn*, **ⲟⲩ-** *u-* is in fact used in conjunction with **ⲟⲩⲱⲧ** *wôt*. **ⲟⲩ-** … **ⲟⲩⲱⲧ** *u-* … *wôt* together form one expression with the meaning 'a single', as in **ⲙ̄ⲛ ⲟⲩⲯⲩⲭⲏ ⲅⲁⲣ ⲛ̄ⲟⲩⲱⲧ** *mn u-psukhê gar n-wôt* 'for there is not a single soul' (Acts 27:22) and **ⲙ̄ⲛ ⲟⲩϥⲱ ⲅⲁⲣ ⲛ̄ⲟⲩⲱⲧ** *mn u-fô gar n-wôt* 'for there is not a single hair' (Acts 27:34). "A single soul" and "a single hair" denote a *complete* class by themselves. In other words, exceptionally, the indefinite article **ⲟⲩ-** *u-* does not refer to *part* of a class. **ⲙ̄ⲛ** *mn* denotes the elimination of an entire class. In conjunction with **ⲟⲩⲱⲧ** *wôt*, **ⲟⲩ-** *u-* denotes an entire class and is for that reason apparently allowed after **ⲙ̄ⲛ** *mn* 'there is no(t)'.

While indefinite articles do now follow **ⲙ̄ⲛ** *mn*, the definite article apparently does when heading "substantivized clauses," as in:

ⲙ̄ⲛ ⲡⲉⲧϩⲟⲃ̄ⲥ (Mark 4:22)
mn p-et-hobs

'There is not anything hidden'.

In this case, however, the definite article does not have its usual function of denoting definiteness. Thus, **ⲡⲉⲧϩⲟⲃ̄ⲥ** *p-et-hobs* 'anything hidden' does not refer to any definite entity. Other examples are **ⲡⲉⲑⲏⲡ** *p-et-hêp* 'what is hidden' (Mark 4:22), **ⲡⲉⲧⲟ ⲛ̄ⲛⲟϭ** *p-et-o n-noc* 'who is greater' (Luke 7:28), **ⲡⲉⲧⲉⲓⲣⲉ** *p-et-eire* 'who does' (Rom 3:12), and **ⲡⲉⲧⲛⲁⲁⲁϥ** *p-et-naaa=f* 'who is greater' (Heb 6:13).

There are nearly two hundred examples of **ⲙ̄ⲛ** *mn* in the New Testament, followed in most instances by an article-less noun. Other expressions following **ⲙ̄ⲛ** *mn* may be considered equivalent in function. **ⲗⲁⲁⲩ** *laaw* is a special case, following both **ⲙ̄ⲛ** *mn* (Matt 10:26, 11:27*bis*, 19:26, 24:36; Mark 3:27, 10:29, 13:20, 32; Luke 1:61; etc.) and **ⲟⲩ̄ⲛ** *wn* 'there is'. **ⲗⲁⲁⲩ** *laaw* means 'any', that is, 'one, no matter which, of two or more'. Unaccompanied by a negation, it denotes part of a class and is therefore compatible with Type 3 and **ⲟⲩ̄ⲛ** *wn*. In conjunction with a negation, **ⲗⲁⲁⲩ** *laaw* means 'not any', that is, 'none', and is then compatible with Type 2 or **ⲙ̄ⲛ** *mn* 'there is not'.

Other expressions after **ⲙ̄ⲛ** *mn* include the following: **ϩⲁϩ** *hah* 'many' (Matt 13:5; Mark 4:5*bis*, 7:15, 9:39; 1 Cor 1:8*ter*; and often); **ⲟⲩⲟⲛ ⲛⲓⲙ** *won nim* (Matt 19:11); and **ϭⲉ** *ce*, **ⲕⲉ** *ke* 'other' both by itself (Mark 12:31, 12:32;

Acts 4:12; Rev 10:6) and in expressions such as **ϭⲉϫⲟⲓ** *ce-joy* (John 6:22) and **ⲕⲉⲟⲩⲁ** *ke-wa* (Gal 1:7).

11.7.2 Explanation of the first component

The function of the indefinite article is to denote *part* of a class. For example, "*a* cow" is a member of the class of cows. On the other hand, **ⲙⲛ̄** *mn* signifies the non-existence of an entire class. Representations of non-existence by means of diagrams reveal that there is no place for a *part* of a class in the expression of non-existence. Non-existence is characteristic of logical propositions of Type 2.

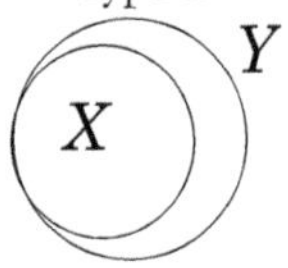

This representation corresponds to this statement of non-existence: "There is no *X* that is not *Y*," which applies to *all*, and not just part, of *X*; the circle marked *X* is entirely encompassed by the circle marked *Y*.

The drawing also corresponds to the two statements "All *X* are *Y*" and "Some *Y* are *X*," but they do not concern us here; they do not express non-existence.

An example of Type 2 is **ⲙⲛ̄ ϩⲙ̄ϩⲁⲗ ⲉϥϫⲟⲥⲉ ⲉⲡⲉϥϫⲟⲉⲓⲥ** *mn hmhal e=f-jose e-pe=f-joys* 'No servant is more elevated than his master' (Matt 10:24). Two classes are being related to one another. They are the class "servant" (*X*) and the class "things no more elevated than their master" (*Y*). Two equivalent statements in unidiomatic English are: "All servants are no more elevated than their master" and "Some of those no more elevated than their master are (all the) servants." In all these equivalent statements, the class *X* is entirely encompassed by *Y*. In the corresponding diagram, one circle entirely encompasses the other. Clearly, the entire class "servant" is involved. There is no reference to some part of it. Such a reference would require an indefinite article.

In Type 2, *two* classes are related to one another. But non-existence can also be stated in relation to just *one* class, as in **ⲥⲉϫⲱ ⲙ̄ⲙⲟⲥ ϫⲉ ⲙⲛ̄ ⲁⲛⲁⲥⲧⲁⲥⲓⲥ ⲟⲩⲇⲉ ⲙⲛ̄ ⲁⲅⲅⲉⲗⲟⲥ ⲟⲩⲇⲉ ⲙⲛ̄ ⲡ̄ⲛ̄ⲁ̄** *se-jô mmo=s je mn anastasis ude mn angelos ude mn pn(ewm)a* 'They say that there is no resurrection, that there are no angels, that there is no spirit' (Acts 23:8). A single class ought to be represented by a single circle. Clearly, the denial of the class involves the denial of the *entire* class and not just *part* of the class. An indefinite article, which would denote *part* of a class, seems therefore out of place. Denying the existence of certain members of a class is perhaps theoretically possible, but it is difficult to imagine a context in which anyone would say such a thing. Consider the statement of non-existence "There are no angels." This statement involves the obliteration of the class "angel." By contrast, it is difficult to see what the logical purport might be of a statement such as "There is not *an* angel" or "A (certain) angel does not exist" without qualification. The statement implies that other angels do exist. What purpose would it serve to state that some members of the class do not exist, without further specification?

11.7.3 Second component of the rule

The second component pertains to **ⲟⲩⲛ̄** *wn* 'there is not' and consists of two complementary parts, which are in a sense opposite to the two parts of the first component of the rule (§11.7.1). On the one hand, the indefinite article (that is, singular **ⲟⲩ**- *u-* or plural **ϩⲉⲛ**- *hen-*) or an equivalent expression is normal after **ⲟⲩⲛ̄** *wn.* But the indefinite article does not occur after **ⲙ̄ⲛ** *mn.* On the other hand, with certain well-defined exceptions (§11.7.5), *absence* of an article or "zero"-article is not possible after **ⲟⲩⲛ̄** *wn.* But "zero" article is mandatory after **ⲙ̄ⲛ** *mn.* Again, the behavior of **ⲙ̄ⲛ** *mn* 'there is not' is in a sense opposite to that of **ⲟⲩⲛ̄** *wn* 'there is'.

> **ⲟⲩⲛ̄ ⲟⲩⲟⲩⲛⲟⲩ ⲛⲏⲩ** (John 4: 21)
> *wn u-unu nêw*
>
> 'An hour is coming'.

***ⲟⲩⲛ̄ ⲟⲩⲛⲟⲩ ⲛⲏⲩ** *wn unu nêw* 'hour is coming' is not a possible combination. It follows that in **ⲟⲩⲛ̄ ⲟⲩⲟⲉⲓϣ ⲅⲁⲣ ⲛⲁϣⲱⲡⲉ** *wn woyš gar na-šôpe* 'for a time will come' (2 Tim 4:3), **ⲟⲩⲟⲉⲓϣ** *woyš* must be an abbreviation for **ⲟⲩⲟⲩⲟⲉⲓϣ** *uwoyš*, and that in **ⲉⲩⲛ̄ ⲟⲩⲟϩ̄ⲥ ⲉϥⲧⲏⲙ ϩ̄ⲛ ⲛⲉϥϭⲓϫ** *e-wn wohs e=f-têm hn ne=f-cij* 'while there is a sharp sickle in his hands' (Rev 14:14), **ⲟⲩⲟϩ̄ⲥ** *wohs* must be an abbreviation for **ⲟⲩⲟⲩⲟϩ̄ⲥ** *u-wohs.* The writing **ⲟⲩ** *u* for **ⲟⲩⲟⲩ** *u-u* (or *uw*) is well documented after **ⲛ** *n.* By contrast, in **ⲙ̄ⲛ ⲟⲩϫⲁⲓ** *mn ujay* 'There is no salvation' (Acts 4:12), **ⲟⲩϫⲁⲓ** *ujay* cannot be a variant of **ⲟⲩⲟⲩϫⲁⲓ** *u-ujay*; nor in **ⲙ̄ⲙ̄ⲛ ⲟⲩϣⲏ ⲛⲁϣⲱⲡⲉ ⲙ̄ⲙⲁⲩ** *mmn uše na-šôpe mmaw* 'There will be no night there' (Rev 21:25) can **ⲟⲩϣⲏ** *uše* be a variant of **ⲟⲩⲟⲩϣⲏ** *u-ušê.* The reason is that the indefinite article does not occur after (**ⲙ̄**)**ⲙ̄ⲛ** *(m)mn* 'there is not'.

There are about 270 instances of **ⲟⲩⲛ̄** *wn* in the New Testament. At John 20:26, **ⲛⲉⲩⲛ̄**- *ne=w-n-* in **ⲛⲉⲩⲛ̄ϩⲟⲩⲛ ⲟⲛ ⲡⲉ** *ne=w-n-hun on pe* 'They were again inside' (Gk. πάλιν ἦσαν ἔσω) consists of three elements:

(1) the imperfect converter **ⲛⲉ** *ne*;

(2) the third person plural suffix pronoun ⲩ *w*; and
(3) the preposition ⲛ̄- *n*- as part of the adverbial phrase ⲛ̄-ϩⲟⲩⲛ *n-hun* 'inside'.

It is not composed, as Wilmet (1957–58: s.v. ⲟⲩⲛ̄) surmises, of *two* elements, to wit:

(1) the imperfect converter ⲛⲉ *ne*;
(2) (ⲟ)ⲩⲛ̄ *wn* 'there is',

which together would mean 'there was'. ⲛⲉⲩⲛ̄- *ne-wn*- 'there was' is not uncommon. In terms of superlineation, one expects ⲛⲉⲩⲛ̄ *ne=w-n*- for 'they were in …' (John 20:26) but ⲛⲉⲩⲛ̄ *ne-wn*- for 'there was'. I have not examined writings of the superlineation in both expressions in good manuscripts.

Indefinite articles are the rule after ⲟⲩⲛ̄ *wn*. Absence of an article is rare, but in some cases, absence is mandatory because other rules apply. Thus, numerals do not take the indefinite article in indefinite expressions, as in ⲟⲩⲛ̄ ⲥⲛⲁⲩ ⲛⲁϣⲱⲡⲉ *wn snaw na-šôpe* 'There will be two (in the field)' (Matt 24:40), ⲥⲛⲁⲩ *snaw* 'two' (Luke 17:34), ⲥ̄ⲛⲧⲉ *snte* 'two' (Luke 17:35), ϥⲧⲟⲟⲩ *ftow* 'four' (Acts 21:23), ϯⲟⲩ *tiu* 'five' (Luke 12:52; John 6:9), ⲥⲟ *so* 'six' (Mark 12:20; John 2:6), ⲥⲁϣ̄ϥ *sašf* 'seven' (Luke 20:29; Rev 1:16), and ⲙ̄ⲛ̄ⲧⲥⲛⲟⲟⲩ̄ⲥ *mntsnows* 'twelve' (Rev 21:12). ⲟⲩⲏⲣ *wêr* 'how many' in ⲟⲩⲛ̄ ⲟⲩⲏⲣ *wn wêr* 'How many are there …?' presumably also belongs here (see Matt 15:34; Luke 15:17, 16:7). The combination *ⲟⲩⲥⲛⲁⲩ *u-snaw* does not exist. Likewise, indefinite articles do not accompany singular ⲕⲉ *ke* 'other', as in ⲟⲩⲛ̄ ⲕⲉⲣ̄ⲣⲟ ϣⲟⲟⲡ *wn ke-rro šo'p* 'Another king exists' (Acts 17:7) and ⲟⲩⲛ̄ ϭⲉ ⲕⲱⲧ ⲉϫⲱⲥ *wn ce kôt ejôs* 'Another one is building on it' (1 Cor 3:12). The combination *ⲟⲩⲕⲉ *u-ke* does not exist. Another special case is the "substantivized" relative clause, in which the definite article has indefinite function, as in ⲟⲩⲛ̄ ⲡⲉⲧⲡⲓⲥⲧⲉⲩⲉ *wn p-et-pistewe* 'There exists one who believes' (Rom 14:2) and ⲟⲩⲛ̄ ⲡⲉⲧ̄ⲕⲣ̄ⲓⲛⲉ *wn p-etkkrine* 'There exists one who judges' (Rom 14:5*bis*). Like ⲗⲁⲁⲩ *laaw* "any," ϩⲁϩ *hah* 'many' appears after both ⲙ̄ⲛ *mn* and ⲟⲩⲛ̄ *wn* (for ⲟⲩⲛ̄ *wn*, see Matt 7:22, 8:11, 9:10, 19:30, 21:8; John 3:23, 10:20, 14:2; and often).

There are some instances, fewer than twenty, in which the absence of any article or "zero"-article cannot be accounted for by any other rule. (§§11.7.4, 11.7.5).

11.7.4 Explanation of the second component

The absence of the indefinite article after ⲙ̄ⲛ *mn* 'there is not' was explained by the fact that the logical structure of negating an existence does not involve *part* of a class (§11.7.2), and denoting part of a class is precisely the function of the indefinite article. Furthermore, it was noted in §11.7.3 that the behavior of ⲞⲨⲚ̄ *wn* is in a sense opposite to that of ⲘⲚ̄ *mn*. After ⲞⲨⲚ̄ *wn*, the indefinite article is the norm. After ⲘⲚ̄ *mn*, the indefinite article never appears. The absence of indefinite articles is explained by the logical structure of ⲘⲚ̄ *mn*, which involves absence of *part* of a class. Accordingly, one expects the logical structure of ⲞⲨⲚ̄ *wn* to involve presence of *part* of a class—as is the case.

Most examples of ⲞⲨⲚ̄ *wn* involve *two* classes. Consider ⲞⲨⲚ̄ ⲞⲨⲞⲨⲚⲞⲨ ⲚⲎⲨ *wn u-unu nêw* 'An hour is coming'. The two classes are "hour" (*X*) and "things that are coming" (*Y*). The example states that these two classes *intersect*. The intersection of *X* and *Y* may be represented by the following diagram.

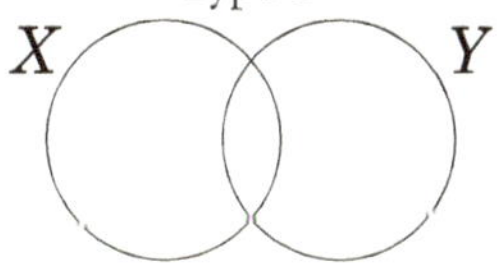

Affirmation of existence means that there are, or there exist, things that belong to both *X* and *Y*. The intersection of *X* and *Y* is an overlap area, and it holds everything that belongs to both the class "hour" (*X*) and the class of "things that are coming" (*Y*)—"an hour (*X*) that is coming (*Y*)." The function of the proposition "An hour is coming" is to declare that this overlap area exists: "(There is) an hour (that) is coming."

How does this overlap area involve *part* of a class? Or, what triggers the presence of the indefinite article with class *X*? Obviously, only *part* of *X* (the class "hour") is found in the overlap area of *X* and *Y*, namely "*an* hour." The indefinite article is therefore needed to denote this fact.

That is, the indefinite article owes its presence after ⲞⲨⲚ̄ *wn* to the specific relationship between the class *X* and the overlap area between *X* and *Y*: only an indefinite *part* of *X* is found in the overlap area between *X* and *Y*. The indefinite article therefore owes its presence *indirectly* to the fact that ⲞⲨⲚ̄ *wn* mostly denotes propositions of Type 3. By contrast, logical propositions of Type 2 exhibit no overlap area between two classes, so the indefinite article does not appear after ⲘⲚ̄ *mn*. The stark contrast between the normal presence of the indefinite article after ⲞⲨⲚ̄ *wn* and its total absence after ⲘⲚ̄ *mn* is therefore a consequence of the fact that ⲞⲨⲚ̄ *wn* and ⲘⲚ̄ *mn* belong to two different types of logical proposition, Types 3 and 2 respectively.

As noted in §11.7.3, there is a small number of instances in which the absence of an indefinite article after ⲞⲨⲚ̄ *wn* cannot be explained by a known rule that overrides the norm. In most of these examples, **ⲟⲩⲛ̄** *wn* plus article-less noun is accompanied by an adverbial phrase; what is striking is the absence of the pattern **ⲟⲩⲛ̄** *wn* plus noun plus infinitive or stative.

ⲛⲉⲧⲉ ⲟⲩⲛ̄ ⲙⲁⲥⲧⲓⲅⲝ ϩⲓⲱⲟⲩ (Mark 3:10)
n-ete wn mastigks hiô=w

'those on whom there are scourges'

ⲡⲉⲧⲉ ⲟⲩⲛ̄ ⲙⲁⲁϫⲉ ϭⲉ ⲙ̄ⲙⲟϥ (Mark 7:16)
p-ete wn ma'je ce mmo=f

'he who has ears, then'

ⲟⲩⲛ̄ ⲛⲟⲃⲉ ⲉⲣⲟⲟⲩ (Luke 13:4)
wn nobe ero=w

'There is sin (reckoned) against them'.

ⲟⲩⲛ̄ ⲥⲟⲩⲟ ϩⲛ̄ ⲕⲏⲙⲉ (Acts 7:12)
wn swo hn kême

'There is grain in Egypt'.

ⲟⲩⲛ̄ ϣⲁϫⲉ ⲛ̄ⲥⲟⲡⲥ ⲛ̄ⲧⲉⲧⲏⲩⲧⲛ̄ (Acts 13:15)
wn šaje n-sops nte-têwtn

'There is a word of exhortation with you'.

ⲟⲩⲛ̄ ϩⲱⲃ ⲉϥϩⲟⲟⲩ ϩⲙ̄ ⲡⲉⲓⲣⲱⲙⲉ (Acts 25:5)
wn hôb e=f-how hm pey-rôme

'There is an evil thing in this man'.

ⲙⲏ ⲟⲩⲛ̄ ϫⲓ ⲛ̄ϭⲟⲛⲥ ⲛ̄ⲛⲁϩⲣⲙ̄ ⲡⲛⲟⲩⲧⲉ (Romans 9:14)
me wn ji n-cons nnahrm p-nute

'Is there injustice in the presence of God?'

ⲟⲩⲛ̄ ⲕⲱϩ ϩⲓ ϯⲧⲱⲛ ⲛ̄ϩⲏⲧⲧⲏⲩⲧⲛ̄ (1 Corinthians 3:3)
wn kôh hi titôn n-hêt-têwtn

'There is jealousy and quarreling among you'.

ϩⲛ̄ ⲟⲩⲛⲟϭ ⲛ̄ⲏⲓ ... ⲟⲩⲛ̄ ϣⲉ ⲟⲛ ϩⲓ ⲃ̄ⲗ̄ϫⲉ (2 Timothy 2:20)
hn u-noc n-êy ... wn še on hi blje

'In a large house there is wood and potsherds'.

ⲡⲉⲧⲉⲩⲛ̄ ⲙⲁⲁϫⲉ ⲙ̄ⲙⲟϥ (Revelation 3:6)
p-ete-wn ma'je mmo=f

'he who has ears'

The small number of examples should be stressed. It is not clear why the indefinite article is absent in these examples while it is so typically present everywhere else. Logically speaking, it is an intersection of two classes that triggers the indefinite article, which refers to part of a class. Is it possible that the adverbial phrase is not considered a second class? The statement **ⲟⲩⲛ̄ ⲥⲟⲩⲟ ϩⲛ̄ ⲕⲏⲙⲉ** *wn swo hn kême* 'There is grain in Egypt' (Acts 7:12) would then not refer to an intersection of the classes "grain" and "things that are in Egypt." In the two examples with **ⲙⲁⲁϫⲉ** *ma'je* 'ear', singular **ⲟⲩⲙⲁⲁϫⲉ** *u-ma'je* 'an ear' seems out of the question because ears occur in pairs and the plural ***ϩⲉⲛⲙⲁⲁϫⲉ** *hen-ma'je* is probably not possible because it would seem to refer to two or more ears and not just to a pair.

Exceptions in which **ⲟⲩⲛ̄** is followed by no more than an article-less noun are discussed in the next section, leaving only four exceptions.

ⲛⲉⲩⲛ̄ ⲓⲟⲩⲇⲁⲓ ⲇⲉ ⲛ̄ⲛ̄ⲁⲣⲭⲓⲉⲣⲉⲩⲥ (Acts 19:14)
newn yuday de n-n-arkhierews …

'There was a Jew, then, of the high priests(?)'. (unclear)

ⲙⲏ ⲉⲓϫⲱ ⲙ̄ⲙⲟⲥ ϫⲉ ⲟⲩⲛ̄ ϣⲱⲱⲧ ⲛ̄ⲉⲓⲇⲱⲗⲟⲛ ϣⲟⲟⲡ
ⲏ ϫⲉ ⲟⲩⲛ̄ ⲉⲓⲇⲱⲗⲟⲛ ϣⲟⲟⲡ (1 Corinthians 10:19)
me e=y-jô mmo=s je wn šôôt n-eydôlon šo'p ê je wn eydôlon šo'p

'Am I saying that there is food sacrificed to idols
or that there are idols?'

ⲉϣϫⲉ ⲟⲩⲛ̄ ⲥⲱⲙⲁ ⲙ̄ⲫⲩⲭⲓⲕⲟⲛ … ⲟⲩⲛ̄ ⲥⲱⲙⲁ
ⲟⲛ ⲙ̄ⲡⲛⲉⲩⲙⲁⲧⲓⲕⲟⲛ (1 Corinthians 15:44)
ešje wn sôma m-psukhikon … wn sôma on m-pnewmatikon

'If there is a psychic body, there is also a spiritual body'.

ⲟⲩⲛ̄ ⲛⲟⲃⲉ ⲉϥϫⲓ ⲉϩⲟⲩⲛ ⲉⲡⲙⲟⲩ (1 John 5:16)
wn nobe e=f-ji ehun e-p-mu

'There exists sin pertaining to death'.

The first example is not entirely clear. The second is the only instance of the pattern **ⲟⲩⲛ̄** *wn* + article-less noun + infinitive or stative. The third is the only instance in which **ⲟⲩⲛ̄** *wn* is followed by article-less noun plus adjective.

The fourth is the only instance in which **ⲟⲩⲛ̄** *wn* is followed by an article-less noun followed by a circumstantial clause functioning as the equivalent of a relative clause.

11.7.5 Third component of the rule

According to the second component of the rule, the indefinite article is usual after **ⲟⲩⲛ̄** *wn* 'there is'. But there are exceptions (most in §11.7.4). In two of them—too few for a definitive conclusion—**ⲟⲩⲛ̄** is followed by no more than an article-less noun, or the following short pattern is not attested: **ⲟⲩⲛ̄** *wn* + **ⲟⲩ-** *u-* + noun. The absence of the indefinite article thus seems mandatory in this pattern.

ⲟⲩⲛ̄ ⲙⲁ (Luke 14:22)
wn ma

'There is room'.

ⲟⲩⲛ̄ ⲁⲛⲁⲥⲧⲁⲥⲓⲥ ⲁⲩⲱ ⲟⲩⲛ̄ ⲁⲅⲅⲉⲗⲟⲥ
ⲁⲩⲱ ⲟⲩⲛ̄ ⲡ̄ⲛ̄ⲁ̄ (Acts 23:8)
wn anastasis awô wn angelos awô wn pn(ewm)a

'There is resurrection and there are angels
and there are spirits'.

An exception is **ⲟⲩⲛ̄ ϩⲉⲛⲁⲛⲑⲩⲡⲁⲧⲟⲥ** *wn hen-anthupatos* 'there are proconsuls' (Acts 19:38). Perhaps **ϩⲉⲛ-** *hen-* is needed to denote the plural number explicitly.

11.7.6 Explanation of the third component

It was mentioned in §11.7.4 that the presence of the indefinite article after **ⲟⲩⲛ̄** is triggered by an overlap of two classes. However, when just a noun follows **ⲟⲩⲛ̄** *wn* as in **ⲟⲩⲛ̄ ⲁⲅⲅⲉⲗⲟⲥ** *wn angelos* 'there are angels' (Acts 23:8) and **ⲟⲩⲛ̄ ⲙⲁ** *wn ma* 'There is room' (Luke 14:22), there is only one class. There is therefore no overlap of two classes. The conditions triggering the presence of an indefinite article **ⲟⲩ-** *u-* are therefore absent. That may explain the necessary absence of an indefinite article after **ⲟⲩⲛ̄** *wn* if only a noun follows. **ⲟⲩⲛ̄ ⲁⲅⲅⲉⲗⲟⲥ** *wn angelos* and **ⲟⲩⲛ̄ ⲙⲁ** *wn ma* declare the existence of the two classes "angel" and "place" without any restrictions to some indefinite part of the classes. Likewise, **ⲙ̄ⲛ̄ ⲁⲅⲅⲉⲗⲟⲥ** *mn angelos* 'there are no angels' (Acts 23:8) (§11.7.2) and ***ⲙ̄ⲛ̄ ⲙⲁ** *mn ma* 'there is no place' declare the non-existence of the same two classes without any restrictions. The indefinite article is absent, as after any other **ⲙ̄ⲛ̄**.

11.8 TWO HISTORICAL TRENDS EVIDENCING GROWTH IN LOGICAL COMPLEXITY

11.8.0 The search for hard evidence

The present search is for hard evidence from Egyptian for an increase in logical complexity or sophistication over time. Two trends may be observed: the disappearance of a pattern belonging to the logical proposition of Type 1, and both an expansion of Type 2 and the rise of the logical proposition of Type 3. These amount to diminution of the less complex Type 1, expansion of the more complex Type 2, and a substantial increase of the even more complex Type 3. The shrinking of a less sophisticated type and the expansions of more sophisticated types may be interpreted as an overall increase in sophistication.

11.8.1 The first trend: Disappearance of Type 1

Type 1 (the balanced sentence) becomes obsolete after Middle Egyptian. Vernus (1981: 77) has called its structure "archaic" even in Old and Middle Egyptian. Possible later examples have on occasion been pointed out, but I believe they can be explained differently—they were simply copied from earlier texts. The fact that the simple Type 1 disappears may be interpreted as an increase in logical sophistication.

11.8.2 The second trend: Expansion of Type 2 and rise of Type 3

11.8.2.1 **ⲟⲩⲛ̄** wn *'there is' and* **ⲙⲛ̄** mn *'there is not' in the present tense system*

One basic rule of Coptic grammar involves the following contrasts. One says **ⲡⲣⲱⲙⲉ ⲥⲱⲧⲡ̄** *p-rôme sôtp* 'the man is choosing' but **ⲟⲩⲛ̄ ⲟⲩⲣⲱⲙⲉ ⲥⲱⲧⲡ̄** *wn u-rôme sôtp* 'a man is choosing' or 'there's a man choosing'. Likewise, in the negative counterpart, one says **ⲙ̄ⲡⲣⲱⲙⲉ ⲥⲱⲧⲡ̄ ⲁⲛ** *m-p-rôme sôtp an* 'the man is not choosing' but **ⲙⲛ̄ ⲣⲱⲙⲉ ⲥⲱⲧⲡ̄** *mn rôme sôtp* 'there is no man choosing'. Coptic grammars dutifully state that an indefinite entity serving as basic component in the present tense system needs to be preceded by existential **ⲟⲩⲛ̄** *wn* 'there is' or existential **ⲙⲛ̄** *mn* 'there is not' (see Layton 2004: 248 §322, 249–50 §324, 385–86 §479; Polotsky 1960: 401 §§19–21, 410–11 §35 [1971: 247, 256–57]; for Coptic **ⲟⲩⲛ̄** *wn* and **ⲙⲛ̄** *mn*, see also Elanskaya 1979 and Nagel 1980, cited by Quecke 1983: 298–302]). For the presence of the indefinite article after **ⲟⲩⲛ̄** *wn* and its absence after **ⲙⲛ̄** *mn*, see §11.7.

ⲟⲩⲛ̄ ⲟⲩⲣⲱⲙⲉ ⲥⲱⲧⲡ̄ *wn u-rôme sôtp* 'there is a man choosing' and **ⲙⲛ̄ ⲣⲱⲙⲉ ⲥⲱⲧⲡ̄** *mn rôme sôtp* 'no man is choosing' look like counterparts, but there is a fundamental difference in logic between them. The latter belongs to the logical proposition of Type 2. **ⲟⲩⲛ̄ ⲟⲩⲣⲱⲙⲉ ⲥⲱⲧⲡ̄** *wn u-rôme sôtp* 'there is a man choosing' belongs to the logical proposition of Type 3.

11.8.2.2 **ⲙⲛ̄ ⲣⲱⲙⲉ ⲥⲱⲧⲡ̄** mn rôme sôtp *'no man is choosing' as a logical proposition of Type 2*

ⲙⲛ̄ ⲣⲱⲙⲉ ⲥⲱⲧⲡ̄ *mn rôme sôtp* 'no man is choosing' belongs to Type 2. There are two classes, "men" and "chooser(s)," and four combination classes, (1) men who are choosers, (2) men who are not choosers, (3) choosers that are not men, and (4) anything that is neither men nor choosers. The proposition negates class (1). "No man is choosing" is therefore the logical equivalent of "All men are non-choosers."

The same Type 2 relation exists on the level of secondary propositions (relations between events, §§4.1, 11.4). Cicero (*Fat.* 8.15) criticized a report by Chrysippus that the Mesopotamians, instead of stating "When someone is born at the rising of Sirius, he will not die in the sea," preferred to say "No one is both born at the rising of Sirius and dies in the sea" (cf. Depuydt 1999a: 106, p. 276 below). It is a fact that the two statements are logically equivalent. Two events are involved: being born at the rising and dying at sea. The result is four combined events: (1) being born at the rising and dying at sea; (2) being born at the rising and not dying at sea; (3) not being born at the rising and dying at sea; and (4) neither being born at the rising nor dying at sea. The statement above denies the existence of one and only one of the four combinations, (1), being born at the rising and dying at sea. The logical purport of the statement is to say that there is no such thing.

Cicero ridicules Chrysippus's preference for the formulation of the Mesopotamians. There is indeed, strictly speaking, no logical difference with the alternative formulation. However, as Venn stresses throughout his *Symbolic Logic* (1894), the negation of exactly one of four possible combination categories is the most definite and categorical part of logical propositions of Type 2, the one that is the most conducive to firm reasoning.

11.8.2.3 **ⲟⲩⲛ̄ ⲟⲩⲣⲱⲙⲉ ⲥⲱⲧⲡ̄** wn u=rôme sôtp *'there is a man choosing' as a logical proposition of Type 3*

ⲟⲩⲛ̄ ⲟⲩⲣⲱⲙⲉ ⲥⲱⲧⲡ̄ *wn u-rôme sôtp* 'there is a man choosing' also belongs to Type 3. There are again two classes, "men" and "chooser(s)," as well as the four combination classes mentioned above. The statement does not deny the existence of any of these four classes. Rather, it *affirms* the existence of one of the four categories: there is, or there exists, such a thing as men who are also choosers.

11.8.2.4 The birth and rise of the indefinite and definite articles as a prerequisite for the coming into existence of the patterns **ⲟⲩⲛ̄ ⲟⲩⲣⲱⲙⲉ ⲥⲱⲧⲡ̄** wn u rôme sôtp *'there is a man choosing' and* **ⲙⲛ̄ ⲣⲱⲙⲉ ⲥⲱⲧⲡ̄** mn rôme sôtp *'no man is choosing'*

Neither the affirmative pattern **ⲟⲩⲛ̄ ⲟⲩⲣⲱⲙⲉ ⲥⲱⲧⲡ̄** *wn u-rôme sôtp* 'there is a man choosing' nor the negative pattern **ⲙⲛ̄ ⲣⲱⲙⲉ ⲥⲱⲧⲡ̄** *mn rôme sôtp* 'there is no man choosing' existed in earlier Egyptian. Winand (1989; see appendix 3.2 below) describes the rise and spread of these patterns in later Egyptian.

The question arises: What caused these two patterns to come into existence? They have to do with the distinction between definiteness and indefiniteness, and so could not have come into existence before the emergence of the article. Old and Middle Egyptian lack definite and indefinite articles, but they become common from Late Egyptian onward. They articulate the logical distinction between a class and *part* of a class. As a result, the logical distinction between Types 2 and 3 also becomes more articulated.

The rise of the definite and indefinite articles was very gradual, a development rooted in logic. "With Late Egyptian," writes Winand (1989: 168), "the definite/indefinite opposition achieves formal expression. This grammaticalization of a logical category set in motion a readjustment of the entire linguistic structure."* This development may be interpreted as an increase in logical complexity. In this evolutionary process, a distinction is necessary between the definite and indefinite articles. Already in Late Egyptian, the definite article (*pꜣ*, *tꜣ*, *nꜣ*) seems needed when a noun needs to be rendered definite. By contrast, the indefinite article (singular *wꜥ*) appears to have spread much more slowly. Only by the time of Coptic is the indefinite article generally required to render a noun indefinite (on the use of the articles in Demotic, see Johnson 1987). Any exceptions in which an article remains absent are well defined and well delineated.

The appearance of the article greatly articulates the opposition between definite and indefinite nouns. This opposition in turn gave shape to another opposition: the one between the association of the *presence* of *wn* or *mn* with *indefinite* nouns, on the one hand, and the association of *absence* of *wn* or *mn* with *definite* nouns, on the other hand.

As regards *wn/mn* 'there is (not)' plus indefinite noun, another distinction is necessary, pertaining to what follows the construction. Either noth-

* Avec le néo-égyptien, l'opposition défini/non défini reçut une expression formelle. Cette grammaticalisation d'une catégorie logique entraîna un réajustement de toute la structure linguistique."

ing, or an adverb, or an adverbial phrase could follow *wn/mn* already in Old and Middle Egyptian. On the other hand, a following infinitive or a stative is a later innovation.

11.8.2.5 Earliest examples of the patterns **ⲟⲩⲛ̄ ⲟⲩⲣⲱⲙⲉ ⲥⲱⲧⲡ̄** wn u-rôme sôtp *'there is a man choosing' and* **ⲙ̄ⲛ ⲣⲱⲙⲉ ⲥⲱⲧⲡ̄** mn rôme sôtp *'no man is choosing'*

11.8.2.5a Dynasty 21

The new constructions **ⲟⲩⲛ̄ ⲟⲩⲣⲱⲙⲉ ⲥⲱⲧⲡ̄** *wn u-rôme sôtp* (*wn wꜥ rmṯ stp*) and **ⲙ̄ⲛ ⲣⲱⲙⲉ ⲥⲱⲧⲡ̄** *mn rôme sôtp* (*mn rmṯ stp*), with an infinitive or stative as third component, first occur with some frequency in Demotic, which is attested from the seventh century B.C.E. onward (for Demotic, see Johnson 1976: 37, 42–45, 85–86 [cf. Quecke 1979: 438]; Johnson 1991: 37 §49; Winand 1989: 165 and passim; and Simpson 1996: 157–59). In a review of the first (1986) edition of Johnson's elementary grammar of Demotic, Quecke (1988: 220–21) made the following central question explicit.

> Why the adverbial sentence requires *wn/mn* before an indefinite subject is another question.*

Earlier examples are scarce. Winand (1989: 166) has located a possible example in Papyrus El-Hibeh 2,6–7 (Spiegelberg 1917: 8), dating to the Third Intermediate Period, Dynasty 21 (ca. 1000 B.C.E.):

hꜣb n.j jn wn ḥtr jw [*n*].*ṯn* {*jw n.ṯn*}

'Write to me as to whether a span (of horses) has come to you'.

The proposed emendations to the text make this example less than perfect for grammatical purposes, although, as Winand (170) points out, the existential auxiliary *wn* 'there is' appears in the same text before an indefinite noun in both the negative perfect and the so-called third future:

* Warum der Adverbialsatz vor unbestimmtem nominalem Subjekt *wn/mn* verlangt, ist eine andere Frage.

bwpw wn ḥtr jj n.n jmy ḥr.k n.n
ḫy-rꜥ jrj wn ḥtr jj mtw.n hꜢb n.k (lines 7–9)

'No span has come to us. Pay attention to us. A span will come and we will write to you (about it)'.

The example of *wn* in lines 6–7 does remain exceptional in being so early. Quack (2001: 128 n. 91) proposes instead that *Wn ḥtr* is a personal name in line 6. This interpretation seems much more probable, especially because it also eliminates the syntactic problems that arise if *wn* is twice interpreted as "there is" in lines 7–9.

11.8.2.5b The New Kingdom (Late Egyptian)

What about the appearance even earlier in the second millennium B.C.E., in the New Kingdom? There seem to be no Late Egyptian examples. "In contradistinction to Co[ptic]," writes Frandsen (1974: 171 §95), "*wn* + first present with an indefinite actor expression and with an infinitive or stative filling the second position does not seem to occur." None appear in Erman's Late Egyptian grammar (1933: 246–47 §§506–7). One possible example is reported by Winand (1989: 170 n. 2):

wn nkt ḫpr r.j (Tale of Two Brothers 8,4 [*LES* 17,13])

'Something has happened to me'.

If this in fact is the pattern *wn* + indefinite noun + infinitive/stative, then *ḫpr* is the stative. This seems to be the sole Late Egyptian example, and extra scrutiny is warranted: an alternative explanation seems equally plausible. *Ḫpr* might well be a past active participle meaning 'that has happened', yielding the literal translation 'There is (*wn*) a thing (*nkt*) that has happened (*ḫpr*) to me (*r.j*)'. According to Winand's detailed description of the behavior of the Late Egyptian active participle (1992: 343–64 §§540–76), there are "many" examples (347 §546) in which the past active participle has no explicit written expression. An instance in the same text is *nꜢ šm* 'those who have gone' (Two Brothers 11,9 [*LES* 21,11], but cf. *nꜢ rmṯ i.šm* at 11,8 [21,10]).

Moreover, if *ḫpr* were a stative referring to feminine *nkt* 'thing', one would expect the form *ḫprt* or *ḫprtj* (cf. Winand 1992: 114–17 §§209–12). Winand has located only seventeen instances dating to Dynasties 19 or 20 in which a stative without an ending seems to be third person feminine singular, and he eliminates some of these. Furthermore, three of them probably represent an entirely different phenomenon. The first example is also found in the Tale of the Two Brothers. This other phenomenon explains the masculine singular form of the stative (see §11.8.2.5c).

To establish a grammatical category, unambiguous examples are needed, so this example does not suffice to establish the existence of the pattern in question for Late Egyptian. *H̱pr* might well be an active participle, not a stative; if it is a stative, it should exhibit a feminine ending as in *ḫprt* or *ḫprtj* (the feminine active participle usually does not exhibit the ending *t*).

For completeness' sake, a related study of expressions denoting existence in Late Egyptian, Vernus 1985, deserves mention here.

11.8.2.5c Excursus

In the following three examples, a form of the stative without a feminine ending accords with the nouns *ḏrt* 'hand' and *jrt* 'eye', which are feminine. The result is a dissonance between the masculine gender of the stative and the feminine gender of the nouns.

jw ḏrt.f wꜣḥ ḥr ḏꜣḏꜣ.f (Tale of Two Brothers 8,7 [*LES* 18,7])

'while his hands are placed on his head'

jw jrt.k wbḫ rꜥ nb (*LEM* 24,15)

'while your eyes are clear every day'

jjꜣ jrt.j kꜣmn (Černý 1939, 30,11)

'Truly, my eyes are blind'.

It is not impossible that the three statives *wꜣḥ*, *wbḫ*, and *kꜣmn* are early genderless forms. By the time of Coptic, the stative form had become entirely genderless. But there is another way to account for the seeming dissonance between feminine nouns and masculine statives. In referring to both members of a pair of body parts, feminine *ḏrt* '(the two) hands' and *jrt* '(the two) eyes' are treated as masculine (see Depuydt 2002a: 813–14; 2002/3; and §A4.2 below).

Thus *ḏrt.f*, *jrt.k*, and *jrt.j* in these examples probably mean 'his hands', 'your eyes', and 'my eyes', denoting both members of a pair of body parts, despite usually being rendered in the singular.

11.8.2.6 The structure of the patterns **ⲟⲩⲛ̄ ⲟⲩⲣⲱⲙⲉ ⲥⲱⲧⲡ̄** wn u-rôme sôtp *'there is a man choosing' and* **ⲙⲛ̄ ⲣⲱⲙⲉ ⲥⲱⲧⲡ̄** mn rôme sôtp *'no man is choosing'*

Certain questions regarding the birth and expansion of the patterns at hand remain unanswered. It is not known, for example, whether the affirmative counterpart evolved at a different pace than the negative. A comprehensive database of examples would be needed to answer this question. It is otherwise clear that there is a difference between the two, such as the absence of an indefinite article after ⲙⲛ̄ *mn* and the normal presence of a definite article after ⲟⲩⲛ̄ *wn*. This difference in behavior in terms of the articles can be explained on the basis of their logical structure (§11.7).

But this leaves the grammatical structure of ⲟⲩⲛ̄ ⲟⲩⲣⲱⲙⲉ ⲥⲱⲧⲡ̄ *wn u-rôme sôtp* and ⲙⲛ̄ ⲣⲱⲙⲉ ⲥⲱⲧⲡ̄ *mn rôme sôtp* without an explanation. At first sight, the patterns look like hybrids of two patterns that existed in earlier Egyptian—in ⲙⲛ̄ ⲣⲱⲙⲉ ⲥⲱⲧⲡ̄ *mn rôme sôtp* it is easy to recognize both Middle Egyptian *nn rmṯ* 'there is no one' and (*jw*) *rmṯ ḥr stp* 'a/the man is choosing'. The preposition *ḥr* 'on' vanished in Late Egyptian. It is not clear whether the disappearance of *ḥr* was a prerequisite for ⲙⲛ̄ ⲣⲱⲙⲉ ⲥⲱⲧⲡ̄ *mn rôme sôtp* to come into existence. Furthermore, it is possible that intermediate constructions need to be postulated, such as *mn rmṯ jw.f* (*ḥr*) *stp* 'there is no one who is (*jw.f*) choosing' (ⲙⲛ̄ ⲣⲱⲙⲉ ⲉϥⲥⲱⲧⲡ̄ *mn rôme e=f-sôtp*). The pattern existed, but is it the intermediate step? Accordingly, ⲙⲛ̄ ⲣⲱⲙⲉ ⲥⲱⲧⲡ̄ *mn rôme sôtp* would then be ⲙⲛ̄ ⲣⲱⲙⲉ ⲉϥⲥⲱⲧⲡ̄ *mn rôme e=f-sôtp* minus ⲉϥ- *e=f-*. However, the existence of intermediate steps would be difficult to prove.

An interesting and important point is that the two patterns in question are completely new in Egyptian, presumably introduced some time in the Demotic phase of the language. Languages change, but the default assumption seems to be that the same means of communication remain overall available. In this case, however, the new patterns allow the expression of a nuance that was not possible before. Moreover, it is difficult to see to what other than logical structure the subtle nuance in question pertains. It is therefore also reasonable to assume that the increase in logical sophistication was the driving force behind the linguistic change. Hybrids were forced into existence, as it were, patched together from constructions that otherwise existed, in order to allow the expression of a higher degree of logical complexity.

11.8.3 Expression of existence and the contingent tenses

It is clear from the preceding sections that expressions of existence such as *wn* 'there is' and *mn* 'there is not' gradually expanded their domain in the Egyptian language as time progressed. The following development can perhaps also be interpreted as part of this same expansion.

The element ϣⲁ *ša* in ϣⲁϥⲥⲱⲧⲙ̄ *ša=f-sôtm* derives from the element *ḫr* in Old and Middle Egyptian *sḏm.ḫr.f*. There is no doubt as to the similarity in function of the two verb forms. Both appear prominently in the apodoses of conditional clauses. In Old and Middle Egyptian, the conditional clause is of the type *jr sḏm.f.* In Coptic, the conditional clause is of the type **ⲉϥϣⲁⲛⲥⲱⲧⲙ̄** *e=f-šan-sôtm*. But there is also a marked difference. When *sḏm.ḫr.f* does not follow a conditional clause, it is still always dependent on a specific condition that can be derived from the context. The form *sḏm.ḫr.f* (and the related *ḫr.f sḏm.f*) therefore means something like '(whenever this is so,) then he hears'. The same cannot be said about **ϣⲁϥⲥⲱⲧⲙ̄** *ša=f-sôtm.* This form 'he (generally) hears' denotes the general present or so-called simple present. In Old and Middle Egyptian, the same general present is denoted by the compound form *jw.f sḏm.f* 'he (generally) hears', which has become obsolete in Late Egyptian, replaced by *ḫr sḏm.f.* The morpheme *ḫr* is now used where Old and Middle Egyptian used both *sḏm.ḫr.f* and *jw.f sḏm.f.* This development sufficiently shows that *ḫr* lost its nuance of specific contingency (Depuydt 1993a: 208–33).

On the one hand, Coptic **ϣⲁϥⲥⲱⲧⲙ̄** *ša=f-sôtm* has lost the characteristic dependency on conditions exhibited by Old and Middle Egyptian *sḏm.ḫr.f.* On the other hand, its frequent appearance following **ⲉϥϣⲁⲛⲥⲱⲧⲙ̄** *e=f-šan-sôtm* clearly indicates that such dependency remains valid. How can it be explained? It is presumably of the following kind. **ϣⲁϥⲥⲱⲧⲙ̄** *ša=f-sôtm* 'he (generally) hears' implies the existence of certain unspecified conditions. Historically, a weakening of meaning has therefore occurred. Dependency on specific, known conditions has evolved into dependence on unspecified conditions. This dependency on unspecified conditions can be further described as follows.

In §0.2 on p. 2, the sentence "When it rains, I stay inside" is represented in the following diagram.

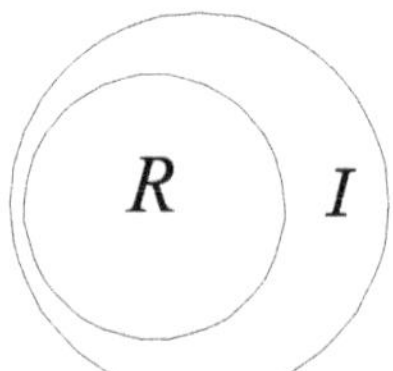

The circle *R* represents the occasions when it rains. The circle *I* represents the occasions when I stay inside. The sentence "When it rains, I stay inside" states that there are no occasions when it rains and I am not inside. The circle *R* is entirely encompassed by the circle *I*.

With reference to the same diagram, let us assume that **ϣⲁϥⲥⲱⲧⲙ̄** *ša=f-sôtm* is represented by the larger circle *I*. **ϣⲁϥⲥⲱⲧⲙ̄** *ša=f-sôtm* indicates that there exists another circle *R*, but in this case, the content of circle *R* is undefined or unknown. By contrast, in the case of Old and Middle Egyptian *sḏm.ḫr.f*, the contents of circle *R* is a specific set of occasions that serves as the condition of *I*.

As regards **ϣⲁϥⲥⲱⲧⲙ̄** *ša=f-sôtm* 'he (generally) hears', the existence of the inner circle *R* guarantees the existence of the larger circle *I* of which *R* is part. That is, because circle *R* has a certain expansion, *I* must have expansion. Consequently, **ϣⲁϥⲥⲱⲧⲙ̄** *ša=f-sôtm* means something like 'He generally hears (when the occasion arises)'. After all, few events are so absolutely universal that there is not some unspecified, unknown condition on which they depend. By contrast, Old and Middle Egyptian *jw.f sḏm.f* presumably exhibits no such nuance of unspecified contingency.

The development just outlined makes it possible to account for a combination of two facts, or how it is that a verb form can exhibit the following two characteristics simultaneously: (1) **ϣⲁϥⲥⲱⲧⲙ̄** *ša=f-sôtm* is an aorist or general present; (2) **ϣⲁϥⲥⲱⲧⲙ̄** *ša=f-sôtm* is typical in the apodosis. The same development can also be interpreted as an expansion of the category of existence.

Like **ϣⲁϥⲥⲱⲧⲙ̄** *ša=f-sôtm*, the French *conditionnel* implies an unspecified condition, but the tense is different—a future tense. Thus, for example, *j'irais* 'I would go' indicates that there exists another event out there that, when it takes place, serves as a condition of my going. It is not necessary to specify the condition, but it can be specified: what matters is the *existence* of the condition

12 Concluding Remarks: Some Possible Ghost Phenomena in Egyptian Grammar

> The voices that once complained that I make Egyptology more difficult with my grammar have long been silenced.*
>
> Adolf Erman, *Mein Werden und mein Wirken*, 264.

12.0 Old and Middle Egyptian Grammar in Flux

The study of Old and Middle Egyptian grammar is a field still very much in flux. Some of its accomplishments are of fairly recent date, such as the establishment, since the 1950s and 1960s, of the master list of generic verb forms. No new ones are likely to be discovered, but it is only for a few decades that we have known them all. Furthermore, there is an important distinction between the master list of generic verb forms found in grammars and the countless specific verb forms found in texts. A majority of the forms in Old and Middle Egyptian texts will never be identified with certainty.

The provisional state of our knowledge of Old and Middle Egyptian brings with it differences of opinion on many points of grammar. In the preceding chapters, positions have been taken on certain issues. Some concern denying the existence of phenomena that have been postulated in other work. The purpose of this chapter is to bring together some of these phenomena, though perhaps, additional evidence will emerge to demonstrate that these phenomena do in fact exist; in that sense, the following presentation is provisional.

12.1 The Term "Prospective"

Prospective verb forms denote future events, but the special term "prospective" suggests to Egyptologists and non-Egyptologists alike that the forms in question denote some special nuance of representing the future. If so, it

* Die Stimmen, die einst darüber klagten, daß ich die Ägyptologie durch meine Grammatik erschwere, sind längst verstummt.

would be nearly impossible to verify its existence and define its precise nature. Thus the term mystifies more than it clarifies.

It entered Egyptian grammar in 1924, in Gunn's foundational *Studies in Egyptian Syntax*, where he presents his discovery of a relative *sḏm.f* form that denotes future events. Gunn originally wanted to call the form simply "future," but (1924: 1 n. 2)

> The term "prospective" was suggested to me by Dr. Gardiner after I had communicated to him my discovery of the new form [and] I have adopted it in preference to "future."

Polotsky (1964: 270–71 [1971: 55–56], with nn. 2–3) describes in some detail the rise of the term in general linguistics in the early twentieth century.

Polotsky also notes that in no field of linguistics has the term "prospective" been as successful as in Egyptology, even if it was originally applied in certain Indo-European languages. He seemed favorably disposed toward it. Yet the circumstances in which the term "prospective" entered Egyptian grammar seem rather casual. In an awkward twist, Gardiner later discredited the very existence of the form for which he had proposed the name "prospective." As a result of Gardiner's denial of the form that he had named, the prospective *sḏm.f* led a kind of shadow existence in Egyptian grammar for several decades (see Depuydt 1993b). Only in the 1980s and 1990s were distinct future *sḏm.f* forms completely rehabilitated. The term "prospective" has been retained, but certain variants of the future *sḏm.f* came to be called the "subjunctive," a term suggested more than a century ago by Erman (1884) for the *sḏm.f* form following *rd*(*y*) 'cause'.

The preference for the term "prospective" over "future" seems to be without basis, and it seems to me that if some grammarians began using "future" instead of "prospective," the confusion would be minimal. The term "prospective" creates a mental bias that the form in question expresses a certain nuance of the future that has not even been shown to exist.

12.2 THE NON-SUBSTANTIVAL *sḏmw.f*

It is still widely accepted that the *sḏmw.f*, that is, the future *sḏm.f* that sometimes exhibits a characteristic *w*, is not always substantival. I am convinced, however, that this is not so. The matter has been discussed at length in §8.0.2 above. It is indicative of the state of Old and Middle Egyptian grammar that two observers contemplating the same facts can still reach such radically different solutions on fundamental points. I am aware that I am expressing the above conviction in opposition to other views. But at the very least, this conviction is stated here in order to prompt and encourage further investigation.

Has something been overlooked when such divergent analyses are extracted by different people from the very same facts?

12.3 THE NON-SUBSTANTIVAL FUTURE *sḏm.f* IN CLEFT SENTENCES

It is generally accepted that the cleft sentence of the type *jn* + Noun + *sḏm.f* 'it is N who will hear', Gardiner's "participial statement," is one of the characteristic contexts in which the *sḏmw.f* appears. I believe that the *sḏm.f* used in that construction is always the same *sḏm.f* form, whether it exhibits *w* or not. Furthermore, I believe that the *sḏm.f* in question is always substantival (§8.0.2.3). The non-substantival *sḏm.f* in cleft sentences thus does not exist. It comes as no surprise that corresponding Late Egyptian constructions have *i͗.jr.f sḏm*. What could be a clearer indication that the verb form in question is substantival? Yet hardly anyone interprets Late Egyptian *i͗.jr.f sḏm* as substantival when it appears in cleft sentences, even if it is everywhere else. And the same reluctance seems to apply to the use of the Old and Middle Egyptian *sḏm.f* in cleft sentences.

12.4 APODOSIS OR *sḏm.ḫr.f* AS EXPRESSIONS OF RESULT OR CONSEQUENCE

12.4.1 Causality

In the sentence "When it rains, I stay inside," the main clause or apodosis "I stay inside" does not denote a result or a consequence. "Result" and "consequence" are not adequate terms when used in this way. Nor does the sentence in question states that the rain causes me to stay inside. Rather, all that the sentence states is that there is no such thing as an occasion when it rains and I do not stay inside (§0.2 above). The proof is that the logically equivalent "When I do not stay inside, it does not rain" does not denote causality. In the same way, but on the first level of thought, the sentence "Swallows are birds" conveys that there is no such thing as swallows that are not birds. The *sḏm.ḫr.f* typically appears in the apodosis (where it may be called a contingent verb form). In that capacity, it does not express a result or a consequence, as is often stated in work on Old and Middle Egyptian grammar.

Without having studied the matter in great depth, I so far do not see any reason to put great substance into the much discussed concept of causality. I consider the concept irrelevant to the analysis of Egyptian verb forms, even if others have given it a prominent role, especially when using terms such as "consequence" and "result." In general, the concept of causality means little if anything. Yet, it seems common to imagine something by the concept of causality. We commonly think of certain events as causing certain other events. I believe such thought to be nothing more than an imagined metaphor. The origin of this metaphor seems to be the repeated observation that entities do things to other entities. For example, one might observe a man washing a car or a boy kicking a ball. The relation between the man and the car and between the boy and the ball is that between an actor and its direct object. Considering the extreme frequency with which such relations between entities are observed, it becomes tempting to think of relations between two events as relations between an actor and its direct object. Causality is when one thinks of one event as doing to another event somewhat similar to what the boy does to the ball. Philosophers have often stressed that is not possible to establish positively that such a relation ever applies between events. But because we so often see entities do things to other entities, it is extremely tempting to keep looking for such relations between events everywhere. It is only natural to see such relations where there is no reason to assume their presence. In this connection, Boole himself (1854: 361, note) recounts the following anecdote.

> A little friend of the author's, on being put to bed, was heard to ask his brother the pertinent question,—"Why does going to sleep at night make it light in the morning?" The brother, who was a year older, was able to reply, that it would be light in the morning even if little boys did not go to sleep at night.

Three minor points regarding three related issues are worth revisitng here.

12.4.2 *Sḏm.ḫr.f* is not modal

I have argued in the past (1988c, 1992b) that *sḏm.ḫr.f* does not convey an exhortation or the like. It is not modal in this sense, as English "he should hear" is. Yet, it is quite common to find statements to this effect about *sḏm.ḫr.f* in grammars. In favor of the nonmodal character of *sḏm.ḫr.f*, *m33* 'see' can be used in the *sḏm.ḫr.f*, as in Papyrus Smith 7,15 and Papyrus Ebers 93,17–18. It is difficult for *m33.ḫr.f* to be an exhortation, as in '(then) he should see'. The reason is that one cannot order or exhort someone to see—we see whether we want to or not. Wordsworth expressed the meaning of "see" much more elegantly:

> The eye—it cannot choose but see;
> We cannot bid the ear be still;
> Our bodies feel, where'er they be,
> Against or with our will.
>
> William Wordsworth, *Expostulation and Reply* (1798)

I have stressed the non-modal character of *sḏm.ḫr.f* (Depuydt 1985a: 52):

> Because transitive "see" precludes an act of the will, it can hardly constitute the substance of an exhortation … (hence the erroneous translation 'you should regard').*

12.4.3 Middle Egyptian *mꜣꜣ* means 'see', not 'look'

In relation to 'see' as the meaning of the verb *mꜣꜣ*, I have proposed that the divine ferryman called *Mꜣ-ḥꜣ.f* has a face at the back of his head and is therefore presumably a Janus-figure (Depuydt 1992b). *Mꜣ-ḥꜣ.f* can hardly mean 'who looks backward (by turning his head)', as has usually been assumed. "Looking" is an act of the will. By contrast, *mꜣꜣ* means 'see', which does not involve volition (§12.4.2). *Mꜣ-ḥꜣ.f* therefore means something like 'who sees at his back' or 'whose backside sees'. The identification of the form *mꜣ* is not clear to me. In this connection, two additional pieces of evidence are as follows. First, Pyr. §497c W mentions a divine being *Mꜣꜣw m ḥrwy* 'who sees with two faces'. Second, a name that is almost identical,  *Mꜣ m ḥrwy.f* 'who sees with his two faces', occurs in a marginal inscription on the unpublished coffin of *Ppjj-jmꜣ*, found in tomb N 4003 at Nag ed-Dêr (see Brovarski 1989: fig. 63; cf. Hussein 2004). The Pyramid text passage occurs close to the mention of a divine entity called *Ḥr.f-m-ḫnt.f-ḥr.f-m-ḥꜣ.f* 'Whose face is at his front and whose face is at his back' (Pyr. §493b WPMN). The context concerns ferrying gods in barks. It is generally assumed that this is the same divine being as the one called *Mꜣ-ḥꜣ.f* and *Ḥr.f-ḥꜣ.f* elsewhere. Presumably the divine ferryman's two faces allow him to look at the same time before and behind himself while steering his barge. This equation seems probable. But even if the equation does not apply, the expression *Mꜣꜣw m ḥrwy* leaves little or no doubt that there is such a thing as a Janus-figure in the Pyramid texts.

* Da das transitive "sehen" einen Willensakt ausschließt, kann es schwerlich Gegenstand einer Aufforderung darstellen … (daher die verfehlte Übersetzung "du sollst *betrachten*").

12.4.4 *M33* as 'look' in earliest Egyptian?

As regards the contrast between 'see' and 'look', I have noted (Depuydt 1985b, 1988c) that verbs meaning 'look' over time change their meaning to 'see'. Most notably, Late Egyptian *nw* means 'look' whereas its Coptic descendant ⲛⲁⲩ *naw* means 'see'. One wonders, therefore, whether *m33* also meant 'look' at some prehistoric time. Perhaps the expression *m33 n*, with preposition *n*, as in Pyr. §489a–d WN, means 'look at' and is a vestige of this earlier meaning.

12.5 SUBSTANTIVAL FORMS DOING DOUBLE DUTY IN BALANCED SENTENCES

A verb form is substantival for the purpose of exercising a certain function. But the form cannot be substantival to perform two functions at the same time. A substantival form whose substantivalness does double duty cannot exist. See §§8.6.1, 9.1, and 10.1 above.

12.6 CLÈRE'S PERFECTIVE RELATIVE VERB FORM

Another notion that I believe to be invalid is the "perfective" relative verb form, as J. J. Clère called it (Schenkel 1987). It is now widely accepted that *jrt.n.f* is past ('that which he has done'), that *jrrt.f* is general present ('that which he does'), and that *jrt.f* is future ('that which he will do'). The theory of the perfective relative claims that the third writing, *jrt.f*, can also be past tense ('that which he has done'). There would then be two forms written *jrt.f* distinguished in pronunciation in some unknown way presumably involving the vowels. The perfective relative may therefore be defined as a relative form without the characteristic past tense affix *n* or gemination in third-weak verbs, yet denoting past events (Depuydt 2003a: 41–45). There is no unequivocal evidence for the relative form in question. I would like to add four observations here.

(1) The fate of the perfective relative in the study of Egyptian grammar resembles that of the prospective (§12.1). Clère announced his discovery at the Twenty-first Congress of Orientalists in Paris. In his brief report in the conference acts (1949), he promised a more elaborate treatment of the new relative verb form, but it never appeared. This circumstance, like the prospective's, is unhappily anticlimactic. Perhaps this non-appearance is related to the fact that two other grammarians had claimed credit for the discovery of the same form (Depuydt 2003a: 43). Still, the lack of a more definitive study of the so-called perfective relative verb form leaves a somewhat unfinished impression.

(2) Why is the perfective relative so prominent in the history of Egyptian grammar? The form owes its popularity to the fact that Gardiner incorporated Clère's proposal in the second edition of his *Egyptian Grammar* (1950). Doubtless Gardiner expected a fuller report. Gardiner was only too glad to embrace the perfective relative. It allowed him to buttress his preference for the two-way contrast between the imperfective *sḏm.f* and the perfective *sḏm.f* over the nascent theory of the emphatic (substantival) verb forms, which involves a three-way contrast between past, present, and future. Owing to the influence of Gardiner's *Egyptian Grammar*, many have jumped on the bandwagon of the perfective relative verb form.

(3) There is a certain logic to postulating a past relative verb form *jrt.f* 'that which he has done'. It is easy to think of *jrt.n.f* 'do', the form with affix *n*, as a later formation. Everyone is familiar with the contrast between the past participles *jr*, active or passive, without gemination and the present participles *jrr*, active or passive, with gemination. Alongside the active present participle *jrr* 'who does' and the passive present participle *jrr* 'which is done', there exists a present relative *jrrt.f* 'that which he does'. By virtue of mere parallelism, one might expect the existence of a past relative verb form *jrt.f* 'that which he has done' alongside the past participle *jr*, presumably as the predecessor of *jrt.n.f* 'that which he has done'. But unambiguous examples seem difficult or impossible to find. The Pyramid texts and other early texts should be searched first; vestiges of past relative *jrt.f* might be found there. The matter awaits further investigation.

(4) The forms *wnt.f jr.f* and *wnt.k jr.k* have been adduced as instances of the purported perfective relative. They are variants of a single expression in the Pyramid texts: *(i.)jr.k wnt.k jr.k m-b3ḥ* 'You will(?) do what you did(?) before' (Pyr. §623c TPMN); *(i.)jr.k wnt.f jr.f* 'You will(?) do what he did(?)' (Pyr. §759c PN); *(i.)jr.f wnt.k jr.k m-b3ḥ* 'He will(?) do what you did(?) before' (Pyr. §760b PMN).

Three arguments favor interpreting *wnt.k jr.k* and *wnt.f jr.f* as past tenses. First, Pyr. §759c M has *i.jr.k wnt.n.f i.jr.f*, with affix *n*. Second, the expression *m-b3ḥ* 'before' refers to past time. Third, the similar expression *(i.)jr.k nw jr.n Wsjr* 'May you do the things that Osiris has done' (Pyr. § 622b TPMN) has the past tense form *jr.n*.

But what kind of form is *wnt.f jr.f*? It seems difficult to take it as a perfective relative *jrt.f*. *Wnt.f jr.f* is different already by virtue of being not simple but compound. It is not clear whether *wnt* is a particle or a form of the auxiliary verb *wnn*. Compound verb forms whose existence seems secure include *jw.f jr.f* 'he (generally) does' and *wn.f jr.f* 'he (generally) did (or: used to do)'. Is *wnt.f jr.f* somehow a relative conversion of the compound verb form *wn.f jr.f*?

12.7 THE COALESCENCE OF THE TWO FUTURE *sḏm.f* FORMS

It has taken some time for the field of Egyptology to accept that there are two future *sḏm.f* forms in addition to the relative (adjectival) future *sḏm.f*. One appears after prepositions and in cleft sentences and occasionally exhibits a characteristic *w*; I call it substantival. The other appears after *rd*(*y*) 'cause' and in wishes and exhibits characteristic forms such as *jwt.f*, *jnt.f*, and *m3n.f*; I call it adverbial. The notion is sometimes expressed that these two forms gradually coalesced in Middle Egyptian (e.g. Vernus 1990). I deny that this happened (§8.0.2.3d). But one argument against coalescence may be added.

Assume that the substantival and adverbial forms did coalesce. If this happened in Middle Egyptian, the process should be complete by the time of Late Egyptian. The future *sḏm.f* form survives in Late Egyptian, but only with a reduced functional scope. Primarily in non-literary Late Egyptian, the adverbial future *sḏm.f* survives in the wish and after *rd*(*y*) 'cause', but the substantival future *sḏm.f* (or *sḏmw.f*) hardly survives, if at all. In fact, *sḏmw.f* has become obsolete—its Middle Egyptian uses are expressed differently in Late Egyptian. One prominent one is in the cleft sentence of the future tense. Late Egyptian has *i͗.jr.f sḏm* where Middle Egyptian has *sḏmw.f*.

What happens after Old and Middle Egyptian teaches us the following. The two trajectories of historical development to which the substantival future *sḏm.f* and the adverbial future *sḏm.f* belong can be kept apart, into Late Egyptian. On the one hand, the substantival future *sḏmw.f* of Old and Middle Egyptian seems to have disappeared. In one prominent case, the cleft sentence, it is replaced by Late Egyptian *i͗.jr.f sḏm*. On the other hand, the adverbial future *sḏm.f* has survived to some extent. The sharp separation of the two historical lines of development supports the strict separation of the two future *sḏm.f* forms in Old and Middle Egyptian and contradicts any scenario of the coalescence of the two.

12.8 NOUN + *sḏm.f*

In recent decades, one occasionally reads about a verb form called Noun + *sḏm.f*. At first sight, one might perhaps be rather inclined to think of Noun + *sḏm.f* as a *construction including* the verb form *sḏm.f*, rather than as a verb form in its own right. Yet, it is as a verb form that Noun + *sḏm.f* is now sometimes understood. I believe that the verb form Noun + *sḏm.f* is a ghost construction. The matter is discussed in more detail in §9.9.2.3 above.

The compound verb form *jw* + Noun + *sḏm.f* does exist. Nothing is more characteristic of Old and Middle Egyptian grammar than the contrast between the presence of *jw* as a marker of independence and the absence of *jw* as a marker of subordination. In that sense, nothing comes easier than to interpret certain instances of the sequence Noun + *sḏm.f* ('while N hears') as nothing else but *jw* + Noun + *sḏm.f* ('N hears') minus *jw*, instead of as a distinct verb form Noun + *sḏm.f*.

APPENDIX 1

CONDITION AND PREMISE IN EGYPTIAN AND ELSEWHERE AND THE LAWS OF THOUGHT IN EXPANDED BOOLEAN ALGEBRA

This is a revision of an article originally published in *Zeitschrift für ägyptische Sprache und Altertumskunde* 126 (1999): 97–111. Minor corrections have been introduced, the phrasing has been changed here and there, and the original footnotes have been incorporated into the main text. The present footnotes comprise additional remarks serving in part as updates. The bibliography is superseded by that of this book.

This article reflects an earlier state of my understanding of modern scientific logic; it has evolved since 1999. I have gained a better understanding of the early history of symbolic logic, how it came about in the mid nineteenth century, and how most of its foundations were established by 1900. I have found John Venn's *Symbolic Logic* (1894) most useful in this respect. Venn's work deserves praise for clarifying and refining Boole's ideas, without replacing or falsifying them. In addition, Venn's book is full of useful notes about the history of logic. The library of the University of Cambridge, where Venn taught, already possessed a rich collection of works in logic in Venn's time, and Venn had studied them like no other.

However, in spite of this development, I have chosen not to rewrite the article. I feel that its overall thesis is still defensible. Furthermore, thought is an organism that evolves. Keeping a record of the earlier stages is desirable, at least if earlier stages seem to lead organically to the later stages without major breaks. Still, here and there, I have deemed it necessary to add remarks that indicate in which direction my own understanding of Boolean thought has evolved.

To the memory of William Dwight Whitney

In what follows, Boolean algebra* is expanded to provide a discrete and complete definition of the basic distinction in meaning between a clause of condition ("When it rains, he stays inside") and a clause of premise ("If it is raining, he is staying inside"). This distinction is grammatically expressed in Egyptian and in most if not all languages.

Boole uses the term "premise" in the sense of traditional logic. But his system does not account for the clear linguistic distinction between clauses of condition and clauses of premise. A "tertiary" type of proposition is therefore proposed—beyond Boole's "primary" and "secondary" types—in order to account for the distinction. But the same laws of thought apply in all three types of propositions.

The "tertiary" type also explains two paramount facts. First, from every statement containing a condition, statements containing a premise can be derived without fail under circumstances strictly controlled by the invariable laws of thought. This process of derivation reveals at once the difference between condition and premise as well as the close affinity between the two. Second, unambiguous linguistic evidence shows that a clause of premise can envelop a statement containing a clause of condition, as in Coptic **ⲉϣϫⲉ ⲉϥϣⲁⲛ–** *ešje e=f-šan* ... 'If it is the case that whenever he ...', but not the other way around.

This is not an exercise in mathematical logic but an examination of every day speech in ancient Egypt. If the minimal algebraic notations cause apprehension, it should be emphasized that hardly any prior knowledge of algebra is required. This article is designed to be self-sufficient. I can otherwise see no way of explaining the never explained distinction between condition and premise as an undeniable empirical fact of Egyptian and other languages than by reducing the distinction to its true origin, the nature of thought and reasoning.

Like most if not all languages, Egyptian exhibits a sharp distinction between two types of conditional clauses ("if"/"when[ever]"–clauses), and this throughout its four thousand year history. The two types will here be called "condition" and "premise." A condition is the italics portion of "*When it rains*, he stays inside." A premise is the italics portion of "*If it is raining*, he is staying inside." The meanings of both types are described with varying degrees of approximation in grammars. The difference between the two

* Boole's algebra is not quite the same as everything that is now called Boolean algebra. Still, the deeper concepts behind all the algebras are the same.

types is empirically more discrete in Egyptian than in most languages. The reason is that Egyptian exhibits special verb forms, contingent verb forms, that depend on conditions only, never on premises.

This paper's design is to provide a clear, solid, comprehensive, and final definition of the basic difference in meaning between condition and premise by tracing these notions to their ultimate origin, the invariable laws of thoughts. The related notion of contingency can then be defined in light of condition and premise. Contingency is dependency on conditions, not on premises. Statements containing conditions can always and without fail be converted into statements containing premises. This underscores both the difference and the affinity between the two.

In his *Investigation of the Laws of Thought on Which Are Founded the Mathematical Theories of Logic and Probabilities* (1854) and in other works, George Boole (1815–1864) first demonstrated that the laws of reasoning are mathematical. Bertrand Russell called pure mathematics the greatest discovery of the nineteenth century and Boole its discoverer in his *Investigation.**

Boolean algebra is now used in computing and in many other areas and forms the basis of the present investigation. It is true that Boole's system has been refined after him. But what is found in the *Investigation* is still valid, and it suffices for our present purposes, except in one respect. In Boole's analysis, a condition is part of a proposition, but a premise is always considered an entire proposition in its own right. However, the linguistic evidence is unequivocal in showing that we do use premises as part of propositions, on the same syntactic level as conditions, namely as subordinate clauses. At one point, Boole (1854: 179–81) examines Cicero's analysis (106–43 B.C.E.) in his

* Laita (1977: 165 n. 8 [reprint, Gasser 2000: 57 n. 8]) gives the source as "Mysticism and Logic (London 1917; repr. 1963), 59." I found it on p. 74 of the 1919 edition, the "third impression." "In mathematics it is often misleading," Laita notes, "to assign a particular discovery to a particular person. Russell himself retracted somewhat his comment on Boole discovering pure mathematics. (I owe the last information to Dr. Ivor Grattan-Guinness, who cites Russell's letter in [the] *Proc[eedings] [of the] Roy[al] Irish Acad[emy]*, 57 (A) (1954–56), 28) [Russell 1955: 64]." I have the impression that Russell's partial retraction may relate to his appreciation of the role of Gottlob Frege in the history of logic. Then again, it also seems to me that Boole's star has risen again in recent years. I feel sympathetic to this development. But a detailed investigation of the respective merits of Boole and Frege would of course require a much more detailed study. My knowledge of Frege's work would not be sufficient for such a task. Still, I sense that Boole's merits deserve something of the quality of the original accolade bestowed upon him by Bertrand Russell. I have my suspicions about Frege but I am not able to substantiate them at this time. [On Frege, see now Grattan-Guinness 2000: 177–99, and elsewhere.]

De Fato ("On Fate") of a discussion by Chrysippus (ca. 280–ca. 206 B.C.E.) of the astrological tenet, *Si quis oriente Canicula natus est, is in mari non morietur* 'when someone is born at the rising of Sirius, he will not die in the sea'. The lack of distinction between clause of condition and clause of premise is a source of confusion in all three discussions.* This paper owes much to Boole's bright light. Any credit of which it may be worthy is merely that of the wayward asteroid accidentally entering an orbit close to the sun and catching fire. Condition and premise well illustrate how we always think in spite of ourselves.

The memory of William Dwight Whitney (1827–1894), Professor of Sanskrit in Yale College, is invoked as that of a contemporary of Boole, as that of a precursor of Ferdinand de Saussure, and as that of someone always first and foremost committed to the inquiry, "Why do we speak as we do?" (Whitney 1867: 10; on Whitney and language, see Silverstein 1971).

* Since this article was first published, I have derived much benefit from studying Venn's *Symbolic Logic* (1894). Venn appears not to have anything to say about the distinction between condition and premise. Above all, like Boole, Venn holds on to the formulation "If x is true, then y is true" as an expression of a conditional sentence, consisting of a conditional clause or protasis and an apodosis, as appears from this passage (Venn 1894: 441).

> [On the two equivalent forms $x\bar{y} = 0$ and $x = xy$] When x and y stand for propositions the same equivalent forms of course demand explanation. The most natural rendering for the negative form [that is, $x\bar{y} = 0$] is the simple denial that x can be true and y false. That which most readily fits the affirmative forms [which include $x = xy$] is the ordinary conditional or hypothetical, "If x is true then y is true."

As will become apparent below, I interpret the same expression "If x is true, …" as a premise and its consequence. Condition and premise are therefore not separated adequately in Venn's work. As a result, the continuation of the above quote seems difficult to understand.

> In offering these renderings, however, we must remember that there is a certain extreme case in which we have to meet a good deal of adverse association. The two interpretations meet the shock at somewhat different points: the former when it has to admit that 'All x is y' covers the case of there being no x or no y, and the latter in sanctioning that 'If x is true y is true' covers the case of y having no connexion whatever with x provided only y is known to be true. That is, the former has a difficulty in carrying the extreme values $x = 0, y = 0$, and the latter in carrying the extreme value $y = 1$. The general property of admitting, or rather the necessity of insisting upon the admission of, both these extreme values, has been fully discussed already (see [chapters] VI., VIII.) and needs no further justification. But the language in which the result is sometimes expressed, upon the special propositional interpretation now under review, needs notice, since it involves a certain violence to popular associations. We have seen already that we must stretch the language of hypothesis in the phrase, 'If x then y', to reach the case of y being certain: we must now claim to do the same with the word *implication*. We must admit that the phrase 'x implies y' does not imply that the facts concerned are known to be connected, or that the one proposition is formally inferrible from the other.

A1.0 INTRODUCTION

In past contributions I have made a number of observations on such notions as contingency, condition, and premise in light of empirical signals conveying these notions in Egyptian and other languages (see Depuydt 1988b, 1989a, 1989b, 1991b, 1993a: 201–55, 1995b; cf. also El-Aguizy 1997). These papers engage most other past research on the topics that are treated in them. Recent bibliographical references not quoted in these articles include two articles by Kruchten (Kruchten 1994 and Kruchten 1997). I hope that the present return to, and expansion on, a topic already treated elsewhere illustrates "that character of steady growth which belongs to science" (Boole 1854: 2).

There is on the whole little in these observations that I would wish to revise in any substantial way. Still, a feeling of dissatisfaction has always accompanied these efforts, especially when it came to defining the *meaning* of contingency, condition, and premise in precise terms. The reader will readily agree that the meaning of these notions is of a different kind than 'house' as a meaning of the word *pr* or reference to past time as the meaning of the verbal morpheme *n* in Middle Egyptian. The exact meaning of the aforementioned notions was also at stake when on occasion I defended the term "contingency" as more adequate to describe the function of certain Middle Egyptian verb forms than "result," "consequence," and other terms. I still believe that "contingency," in its accepted meaning of "dependency on a condition," is preferable. But without anchoring these terms in a conceptual framework that is fully transparent and independent of anyone's will, a discussion about terminology can easily turn into an endless semantic quibble. The need is for a tool that allows one to establish the meaning of notions such as contingency, condition, and premise in terms that are independent of individual choice.

I would provisionally suggest the following terminology:

SUBORDINATE CLAUSE	MAIN SENTENCE
condition	contingent statement
premise	result

A simple test may help to clarify the difference between condition and premise in a non-technical way. From "When it rains, I stay at home," consisting of a condition and a contingent statement, it is not possible to derive the two independent statements (1) "It rains" and (2) "I stay at home." By contrast, from "If it is raining, I am staying at home," consisting of a premise and its result, one *can* derive the two independent statements sentences (1) "It is raining" and (2) "I am staying at home."

A1.1 IN SEARCH OF A TOOL FOR DEFINING THE MEANING OF CONTINGENCY, CONDITION, AND PREMISE

A1.1.1 Insufficiency of induction

In the quest for an objective tool, it may be useful to point out a realm that obviously and readily comes to mind as worthy of exploration but where this tool will not be found. It is the realm of observation of material fact. Consider the generic problem of the precise meaning of an Egyptian word denoting, say, a vessel. An archaeological discovery might bring to light a vessel on which that word is inscribed. In this case, the meaning of the word would acquire clarity and solidity by observation of material evidence. However, no kind of material evidence, whether a new artifact or a new text, will allow refining the meaning of notions such as contingency, condition, and premise.

A1.1.2 Maximizing induction

What is said in §A2.1.1 should not discourage attempts to produce definitions that maximize the empirical in the phenomena under investigation. Pending a more comprehensive definition, an empirical definition may at least serve to avoid conceptual quagmire and confused prose. For example, contingent verb forms may be defined as follows. The available evidence suggests that events denoted by Old and Middle Egyptian *stp.ḫr.f* and *stp.k3.f* (and by other verb forms featuring the morphemes *ḫr* or *k3*) depend on an *jr*–clause. It is an undeniable fact that *jr*–clauses often precede these verb forms. When no *jr*–clause is present, it is easy to infer one. The word *jr* is for all practical purposes an equivalent of "if," the conjunction introducing so-called clauses of condition. Since dependency on conditions is an accepted meaning of the term "contingency," the verb forms may be called contingent. The future contingent verb form *stp.k3.f* then means 'if so, he will choose' or the like. One may also invoke the sense that modern observers have of how the word "if" is used in clauses of condition in their own language.

This definition of contingent verb forms is as empirical as circumstances will allow. It relies on the association in fact of these verb forms with *jr*–clauses and on what native speakers sense the equivalent of *jr* in their own language to mean. This definition suffices for practical purposes but it also has shortcomings. First, it dodges the responsibility of producing an unequivocal and final explanation of the concepts of contingency and condition.

Second, it disregards the undeniable fact that a well-defined group of *jr*–clauses *cannot* precede *stp.ḫr.f* and *stp.k3.f*. In this group, *jr* is followed by the element *wnn* and a tense (past, present, future, or the aorist or general tense) and means something like 'if it is the case that'. By contrast, in *jr*–clauses that do precede *stp.ḫr.f* or *stp.k3.f*, the verbal formation *stp.f* immediately follows *jr*.

The verbal formation *stp.f* encompasses several verb forms. Because of the nature of hieroglyphic writing, it is not fully certain which *stp.f* verb form follows *jr*. Several unambiguous instances of the future verb form *stp(w).f* are on record. But this verb form only rarely exhibits the diagnostically distinctive *w*. However, it is reasonable to assume that the nondistinctive instances are also instances of *stp(w).f*. I suspect that it may be possible to discredit instances to the contrary. But this matter exceeds the scope of the present paper. The *stp(w).f* is in all probability a substantival verb form. It refers to "the fact that" a certain event will happen. A substantival verb form is what one expects after *jr*. The conjunction *jr* literally means something like 'as for (the fact that …)'.

Traditionally, both classes of *jr*–clauses are indiscriminately called clauses of condition. The question arises: What does *jr stp.f* have that *jr wnn*–plus–any–tense does not, and vice versa? And if we limit the term "condition" to *jr stp.f* because it can be followed by *stp.ḫr.f* and *stp.k3.f*, what should we call *jr wnn*–plus–any–tense? I have elsewhere suggested the term "premise."

In English, conditions are expressed by "when(ever)" or "if," as in "When(ever) it rains, he stays inside." Premises are expressed by "if," as in "If it is raining, he is staying inside."

The terms "condition" and "premise" do not need to be more than arbitrary names for two distinct phenomena. Names do not necessarily matter. Definitions do. In fact, I have in the past on occasion used the random terms "Type A" and "Type B" instead of "condition" and "premise." But as it happens, there is a long tradition of understanding the terms "condition" and "premise" in a sense that roughly matches both the distinction between *jr stp.f* and *jr wnn* in Middle Egyptian and the definitions provided below. Therefore, to the extent that the two terms are not arbitrary, they are not entirely inappropriate.

Contingency, condition, and premise involve a remarkable constellation of undeniable facts. The distinction between condition and premise emerges as the core of the problem. Contingency can be defined in light of it: contingency is compatible with conditions and incompatible with premises.

A1.1.3 Deduction and the laws of thought

The need is still for a tool that allows a definition of the meanings of condition and premise that is final and independent of individual choice. In this abstract domain, the temptation should be resisted to produce an impressionistic account that relies on persuasion and sympathy. The truth is not negotiable. It was noted earlier that the observable facts will not provide the needed tool. In which realm can this tool then possibly be found?

The nature of this realm is suggested by the type of meaning conveyed by the terms "condition" and "premise." As noted, the meanings of these terms are of a different type than 'house' as a meaning of *pr*. Condition and premise involve the relation between propositions. The sought realm may therefore be identified as the faculty of reasoning. The final definition about the difference between the two notions resides in the nature of thought. Contrastive emphasis is defined along the same lines in Depuydt 1999b (appendix 2 below).

Boole's *Investigation of the Laws of Thought* (1854), in which it is shown that reasoning can be represented mathematically, forms the basis of the following line of argument. His principles are expanded to include the difference between condition and premise by introducing a "tertiary" type of proposition beyond his distinction between "primary" and "secondary" propositions.

What follows is not in the first place an exercise in mathematical logic. It is an analysis of ordinary speech. Nor does what follows in any way mean to suggest how one should think properly. Like all the laws of science, the laws of thought are not made. They are discovered. It is not necessary to learn logic to think properly. We think logically in spite of ourselves. What matters most to the present investigation is that the invariable patterns of human thought find expression in certain fixed patterns of human speech. These patterns of thought are therefore the origin of the meaning of the corresponding patterns of speech.

The solidity of the following definitions lies not in empirical observation as is mostly the case, and should be the case, in science. The laws of the mind, as opposed to the laws of the individual sciences to which the mind is applied, have a different, deductive, quality. Boole describes the difference as follows (1854: 4).

> The general laws of Nature ... are either inductive inferences from a large body of facts, the common truth in which they express, or, in their origin at least, physical hypotheses of a causal nature serving to explain

> phænomena with undeviating precision, and to enable us to predict new combinations of them. They are in all cases, and in the strictest sense of the term, *probable* conclusions, approaching, indeed, ever and ever nearer to certainty, as they receive more and more of the confirmation of experience .... On the other hand, the knowledge of the laws of the mind does not require as its basis any extensive collection of observations. The general truth is seen in the particular instance, and it is not confirmed by the repetition of instances .... The perception of such general truths is not derived from an induction from many instances, but is involved in the clear apprehension of a single instance .... [They are] not probable knowledge. For we not only see in the particular example the general truth, but we see it also as a certain truth,—a truth, our confidence in which will not continue to increase with increasing experience of its practical verifications.

The cornerstone of the following argument is the demonstration that premises can be derived from conditions without fail. This proves three things. First is the fact that conditions and premises differ. Second is the precise ways in which they differ. Third is the fact that, in spite of their differences, statements containing conditions and those containing premises are members of the same species because one can always be resolved into the other by an invariable and fully transparent process. After all, the close affinity between statements containing conditions and statements containing premises as members of a single species is obvious from the fact that both can be expressed by *jr* in Middle Egyptian and by "if" in English.

A1.2 FACTS ABOUT CONDITION AND PREMISE AND ABOUT CONTINGENCY

What follows is a set of unambiguous facts of the Egyptian language. This selection is sufficient for our present purposes. More facts are found in the contributions listed in Depuydt 1988b, 1989a, 1989b, 1991b, 1993a: 201–55, and 1995b. To consider all facts in every stage and dialect of Egyptian would hide the forest for the trees. The present facts are restricted to Middle Egyptian, spoken some time before and after 2000 B.C.E., and the Sahidic dialect of Coptic, spoken some time before and after 400 B.C.E. Three kinds of empirical signals are listed:

(1) markers of "if"-clauses, which are subordinate clauses;
(2) markers of "then"-clauses, which are main clauses;
(3) compatibilities and incompatibilities between "if"-clauses and "then"-clauses.

The compatibilities and incompatibilities in (3) are perhaps the most striking presence in the linguistic flesh of the difference between condition and premise. The use of different tenses (past, present, future) in premises is another.

(1) "If"–clauses
  Type A ("condition")
    Middle Egyptian *jr* 'if, when' + *stp*(*w*).*f*
    Sahidic Coptic **ⲉϥϣⲁⲛⲥⲱⲧ̄ⲡ̄** *e=f-šan-sôtp* 'when(ever) he chooses'
    Greek ἐάν
  Type B ("premise")
    Middle Egyptian *jr* 'if' plus *wnn* plus a tense
    Sahidic Coptic **ⲉϣϫⲉ** *ešje* 'if'
    Greek εἰ
(2) "Then"–clauses
  Type I (contingent, that is, dependent on a condition)
    Middle Egyptian *stp.ḫr.f* '(If or when so,) he chooses'
      stp.k3.f '(If or when so,) he will choose'
    Sahidic Coptic **ϣⲁϥⲥⲱⲧ̄ⲡ̄** *ša=f-sôtp* 'he chooses' (the "aorist")
  Type II (result or consequence)
    Sahidic Coptic **ⲉⲓⲉ** *eye* 'then'
(3) Compatibilities and incompatibilities
  Type I can follow Type A but not Type B
  Type II can follow Type B but not Type A

A remark on contingency as a legitimate grammatical category of Egyptian is in order. A first undeniable fact, namely that the Middle Egyptian verb forms *stp.ḫr.f* and *stp.k3.f* often appear after *jr*–clauses, empirically suggests that contingency is their function. A second undeniable fact, namely that these verb forms occur only after *jr*–clauses of Type A (condition), not after Type B (premise), empirically much reinforces the notion that contingency, or dependency on conditions, is what these verb forms are very specifically all about. This second fact more than anything focuses the function of Middle Egyptian *stp.ḫr.f* and *stp.k3.f* and related verb forms narrowly and sharply on condition. Any theory about these verb forms should account fully for both these unequivocal facts. The clear distinction between condition and premise proposed below provides the meaning of contingency with its ultimate definition on a more abstract level.

To illustrate the facts listed above, a genuine Middle Egyptian example of Type A (condition) is cited and a generic example of Type B (premise) is modeled after it in order to highlight the difference. Two genuine Sahidic statements describing the same idea in both Type A and Type B follow.

Type A *jr h3b.ṯn wpwtyw.ṯn r.j n jw.j n.ṯn* (CT 7.168b)

'When(ever) you send your messengers against me, I will not come to you'.

Type B **jr wnn.ṯn ḥr h3b wpwtyw.ṯn r.j n jw.j n.ṯn*

'If (it is the case that) you are sending your messengers against me, I will not come to you'.

Type A **ⲉⲣϣⲁⲛ ⲡⲉⲕⲃⲁⲗ ⲥⲕⲁⲛⲇⲁⲗⲓⲍⲉ ⲙ̄ⲙⲟⲕ ⲡⲟⲣⲕ̄ϥ** (Mark 9:47)

eršan (Gk. ἐὰν) *pe=k-bal skandalize mmo=k pork=f*

'When(ever) your eye causes you to sin, pluck it out'.

Type B **ⲉϣϫⲉ ⲡⲉⲕⲃⲁⲗ ⲥⲕⲁⲛⲇⲁⲗⲓⲍⲉ ⲙ̄ⲙⲟⲕ ⲡⲟⲣⲕ̄ϥ** (Matthew 18:9)

ešje (Gk. εἰ) *pe=k-bal skandalize mmo=k pork=f*

'If (it is the case that) your eye is causing you to sin, pluck it out'.

A1.3 THE MEANING OF CONDITION AND PREMISE IN TERMS OF BOOLEAN ALGEBRA

A1.3.1 Elements of Boolean algebra

A1.3.1.0. The symbols found in Boole's *Investigation* (1854) are used here, even if somewhat different ones are now common. I will try to make what follows fully self-sufficient to afford the reader the best possible opportunity for criticism. Hardly any prior knowledge of algebra is required.

A1.3.1.1. Classes of objects. In *primary* propositions, letters are used to represent the classes of *objects* that form the subjects of our conception. For example, *s* may represent "sheep" and *w* may represent "white objects." In *secondary* propositions, letters represent *primary* propositions. For example, *i* may represent "He stays inside" and *r* may represent "It rains." It is perhaps slightly confusing that italic lower case letters are used for both objects and propositions, but such are Boole's conventions.

A1.3.1.2. Combinations of classes. On the primary level, the combination *sw* denotes objects that belong to both classes *s* and *w*, that is, to "white sheep." The combination *s* + *w* denotes "sheep and white objects" or "either sheep or white objects." On the secondary level, the combination *ir* represents "(the occasions or events in which) he stays inside and it rains" (or more precisely, in Boolean terms, "the time* during which 'He stays inside' and 'It rains' are both valid"). *i* + *r* represents "Either he stays inside or it rains," that is, "the occasions when he stays inside and the occasions when it rains."

A1.3.1.3. The numbers 1 and 0. Only two numbers are employed in Boole's algebra, 1 and 0. On the primary level, 1 denotes "everything" and 0 denotes "nothing." On the secondary level, 1 denotes "true" and 0 denotes "false," or more precisely, "the time during which a proposition is true," that is, either all the time (1) or never (0).

A1.3.1.4. Complementary classes or denials. Complementary classes are denoted in the form 1 – *x*. Thus, on the primary level, 1 – *s* represents "all objects except sheep," or "the universe minus sheep," or "not-sheep." On the secondary level, 1 – *i* represents "he does not stay inside" (more precisely, "all the time minus the time during which he stays inside").

A1.3.1.5. The indefinite element v. The symbol *v* denotes the indefinite class. On the primary level, it stands for something like "none, some, or all." Thus, *vs* would be an indefinite amount of the class "sheep," or a class of objects that belongs both to the class *s* "sheep" and to the class *v* "indefinite." On the secondary level, *v* means "true during an indefinite amount of time." Thus, *vr* means "none, some, or all of the time that it rains."†

* Boole idiosyncratically stresses the notion of time in relation to secondary propositions. Boole's emphasis on time has not been widely accepted after him. This is not the place to discuss Boole's use of the concept of time and the objections that have been made against it. What matters here is that the controversy does not affect the coherence of Boole's algebra negatively in any way. Nor does it affect the points that will be made in the following argument.

† The function of Boole's symbol for indefiniteness, *v*, has been much discussed. I have become better acquainted with aspects of the problem since writing the above lines. Venn's *Symbolic Logic* (1894) has again been very helpful. A detailed discussion exceeds the scope of the present argument. Two general observations may suffice. First of all, there seem to be some slight internal inconsistencies in Boole's presentation of *v*. Second, there are ways of avoiding *v*. Nevertheless, within Boole's system, the symbol *v* appears to be functional and workable. At all events, the debate concerning *v* does not essentially affect the points that will be made in the following argument.

A1.3.1.6. Primary and secondary propositions. An example of a primary proposition is as follows. If a is "Aristotle" and p is the class of "philosophers," then "Aristotle is a philosopher" can be represented as

$$a = vp.$$

The indefinite element v indicates that there are other philosophers besides Aristotle. The equation $a = p$ would mean that the class "Aristotle" and the class "philosophers" completely overlap, in other words, that there are no other philosophers but Aristotle.

An example of a secondary proposition is as follows. If i is "he stays inside" and r is "it rains," then "When it rains, he stays inside" can be represented as

$$r = vi.$$

This representation means that the time during which "It rains" (r) is valid completely overlaps with an indefinite amount (v) of the time during which "He stays inside" (i) is valid. He may or may not stay inside when it does not rain but whenever it rains he does.

On the secondary level, the equation $r = 1$ means "It always rains": the time that "It rains" is valid (r) equals all the time (1). The equation $r = 0$ means "It never rains" and the equation $r = v$ "Sometimes it rains." On the primary level, $p = 1$ conveys that the universe consists entirely of philosophers.

A1.3.1.7. The fundamental law of thought. If the mind contemplates the class of sheep, s, and then again s, then the result, ss, which is the class of objects belonging both to "sheep" and again to "sheep," is not larger than the class s. Thus,

$$s = ss \text{ or } s = s^2.$$

If one subtracts s^2 from both sides, we obtain

$$s - s^2 = s^2 - s^2,$$

and therefore also

$$s - s^2 = 0.$$

Isolating s results in $s\,(1 - s) = 0$. The interpretation of this equation is that the objects belonging to both "sheep" and "not-sheep" are zero or do not exist. This amounts to the basic axiom of philosophy according to which a thing cannot be itself and something else.

A1.3.1.8. The laws of thought exemplified by elimination. The primary proposition

$a = vp$ "Aristotle is a philosopher"

has the same form as the secondary proposition

$r = vi$ "Whenever it rains, he stays inside."

In fact, the same laws of thought are valid on both levels. (As will be shown below, they are also valid on a tertiary level.) This may be illustrated by applying the procedure which Boole calls "elimination" to the two levels. The procedure of elimination is here first performed abstractly upon $y = vx$, regardless of level. It is then applied, in §A2.3.1.9, to specific examples on both the primary and the secondary levels.

The theorems proven by Boole on which the procedure of elimination rests cannot be repeated here. It should also be noted that some of the following steps and equations are purely mathematical. They cannot be interpreted in terms of the real world as we know it. Nevertheless, every step obeys rigorously proven theorems and is therefore true even when not interpretable.*

To eliminate a symbol, say v in the equation $y = vx$, Boole proceeds as follows. If vx is subtracted from both sides, we have

$$y - vx = vx - vx,$$

and then also

$$y - vx = 0.$$

One then equates v successively with 1 and with 0 and multiplies the results. This means that $y - 1x$ is to be multiplied with $y - 0x$. Because $1x = x$ and $0x = 0$, we obtain

$$(y - x)\, y = 0,$$

and further

$$yy - xy = 0.$$

* In his *Symbolic Logic* (1894), Venn showed that many steps that Boole considered uninterpretable can in fact be interpreted in terms of the real world. Then again, Venn's observation does not really affect the soundness of Boole's system. Logic is like mathematics. It is imperative that each step of an argument follow with necessity from a preceding step according to an internal consistency. It does therefore not matter to the wholesomeness of Boole's system whether the steps are interpretable in terms of the real world or not. If the steps exhibit perfect consistency, then they will be effective.

Since, according to the fundamental law of thought (see §A2.3.1.7), $yy = y$, we have

$$y - xy = 0,$$

and therefore

$$y(1 - x) = 0.$$

If $y(1 - x)$ is zero, it follows that all the other possible constituents of both x and y combined are together 1, as follows:

$$1 = yx + (1 - y)x + (1 - y)(1 - x).$$

On the primary level, 1 is Everything or the universe. On the secondary level, 1 is the total amount of time.

A1.3.1.9. Elimination applied to primary and secondary propositions. The derivations presented at the end of §A2.3.1.8 may now be applied to the two propositions mentioned in §A2.3.1.6. The laws of thought allow deriving from the primary proposition

$a = vp$ "Aristotle is a philosopher"

the following conclusions:

$$a(1 - p) = 0;$$

$$1 = ap + (1 - a)p + (1 - a)(1 - p).$$

These conclusions are to be interpreted as follows. The class of objects that are both Aristotle and not philosophers ($a[1 - p]$) is zero (0) or does not exist: "One is either Aristotle or not a philosopher." Furthermore, the universe (1) consists entirely of objects that are both Aristotle and philosophers (ap), objects that are not Aristotle but philosophers ($[1 - a]p$), and objects that are neither Aristotle nor philosophers ($[1 - a][1 - p]$). As one can see, this conclusion is true.

Likewise, the same laws of thought allow deriving from the secondary proposition

$r = vi$ "Whenever it rains, he stays inside"

the following conclusions:

$$r(1 - i) = 0;$$

$$1 = ri + (1 - r)i + (1 - r)(1 - i).$$

These conclusions are to be interpreted as follows. The overlap between the time during which it rains (*r*) and the time during which he does not stay inside ($1 - i$) is zero (0). That is, the two propositions are never true at the same time: "Either it rains or he is not inside." Furthermore, the total amount of time (1) consists of three elements: first, the time in which both "It rains" and "He stays inside" (*ri*) overlap; second, the time during which it does not rain and he stays inside ($[1 - r]\ i$); and third, the time during which it does not rain and he does not stay inside ($[1 - r]\ [1 - i]$). As one can see, this conclusion is true.*

The operations of Boolean algebra always result in statements that are unequivocally true. Their power only becomes fully apparent when they are applied to a very complex example. But producing such an example here would take up too much space.

A1.3.1.10. Statements containing conditions and truth or falsehood. In the equation $r = vi$ "Whenever it rains, he stays inside," no commitment is made regarding the truth or falsehood of the propositions involved, *r* and *i*. It is not stated that it is actually raining or that someone is actually staying inside. Only the times during which these two propositions are true are related to one another. It is stated that the time when it rains always overlaps with the time when he stays inside.

By elimination of *v*, the criterion of truth (1) versus falsehood (0) can be entered. The equivalent equation $r\ (1 - i) = 0$ "Either it rains or he stays inside" states that it is not true (0) that it both rains and he does not stay inside. The element of uncertainty of course remains but is now cast in a different mode.

According to Cicero in his *De Fato* 8.15, the Stoic philosopher Chrysippus believed that the Chaldaean astrologers, instead of stating "If someone is born at the rising of Sirius, he will not die in the sea," preferred to say, "No one is at the same time born at the rising of Sirius and dies in the sea." However, the two statements are logically fully equivalent. The former statement is an instance of $y = v\ (1 - x)$; the latter, of $yx = 0$. The two equations can readily be converted into one another according the unfailing procedure described above.

Cicero therefore ridicules Chrysippus's report (*o licentiam iocularem*), in addition to casting doubt on his knowledge of ancient Mesopotamia. None-

* At this point, the original text has a paragraph with an incidental note on the equation $r = v$, which is taken to mean "Sometimes it rains." This paragraph is here omitted because the original text contains an error. The result of eliminating *v* in $r = v$ is not $r\ (1 - r)$, but rather the inconclusive equation $0 = 0$.

theless, the equation $yx = 0$ does introduce the truth criterion in the form of 0. The basic uncertainty of course remains, but it shifts elsewhere. If one cannot condone the faulty logic underlying Chrysippus's report, the element 0 at least makes it possible to understand the intuitive desire to escape from statements containing conditions because they exhibit total absence of 1 or 0, that is, the truth criterion.*

A1.3.1.11. Contingent tenses and truth or falsehood. As contingent verb forms, *stp.ḫr.f* and *stp.k3.f* refer to events that involve no commitment to either truth or falsehood. If x is the event "He chooses," then the meaning of the aorist contingent verb form *stp.ḫr.f* 'if so, then he (generally) chooses' may tentatively be represented as

$$c = vx.$$

The symbol c represents a certain specific unspoken condition ("in that case"). After Middle Egyptian, the morpheme *ḫr* seems to have shifted from *specific* contingency to *general* contingency. The Coptic verb form **ϣⲁϥⲥⲱⲧⲡ̄** *ša=f-sôtp* 'he chooses', which derives from *ḫr jr.f stp*, retains its close association with "if"–clauses. It may perhaps also be represented by $c = vx$. But c has then weakened from a specific condition ("in that case") to the most general of conditions ("Whenever the case arises [whatever that case may be]").

If the components *ḫr* and *k3* do indeed derive from verbs meaning 'say' and 'think', as is commonly assumed, perhaps the meaning of the verb forms containing them was originally something like 'he chooses, they say' and 'he will choose, it is thought'. In other words, these are not events about which I have personal certainty. I have been told about them or it is thought that they will happen. I am not the source. I am therefore not certain whether they are true or false.

A1.3.2 Deriving clauses of premise from clauses of condition in accordance with the laws of thought

Without further ado, the process by which clauses of premise can be developed unfailingly from every clause of condition may be demonstrated. The secondary proposition mentioned above is repeated here along with the two logically equivalent statements mentioned at the end of §A2.3.1.9.

$r = vi$ "Whenever it rains, he stays inside";

* Throughout his *Symbolic Logic* (1894), Venn hails the logical advantages of absolute denials such as $xy = 0$, especially their categorical character.

$r(1 - i) = 0$ "It is never true that it rains and he does not stay inside"; (1)

$1 = ri + (1 - r)\, i + (1 - r)(1 - i)$
"Either it rains and he stays inside, or it does not rain and he stays inside, or it does not rain and he does not stay inside." (2)

In Boolean algebra, every symbol except v is susceptible of two interpretations. It stands for either 1 or for 0. Because the equations above do not state whether r and i are 1 or 0, one has to *assume* it. There are four possible assumptions: $r = 1$; $r = 0$; $i = 1$; and $i = 0$. These four equations are themselves secondary propositions. The meaning of these four secondary propositions is as follows: "It is true that it is raining"; "It is not true that it is raining"; "It is true that he is staying inside"; and "It is not true that he is staying inside."

By stating that "It rains" is a real event, it becomes necessary to add *tense* (past, present, future, and aorist or general tense). Depending on the context, $r = 1$ may appear in one of the following main options: "It is raining"; "It was raining"; "It will rain"; or "It (generally) rains." Any event that we think of as an actual occurrence has to happen at some time in relation to our own point in time. From the linguistic point of view, tense indirectly becomes the marker of the truth value that we attribute to an event. As will be seen below, it is also tense that quintessentially distinguishes premise from condition in language.

Let us assume, then, that $r = 1$ and also that the event is occurring in the present tense. Thus, we are assuming that "It is true that it is raining." On this assumption, the two equations (1) and (2) above assume the following forms:

$$1 \times (1 - i) = 0; \qquad (1')$$

$$1 = 1 \times i + (1 - 1)\, i + (1 - 1)(1 - i). \qquad (2')$$

These two equations can readily be simplified as follows:

$$1 - i = 0; \qquad (1'')$$

$$1 = i + 0 \times i + 0 \times (1 - i). \qquad (2'')$$

And then to:

$$i = 1; \text{ or}$$

$$1 = i.$$

The unambiguous interpretation of $i = 1$ is that "It is true that it is raining." This inevitably follows from assuming that $r = 1$.

The three other assumptions are: $r = 0$; $i = 1$; and $i = 0$. If one enters these values into equations (1) and (2) above, the following conclusions will be obtained.

If $r = 0$ then (1) $0 = 0$ and (2) $1 = 1$

If $i = 1$ then (1) $0 = 0$ and (2) $1 = 1$

If $i = 0$ then (1) $r = 0$ and (2) $r = 0$

It appears that no conclusion is possible when $r = 0$ ("It is true that it does not rain") and when $i = 1$ ("It is true that he stays inside") because the result is either $0 = 0$ or $1 = 1$. Upon slight reflection, it becomes obvious that this is true. The two cases in which a conclusion is possible are as follows:

If r = 1 then $i = 1$ "**If it is raining**, then he is staying inside."

If i = 0 then $r = 0$ "**If he is not staying inside**, then it is not raining."

The portions in bold are **premises**. Thus it has been shown with the utmost parsimony how premises differ from conditions at the very core, that is, in relation to the invariable laws of thought.

Although language is an "imperfect instrument of thought," as Boole calls it, and although thought often ends up in fragmented forms in acts of speech, and although certain semantic nuances in speech do not pertain to strict reasoning, nevertheless the basic structure of language must reflect the constant laws of the mind. We always think in certain fixed ways in spite of ourselves, and cannot be understood if we do not. The distinction between condition and premise illustrates this.

A1.3.3 Three properties of premises: Truth value, assumption, and tense

Premises clearly have three properties that set them apart from conditions. First, they involve an element of *truth* or *falsehood*. This is obvious from the elements … = 1 and … = 0 in the equations above. Secondly, they involve an element of *assumption*. This is immediately apparent from the fact that $r = 1$ involves a choice of one possibility from a total of four. The other three are $r = 0$, $i = 1$, and $i = 0$. Thirdly, on the linguistic level, as already noted, a premise involves *tense*. Indeed, an event cannot be actual and real to us without occurring at some time in relation to the time when we speak.

The commitment to true or false in a clause of premise and its main clause explains why Middle Egyptian *stp.ḫr.f* and *stp.kꜣ.f* cannot follow a premise (cf. §§A2.2, A1.3.1.11). The Sahidic Coptic verb form **ϣⲁϥⲥⲱⲧⲡ̄** *ša=f-sôtp* 'he (generally) chooses', also called aorist, has retained this property.

A1.3.4 Derivation by logical equivalence and derivation by assumption

A premise is not logically equivalent to a condition. The equation $r = vi$ "Whenever it rains, he stays inside," which contains a condition, cannot be manipulated so that it yields a premise. Deriving a premise from a condition at some point involves making an assumption among assumptions, for example, assuming that $r = 1$. But then, the choices of assumption regarding the truth (1) or falsehood (0) of r and i are strictly limited and numbered. There are exactly four choices with two propositions r and i. Deriving a premise from a condition therefore exhibits no random or unknown factor. It is a fully controllable and unerring derivation.

Then again, because something has to be assumed, the derivation of a premise from a condition is not necessary or inevitable. It involves *initiative*.* The conscious decision of someone choosing an assumption among assumptions, say that $r = 1$, is required, on whatever prompting from the circumstances. By contrast, the equations $r = vi$ "When it rains, he stays inside" and $r(1 - i) = 0$ "It is not true that it rains *and* he does not stay inside" are logically equivalent. The equation $r(1 - i) = 0$ is latently present in the equation $r = vi$. No choices are necessary. Latent presence means logical equivalence.

A1.3.5 Deriving conditions from premises

The discreteness of condition and premise has already been sufficiently demonstrated above by deriving the latter from the former. But is it also possible to derive a condition from a premise? If the derivation of premises from conditions requires assuming something, then the path back from premises to conditions must also be indirect and lack logical equivalence.†

* I have attempted to develop the role of initiative and choice in logic at length in Depuydt 2007, by proposing an area of thought called "incipient thought." It is described there as "a zone of intellectual activity, not part of logic yet describable partly in Boolean terms, not involving the process of thought but rather the formation of thought as resulting from outside impulses."

† This §3.5 may now be considered obsolete. The section considers the possibility

From "If it is raining, he is staying inside," a premise plus its main clause, one knows for certain that *r* is being equated with 1, and not with 0, and that, by doing so, $i = 1$ is obtained. At the same time, one knows that an equation is implied. Only through an equation can *i* assume a value of 1 or 0 depending on a value of 1 or 0 for *r*. Two most obvious candidates as implied equations are $r = vi$ "When it rains, he stays inside" and $r = i$ "Whenever it rains and only then does he stay inside." This includes equations that have either of these two equations as a logically equivalent development, such as "When it rains or when he is sick, he stays inside." I conclude provisionally that implying either of these equations is extremely tempting. I have not ascertained that it is binding.*

I further refrain from making assumptions about what different speakers know at different times based on their collective lifetime experiences when they make statements. This is variable. The present concern is with the invariable laws of thought.

A1.3.6 Premises and tertiary propositions

Premises do have a place in Boolean algebra, but not one that reflects the place of premise in language. From $r = vi$ "When it rains, he stays inside," the proposition $i = 1$ "It is true that he is staying inside" was derived by assuming that $r = 1$ "It is true that it is raining." In Boole's system, this process is part of the larger phenomenon of "reduction." Reducing involves making one equation out of two or more according to the laws of thought. Indeed, the two equations $r = vi$ and $r = 1$ can easily be reduced to $i = 1$. However, this is not how language operates. All that is *stated* in "If it is raining, he is staying inside" is $r = 1$ and $i = 1$. The challenge, then, is to represent "If it is raining, he is staying inside" in Boolean algebra. The desired equation would seem to be

$$[r{=}1] = v\,[i{=}1].$$

Secondary propositions such $r = 1$ have themselves become terms of a higher order of proposition. In this capacity, their corresponding equations are represented here without spaces (as in $r{=}1$). If secondary propositions concern the truth or falsehood of propositions, then the tertiary level concerns the *reasonable compatibility* of secondary propositions. Already by mere

of deriving conditions from premises. Proof that such derivations are possible have now been presented in chapter 6 above.

* As was observed in the previous note, it has now been proven in chapter 6 that the whole process is indeed binding.

deduction, in pure mathematical terms, the postulation of a tertiary level is allowed. The question is whether it can be interpreted in terms of the real world. I believe it can.*

A1.3.7 Secondary propositions as terms of tertiary propositions

The difference between secondary and tertiary propositions may be further illustrated as follows. From $r = vi$ "When it rains, he stays inside," it follows inevitably that

$1 - i = v\,(1 - r)$ 'When he does not stay inside, it does not rain'.

On the other hand, from

$[r{=}1] = v\,[i{=}1]$ 'If it is raining, he is staying inside', (3)

it does not follow logically that

$[i{=}0] = v\,[r{=}0]$ 'If he is not staying inside, it is not raining'. (4)

After all, how could declaring that he is actually staying inside ($[i{=}1]$) in (3) be logically equivalent to declaring that he is actually not staying inside $[i{=}0]$ in (4)? It can hardly follow from the fact that something is true that it is not true.

What does follow from (3) if one applies the invariable laws of thought consistently to the tertiary level is the equation

$$1 - [i{=}1] = v\,\{1 - [r{=}1]\}. \quad (5)$$

This equation must be interpreted as follows: "What follows from any true statement other than that–he–is–inside ($1 - [i{=}1]$) is any true statement other than that–it–is–raining" ($1 - [r{=}1]$). What does this mean?

Equation (3) states that every instance of assuming that it is actually raining ($[r{=}1]$) is accompanied by the assumption that he is actually staying inside ($[i{=}1]$). This means that if one assumes anything else than that he is actually staying inside ($1 - [i{=}1]$), then this assumption can only be accompanied by an assumption other than the assumption that it is actually raining. Upon slight reflection, it appears that this is true.

* In developing Boole's algebra beyond Boole, one can derive some encouragement from the following statement by Venn (1894: 436 n. 1):

> My only complaint against Boole on this point [that is, using class symbols as propositions on the second level of thought, the level of the events] is that even he did not generalize sufficiently. Doubtless he directed attention to what are far the most important interpretations, but I prefer to regard even these as constituting only a portion of what is potentially an indefinitely wide field of application.

A1.3.8 Clauses of condition as part of clauses of premise

Primary propositions relate objects. Secondary propositions relate propositions to the truth criterion, either absolutely as in $x = 1$ or one proposition in relation to another as in $y = vx$. Tertiary propositions relate true or false propositions to one another.

In the tertiary proposition "If it is raining, he is staying inside," the true or false propositions related to one another are of the type $x = 1$. Linguistically speaking, it is not always clear whether "It rains" means just that or 'It is true that it rains'. A linguistically less ambiguous type of secondary proposition is $y = vx$ ("Whenever … then …"). This type can indeed be found in a clause of premise or in the main clause that follows it. Examples from Coptic have been cited elsewhere (Depuydt 1995b: 87–88 n. 2). Such examples prove the existence of tertiary propositions. One example is as follows: 'If it is the case that (**ⲉϣϫⲉ** *ešǰe*) when someone is caught (**ⲉⲩϣⲁⲛⲧⲁϩⲟ ⲟⲩⲁ** *e=w-šan-taho wa*) stealing he is put to shame before those who see him, then …'. Years ago, H. J. Polotsky once pointed out this phenomenon to me in a Coptic class, in a way, not altogether uncharacteristic of him, that students strongly sensed that something needed to be done about it. Let others judge whether this paper does.

The above also accounts for why the opposite is impossible: a clause of condition cannot encompass a clause of premise. This is already apparent from the structure of Coptic. The verb form **ⲉϥϣⲁⲛⲥⲱⲧⲡ̄** *e=f-šan-sôtp* obviously cannot contain a clause beginning with *ešǰe* 'if'. But it is the laws of thought, including the hierarchy of three levels, that provide the ultimate reason for this incompatibility. A subordinate clause of condition is *part* of a secondary proposition. A subordinate clause of premise is a *complete* secondary proposition. A premise is hence larger, as it were, than a condition. A condition cannot therefore encompass something larger than itself.

In sum, a premise is unthinkable as part of a condition. If something cannot be thought, it cannot normally be said. And if it is somehow said nonetheless, it cannot be understood.

APPENDIX 2

CONTRAST IN EGYPTIAN AND IN GENERAL AND THE LAWS OF THOUGHT IN BOOLEAN ALGEBRA

This is a revision of an article first published in *Göttinger Beiträge zur Sprachwissenschaft* 2 (1999): 37–60. Minor corrections have been introduced and the phrasing has been changed here and there. My understanding of symbolic logic has evolved since 1999. I have been able to gain a better understanding of the early history of symbolic logic, how it came about in the mid nineteenth century and how most of its foundations were established already by 1900. Here and there, I have deemed it necessary to add remarks that indicate in which direction my own understanding of Boole's theories have evolved. Such remarks are placed between square brackets or are in footnotes.

> It may, perhaps, be permitted to the mind to attain a knowledge of the laws to which it is itself subject, without it being also given to understand their ground and origin, or even, except in a very limited degree, to comprehend their fitness to their end, as compared with other and conceivable systems of law ....
>
> It is the business of science not to create laws, but to discover them.
>
> We do not originate the constitution of our own minds, greatly as it may be in our power to modify their character.
>
> George Boole, *An Investigation of the Laws of Thought* (1854): 11

Emphasis. Few terms are heard more often wherever Egyptian is taught in recent times. Most if not all languages appear to exhibit the phenomenon. But what is emphasis? What kind of discrete message is a speaker conveying and a hearer understanding when an empirically observable marker of emphasis is sent from the former to the latter. It seems obvious that the meaning of emphasis is of a different kind than 'house' as the meaning of the Middle Egyptian word *pr* or 'reference to past time' as the meaning of the verbal morpheme *n* in Middle Egyptian.

In what follows, a theory of emphasis is proposed that I believe to be sharp and clear while also eminently simple. What follows is not in the first place an exercise in mathematical logic, but an examination of everyday speech in ancient Egypt. If the minimal algebraic notations cause apprehension, it should be emphasized that hardly any prior knowledge of algebra or of any system of logic is required. This article is designed to be completely self-sufficient. The principles that are needed are fully described and explained before they are applied.

I personally can see no other way of accounting for "emphasis" as an undeniable empirical fact of Egyptian and other languages than by reducing the seemingly elusive concept transmitted by certain empirical markers to what I believe to be its true origin, the nature of thought.

Since the 1970s or so, "emphasis" has become one of the most prominent and pivotal grammatical terms heard in classrooms in which Egyptian is taught. This observation applies equally to all stages of Egyptian, from Old Egyptian to Coptic, spanning a period of some four thousand years, from sometime between 3000 B.C.E. and 2500 B.C.E. to sometime between 1000 C.E. and 1500 C.E. It is now generally agreed which empirical signals convey emphasis. It is now also generally agreed what emphasis in Egyptian roughly corresponds to in languages spoken by modern observers of Egyptian. But there is no agreement whatsoever as to what emphasis is or what it exactly means. The question arises: What mode of existence does something lead that we all accept exists (regardless of what we call it) yet no two independent observers are able to define in the same way?

The purpose of what follows is to propose a complete definition of emphasis. Two types of emphasis will be distinguished. I have called the two types "isolating emphasis" and "distinctive emphasis" in Depuydt 1991a. The definitions of the two types are repeated below. In the present paper, the terms "isolating contrast" and "distinctive contrast" are preferred.

The definitions proposed in §A2.3 for the meaning of the two types of emphasis or contrast are, I believe, at once securely based in material fact, conceptually parsimonious, and comprehensive of the whole meaning of both types. What is more, these empirical and simple definitions involve what Boole calls, at the outset of his *An Investigation of the Laws of Thought* (1854), "the fundamental laws of those operations of the mind by which reasoning is performed." These laws make it possible to define either type of contrast *in a single statement.*

The two statements are brief answers to the question: What meaning do empirical signals of isolating or distinctive contrast convey when they are applied to an element x? The answers, expressed in Boolean algebra, are not any longer than as follows. Isolating contrast applied to x states (in addition to stating x itself) also $1 - (1 - x)$ "not not x." Distinctive contrast applied to x states (in addition to stating x itself) also $1 - x = \ldots$ "as opposed to not x (who/which is/does or not)."

ISOLATING CONTRAST applied to x states	$1 - (1 - x)$ in addition to x
DISTINCTIVE CONTRAST applied to x states	$1 - x = \ldots$ in addition to x

As suggested in §2.4.6 below, these two definitions perfectly complement each other. The definitions show how close and yet distinct the two types of contrast are. The two types are just slightly different ways of doing the same basic thing.

Contrast exposes the fiber of the most fundamental properties of human thought. The phenomenon deserves to be a showcase illustration of how we think in spite of ourselves.

A2.0 INTRODUCTION

Students of Egyptian learn at some point that, in a simple and well-known sentence such as *sḳdd tꜣ ḫft wḏ.k* 'The land sails according to your command' (Eloquent Peasant B1, 298–99 [old numbering: 267–68]) the adverbial phrase *ḫft wḏ.k* 'according to your commandment' receives emphasis, is emphasized, bears emphasis, or the like. Other technical terms are used to describe the same property of *ḫft wḏ.k*. "Emphasis" has remained the most common term, the *jus et norma loquendi*. As noted above, the term "contrast" is preferred here. There are two types of contrast, isolating and distinctive. The phrase *ḫft wḏ.k* exhibits isolating contrast.

Students also learn that the gemination of the last consonant *d* in the verb form *sḳdd* is the crucial empirical moment in diagnosing the isolating contrast affecting *ḫft wḏ.k*.

There is now general agreement about what kinds of empirically observable signals can produce the effect called contrast or emphasis. Another instance of isolating contrast is found in *jn ḥm.f rd jr.tw.f* 'It is his majesty who caused it to be done' (Sinuhe B308). The combination of *jn* and the participle form of *rd(y)* 'give, cause' indicates that *ḥm.f* is emphasized. Other markers of the two types of contrast, both isolating and distinctive, are listed in §A2.2.

It is now also generally agreed to which expressions in modern languages markers of contrast in Egyptian roughly correspond. For example, *sḳdd tꜣ ḫft wḏ.k* can be rendered by a cleft sentence: 'It is according to your commandment that the land sails'. Cleft sentences are only rarely used in English. In certain contexts, they are exactly the right thing to say. Everywhere else, they do produce isolating contrast but sound needlessly wooden or stiff.

Another means of marking isolating emphasis or contrast, at least in writing, is italics: "The land sails *according to your commandment*." Italics is in writing what contrastive intonation is in speech. "Contrastive intonation" is a widely used term in modern linguistics. Parallel to isolating and distinctive contrast, two types of contrastive intonation will be distinguished, isolating and distinctive. However, like cleft sentences, italics are better avoided, but for a different reason. As distinct from cleft sentences, isolating intonation is natural and common in speech. Yet modern manuals of style discourage the use of italics to represent intonation, perhaps because italics look busy on the printed page. Instead, one is advised to write in such a way that it is clear which words receive contrastive intonation. In sum, markers of contrast in Egyptian are mostly better left without explicit equivalent in English translation. This practice is not likely to boost the student's confidence in the importance of the phenomenon of contrast.

The general agreement about the empirical body of signals marking emphasis guarantees that emphasis is a legitimate topic of investigation. The general agreement about the equivalents of emphasis in modern languages affords a certain indirect grasp of the meaning of emphasis. By contrast, there is total lack of agreement as to what emphasis really means. No two serious students of the phenomenon have defined its meaning in the same way.

Meaning is the ultimate aim of communication. But meaning is not directly accessible to observation. Only the forms that convey it are. Students of Middle Egyptian learn that the form *pr* conveys the meaning 'house', or that the affix *n* in a verb form indicates that the event expressed by that verb form happened before the present moment in time. They are entitled to a similar straightforward definition of the meaning of emphasis or contrast. After all, contrast is more than just a phenomenon found in ancient texts. The slight yet distinct intonation that marks contrast is ubiquitous in conversation in modern English and in most if not all other languages. It occurs less often in narration. Conversation is full of specific questions ("who?", "what?", "where?", and so on) and of the answers to these questions. Question words by definition always exhibit contrast. So do the answers to them. In Egyptian, this is obvious from the fact that question words typically appear in cleft sentences or are accompanied by other explicit markers of emphasis or contrast. Reactions to statements exhibiting contrastive intonation, as in conversation, trigger more contrastive intonation. In sum, establishing the meaning of something that we are doing all the time seems worthwhile.

What follows is an attempt to define the two types of contrast *in one statement each*. For this purpose, I will make use of the analysis of everyday speech in Boole's *An Investigation of the Laws of Thought on Which Are Founded the Mathematical Theories of Logic and Probabilities.* George Boole (1815–1864) was by all accounts one of the greatest thinkers of all time. Bertrand Russell went as far as calling pure mathematics the greatest discovery of the nineteenth century and Boole its discoverer in his *Investigation*. One biographical detail is perhaps permitted. Boole's wife, Mary, with whom he had five daughters, was herself the niece of Sir George Everest, after whom the mountain is named.

I am aware that the type of work done by Boole was carried on by others after him. I cannot claim to be conversant with all this work. But what is said in Boole's *Investigation* amply fulfils the limited needs of the present paper. One guarantee of the enduring validity of the ideas in Boole's *Investigation* is their application in the modern analysis of relay and switching circuits. In this computer age, we all owe much to Boole. What is more, the human mind, which produces the phenomenon of emphasis or contrast, resembles electric circuits and computers in many ways, even if the details of this resemblance will remain obscure for some time due to a lack of knowledge about the brain.*

In what follows, I will make some use of algebraic notation. But hardly any prior knowledge of algebra is required. To ensure the continuity of the argument, I will not use footnotes. Perhaps I may thus also "avoid anything that may be construed into the parade of learning," as Boole aptly puts it in his Preface. Any references needed for verification will be placed between parentheses.

Students of other ancient languages such as Akkadian, Greek, Hebrew, and Latin may wonder what the fuss is all about. Indeed, contrast has much

* A pivotal contributor was Claude E. Shannon (1916–2001). For some notes and bibliography on Shannon's contribution and on the role of Boolean algebra in the development of modern computing, see §0.9 above; see also Depuydt 2007.

less of a presence in the scientific study of those other dead languages. The reason is that texts written in those languages do not nearly contain the number of explicit markers of contrast that can be observed in Egyptian texts. One likes to assume that, in those other languages, intonation was used mostly to convey contrast in speech, as they are in most living languages.

A2.1 EXISTING DEFINITIONS OF EMPHASIS OR CONTRAST

Many definitions of emphasis or contrast are available. These definitions are usually part of large theoretical frameworks that far exceed the discussion of the unambiguous empirical markers of emphasis. A comprehensive discussion of the many existing definitions exceeds the scope of this paper. Two important recent surveys conveniently lead one to other work. The first survey is a chapter is Ritter's (1995: 17–43; reviewed by Peust 1997). Ritter notes (18) that there are as many opinions on the matter as there are authors. The second survey is Satzinger's (1993).

As is well known, Polotsky first identified many features of Egyptian as markers of emphasis, associating them with their equivalents in modern languages. This is not the place to recall in detail the revolutionary impact of this discovery on Egyptian grammar. As for meaning, Polotsky set the stage for almost all subsequent definitions by describing the meaning of emphasis in terms of the notions subject and predicate. I have joined Ferdinand de Saussure, Hugo Schuchardt, Lucien Tesnière, and many others in doubting the usefulness of the Aristotelian logical concepts of subject and predicate for the purposes of grammar. Subject and predicate are not even valid in logic anymore. A more detailed account of my objections is found in the works listed in the Preface to Depuydt 1999d.

My basic objection is that subject and predicate have no demonstrable presence in material fact. Satzinger (1993: 202) criticizes this objection because it gives priority to method over "results," though he does not call it "method" but "methodical delicacies." Yet, all there is to my method is two tenets: first, that science is rooted in observation (a "methodical delicacy" for which I can claim no credit); and second, that the notions of subject and predicate offend this first tenet by not being accessible to observation. However, Satzinger and I do agree on what, in the many discussions on emphasis, are the bare facts around which the discussion revolves.

A2.2 THE FUNDAMENTAL LAW OF THOUGHT AND THE ROLE OF SUPPLEMENTARY CLASSES IN BOOLEAN LOGIC

A2.2.0 Outline

Without further ado, I adduce the small number of successful elements of Boolean thought needed to produce brief, simple, complete, and mutually complementary definitions of isolating and distinctive contrast. No mention of contrast is made in Boole's work (a probable reason is noted in §A2.3).

§A2.2.1 describes what Boole calls the fundamental law of thought. §A2.2.2 concerns an idea that permeates all reasoning according to Boole's analysis of thought and can also be observed in the law of thought described in §A2.2.1. This idea is that, in every process of reasoning about any class of objects ("people," "cows," "black objects," "black cows," and so on), the supplementary classes (that is, "not people," "not cows," "not black objects," "not black cows," and so on) are always immediately present to the mind. The supplementary classes are used in every possible logical derivation from any statement about those classes. §A2.2.3 presents an example illustrating the idea described in §A2.2.2. Reproducing all the logical steps required for the development of the example exceeds the scope of this paper.

A2.2.1 The fundamental law of thought

Boole was ambitious. He endeavored to define the laws that govern intellectual operations. Boole set out to explain how everyone always inevitably thinks, regardless of whether logical thought is applied to science or to everyday speech. The purpose of this paper is to show that the phenomenon of contrast confirms the basic laws of intellectual operations.

What follows hardly does justice to Boole's treatise. But what we need is in fact rather basic. No single law is more often mentioned in the *Investigation* than what Boole calls the fundamental law of duality or the fundamental law of thought. It is obtained as follows.

The mind is able to select, through intellectual operations, classes of objects from the universe. Consider the intellectual operation involved in the word "cows." Let the class of "cows" be represented by the symbol x.

Boole represents the contrary or supplementary class as $1 - x$. The number 1 represents the universe. $1 - x$ is the class comprehending all the objects not included in the class x, that is, all the objects that are not cows, or the universe minus "cows."

Consider, then, the intellectual operation involved in the word "black." Let the class of "black" objects be presented by y. The class of "black cows" may then be written as xy. The supplementary class is $1 - xy$, that is, all the objects that are "not black cows," or the universe minus "black cows."

While standard algebra often agrees with Boolean algebra, Boolean algebra differs in the following respect. If I think of "cows" and then again of "cows," the class of "cows" is not augmented. Thus the class that exhibits the attribute "cows" is the same as the class that exhibits the two attributes "cows" and "cows." Thus we have

$$x = xx.$$

By borrowing the symbol of powers from algebra, this fundamental law can be expressed as

$$x = x^2.$$

By subtracting x^2 from both sides of the equation, we obtain

$$x - x^2 = x^2 - x^2,$$

and because $x^2 - x^2 = 0$, also

$$x - x^2 = 0.$$

By separating the common element x from both x and x^2 and placing x outside parentheses, we have the following variant formulation, preferred by Boole, of the fundamental law of thought:

$$x\,(1 - x) = 0.$$

The interpretation of this last equation is that the class of objects that are both "cows" (x) and "not cows" ($1 - x$) is zero. In other words, the class does not exist. The equation $x\,(1 - x) = 0$ is equivalent to the fundamental axiom of philosophy according to Aristotle (*Metaphysica* 3.3): "A single thing cannot both possess and not possess the same attribute in the same relation. … That is the firmest of all axioms" (my translation). Indeed, nothing is more basic to the way we think.

A2.2.2 Omnipresence of supplementary classes in thought

In the fundamental law of thought expressed by the equation $x\,(1 - x) = 0$, assuming x represents "cows," the class of "cows" and the class of "not cows" both appear. The fundamental law of thought thus involves the contemplation of both a class, namely x ("cows"), and its supplement or denial, namely $1 - x$ ("not cows"). This observation may now be generalized. It is in this general form that the observation will be used below to define isolating contrast and distinctive contrast comprehensively.

The general observation is that the class $1 - x$ is present in all processes of reasoning involving the class x. Merely thumbing through Boole's *Investigation* should readily convince anyone of the veracity of this general observation. For example, Boole's crucial formula of "development," which cannot be analyzed here in detail, is

$$f(x) = f(1)\,x + f(0)\,(1 - x).$$

The formula contains both x and $1 - x$.

It appears that there is something very dichotomous about human thought. One might perhaps want to ponder the possibility of a type of thought in which something can both belong *and* not belong to a class, that is, a type of trichotomous thought. According to Boole, it would perhaps be possible to investigate the laws of such types of thought as objects of mere speculation. But one could never conceive their real nature adequately with our existing faculties. We are unable to think beyond the categories that the brain is equipped with.

A2.2.3 An example illustrating the omnipresence of the supplementary class in reasoning about a class

Let us consider an example Boole uses, namely the Jewish law,

> Clean beasts are those which both divide the hoof and chew the cud.

Let us further assume that

x = clean beasts;
y = beasts dividing the hoof;
z = beasts chewing the cud.

The above proposition may be represented by the equation

$$x = yz.$$

From this proposition, Boole derives the propositions in (i), (ii), (iii), (iv), and (v) below. These propositions include the supplementary classes $1 - x$ ("unclean beasts"), $1 - y$ ("beasts not dividing the hoof"), and $1 - z$ ("beasts not chewing the cud"). For the sake of brevity, only the final steps of the logical developments are listed below. But the supplementary classes are also fully operative, if not more so, in the intermediate steps.

Some steps are not interpretable in terms of the real world. But all the steps follow Boole's rigorous laws. When no logical interpretation is possible, Boole's line of argument does not become invalid. It is then purely mathematical.*

In what follows, each algebraic equation derived from $x = yz$ is followed by its logical interpretation. The number 1 represents the universe. The number 0 represents the class of nothing. The symbol v denotes the indefinite class. The following logical inferences show that Boole's algebra is true. That is, all the propositions listed below, derived from the original proposition above through derivations of the Boolean kind, are undeniably true statements.

(i) $xy(1 - z) = 0$, which is equivalent to:

Clean beasts (x) dividing the hoof (y) but not chewing the cud ($1 - z$) do not exist ($= 0$).

$xz(1 - y) = 0$, which is equivalent to:

Clean beasts (x) chewing the cud (z) but not dividing the hoof ($1 - y$) do not exist ($= 0$).

$x(1 - y)(1 - z) = 0$, which is equivalent to:

Clean beasts (x) neither dividing the hoof ($1 - y$) nor chewing the cud ($1 - z$) do not exist ($= 0$).

$(1 - x)yz = 0$, which is equivalent to:

Unclean beasts ($1 - x$) dividing the hoof (y) and chewing the cud (z) do not exist ($= 0$).

(ii) $1 = xyz + (1 - x)y(1 - z) + (1 - x)(1 - y)z + (1 - x)(1 - y)(1 - z)$, which is equivalent to:

All existing things or the entire universe (1) consists of (=) either

clean beasts (x) dividing the hoof (y) and chewing the cud (z), or
unclean beasts ($1 - x$) dividing the hoof (y)
but not chewing the cud ($1 - z$), or
unclean beasts ($1 - x$) not dividing the hoof ($1 - y$)
but chewing the cud (z), or
things that are neither clean beasts ($1 - x$) nor dividers of the hoof ($1 - y$) nor chewers of the cud ($1 - z$).

* In his *Symbolic Logic* (1894), Venn showed that most if not all of the steps that Boole considered uninterpretable in the real world can in fact be related to the real world. But that does not affect the rigor of Boole's arguments.

(This writer belongs to the fourth and last class.)

(iii) $z = xy + v\,(1 - x)\,(1 - y)$,
which is equivalent to:

Chewers of the cud (z) are (=) either

clean beasts (x) dividing the hoof (y), or
belong to an indefinite remainder (v) of objects that are
neither clean beasts ($1 - x$)
nor dividers of the hoof ($1 - y$).

Furthermore: $x\,(1 - y) = 0$, which is equivalent to:

Clean beasts (x) not dividing the hoof ($1 - y$) do not exist (= 0).

(iv) $1 - x = y\,(1 - z) + z\,(1 - y) + (1 - y)\,(1 - z)$
which is equivalent to:

Unclean beasts ($1 - x$) are (=) those that either

divide the hoof (y) but do not chew the cud ($1 - z$), or
chew the cud (z) but do not divide the hoof ($1 - y$), or
neither divide the hoof ($1 - y$) nor chew the cud ($1 - z$).

(v) $1 - y = (1 - x)\,z + v\,(1 - x)\,(1 - z)$,
which is equivalent to:

Those that do not divide the hoof ($1 - y$) are (=) either

unclean beasts ($1 - x$) that chew the cud (z),
or belong to an indefinite remainder (v) of objects that are
neither clean beasts ($1 - x$)
nor chewers of the cud ($1 - z$).

Furthermore: $x\,(1 - z) = 0$, which is equivalent to:

Clean beasts (x) not chewing the cud ($1 - z$) do not exist (= 0).

What matters to the present argument is the ubiquitous presence of the class $1 - x$ in any thought involving the class x. Hardly anything is more typical of the way we think.

A2.3 ISOLATING CONTRAST AND DISTINCTIVE CONTRAST: PROVISIONAL DEFINITIONS

If one surgically removes from the vast debate around and about subject and predicate the entities that positively present themselves to the senses, one is left with a number of empirical signals that may be divided into two groups. The first group denotes isolating emphasis or contrast. The second group denotes distinctive emphasis or contrast. A list of members of both groups follows. "Emphasis" is a time-honored term. But it is a little vague. "Contrast" better defines the meaning of the phenomenon.

I have performed the invasive procedure just mentioned in an earlier article, Depuydt 1991a. The patient appears to me alive and well.

EMPIRICAL MARKERS OF ISOLATING CONTRAST

(1) Cleft sentences
(2) Substantival verb forms in Old and Middle Egyptian and so-called emphatic verb forms or second tenses in Late Egyptian, Demotic, and Coptic
(3) A distinct type of intonation (observable in living languages only)

EMPIRICAL MARKERS OF DISTINCTIVE CONTRAST

(1) So-called "anticipatory" emphasis, of the type found in "My mother, she ...," or more elaborately, "As for my mother, she ...," expressed for example by *jr* in Middle Egyptian
(2) The so-called *Verstärker* in later Egyptian, especially in Coptic
(3) Greek μέν ... δέ ..., also used in Coptic
(4) A distinct type of intonation (observable in living languages only)

Incidentally, in a discussion of these lists, Satzinger (1993: 201) attempts to equate what I call distinctive contrast with topicalization, also called extraposition or anticipatory emphasis. Topicalization is empirical. An element is put at the head of a sentence and resumed by a pronoun, as in "My mother, she ...." What follows the topicalized or extraposed element can stand as a sentence in its own right. "Extrapose" means after all "place outside." One of my examples of distinctive contrast was "*I* will" (in the sense of: "And what about you?"). Since distinctive contrast is topicalization according to Satzinger, he splits "*I* will" in the middle, calling "*I*" the extraposed element. But it is clear that "*I*" is not extraposed, in the way that "as for me" would be in "As for me, I will." Clearly, in this instance, distinctive contrast is not extraposition.

As background for the following definitions of isolating and distinctive contrast, let us consider one example of each. I believe that "*Jack* won" can be pronounced in such a way that it unambiguously means "It was Jack who won, no one else." This is isolating contrast. But then, "*Jack* won" can also be pronounced to mean "Jack, for his part, won (implying: and what did someone else, say Jeff, do?"). This is distinctive contrast.

"*Jack*" is twice contrasted with someone else, but with two different types of contrastive intonation. Both types have been represented by italics. I have proposed elsewhere (1991a: 38) the following provisional definitions for both types of contrast.

PROVISIONAL DEFINITION OF DISTINCTIVE CONTRAST

> If distinctive contrast is applied to *x*, then *x* is presented *as distinct from* someone or something else.

PROVISIONAL DEFINITION OF ISOLATING CONTRAST

> If isolating contrast is applied to *x*, then *x* is presented *as distinct from and at the exclusion of* someone or something else.

A2.4 ISOLATING AND DISTINCTIVE CONTRAST: DEFINITIONS ACCORDING TO THE LAWS OF THOUGHT IN BOOLEAN ALGEBRA

A2.4.1 Generic examples of isolating and distinctive contrast

Students of Egyptian are entitled to a simple yet comprehensive definition of the difference in meaning between distinctive contrast in *jr rḫ r pn jw.f ꜥḳ.f* 'As for him who knows this spell, he enters' ("as opposed to someone else who does or does not") and isolating contrast in *jn rḫ r pn ꜥḳ* 'It is he who knows this spell that enters' (implying "not someone else").

The same difference in meaning sets apart the distinctive contrast in Coptic ϯⲥⲟⲟⲩⲛ̄ ⲁⲛⲟⲕ *ti-sown anok* '*I* know' (in the sense of "I, for my part, know"; that is, "as opposed to someone else who does or does not") from the isolating contrast in ⲁⲛⲟⲕ ⲉⲧⲥⲟⲟⲩⲛ̄ *anok et–sown* '*I* know' (in the sense of "It is I who know," "*I* am the one who knows"; that is, "not someone else").

The same difference is found between "*Jack* won" intoned for distinctive contrast ("as opposed to someone else who did something else"; implying, say, "and what about Jeff?") and "*Jack* won" intoned for isolating contrast ("not someone else").

In each example, a class of objects ("the one who knows the spell," "I," "Jack") is contrasted in two different ways with the rest of the universe by means of empirical signals. If the class itself is x, then the supplementary class is $1 - x$ (all that is not x).

How is x contrasted differently with $1 - x$ in isolating contrast and distinctive contrast? It will be useful to present both types in elementary Boolean algebra.

A2.4.2 Absence of contrast

If Jack is x and winners are y, then "Jack won," without contrast, may be rendered by the equation

$$x = vy. \tag{1}$$

This can be paraphrased as "Jack is one or all of those who won." The symbol v denotes an indefinite class. The class vy therefore encompasses "one, some, or all (it is not specified which) of those who won."

A2.4.3 Isolating contrast

When properly intoned for isolating contrast, "*Jack* won" implies "not someone else," that is, "no one that is not Jack." If Jack is x, then "not Jack" is $1 - x$. Furthermore, "no one that is not Jack" ("not not Jack") is unambiguously $1 - (1 - x)$, which can be entered in equation (1) above. The Boolean equivalent of "*Jack* won," with isolating contrast, is then

EQUATION EXHIBITING ISOLATING CONTRAST $$x\,[1 - (1 - x)] = vy. \tag{2}$$

Simply put, "*Jack* won," with isolating contrast, means "Jack, not not Jack, won." "Not Jack" is obviously the same as "someone else." Thus one can also paraphrase "Jack won, not someone else."

From a logical point of view "Jack won," without isolating contrast, and "*Jack* won," with isolating contrast, appear to be equivalent. Indeed, these two propositions basically convey the same message. It is easy to see that they are logically equivalent. Boolean algebra readily confirms this. By removing the parentheses from (2) above, we obtain

$$x\,(1 - 1 + x) = vy.$$

But since $1 - 1 = 0$, we have

$$x\,(0 + x) = vy,$$

and therefore

$$xx = vy. \tag{3}$$

Because of the fundamental law of thought (see §A2.2.1), $xx = x$ or $x^2 = x$, it follows from (3) that

$$x = vy. \tag{4}$$

The class that is both "Jack" and "Jack" (xx) is the same as the class "Jack" (x). (4) is again equation (1), "Jack won," without isolating contrast. In other words, equation (2) can be inferred logically from equation (1) = (4) and (1) = (4) from (2).

If (1) and (2) are logically the same, how can one best describe the difference that appears to exist between them? It will be remembered that the supplementary class $1 - x$ is involved in all logical inferences derived from statements about x. The class $1 - x$ is lurking just under the surface, as it were. Such are the laws of thought. What isolating contrast does, then, is to *produce empirical and explicit reference to* $1 - x$ *in a statement about* x.

Boolean algebra shows how easily the class x ("Jack") expands into $x [1 - (1 - x)]$ ("Jack and not not Jack") and shrinks back into x.

A2.4.4 Distinctive contrast

[*New version of §A2.1.1, written in 2004. The old version, which was published in 1999, follows below.*]

The main focus of the present article is isolating contrast, not distinctive contrast. It will be useful, however, to present a definition of distinctive contrast as well. The aim is to consolidate the notion that isolating contrast and distinctive contrast are both closely related and distinctly different. The present analysis approaches the problem of distinctive contrast from a different angle than the original version published in 1999 and reproduced in full below. The main characteristic of the present analysis is that Boole's indefinite symbol v is disregarded. Boole uses v in statements like "Swallows are birds," that is, "(All) swallows are (some of the) birds." Boole would represent this statement by the equation

$$s = vb.$$

This equation indicates that swallows (s) are some of (v) the birds (b). It is possible to work with v within Boole's system. Then again, certain aspects of the symbol v are controversial. Presenting an alternative analysis of distinctive contrast without using v is therefore desirable. As Venn often notes in his *Symbolic Logic* (1894), an equation that is equivalent to $s = vb$ while not involving v is

$$s = sb.$$

This alternative equation denotes that the class of swallows (s) is the same as the class of things that are both swallows and birds (sb). This is indeed equivalent to stating that swallows are some of the birds. Consider the following diagram. As one can see, SB is the same as S because all members of S are also members of B.

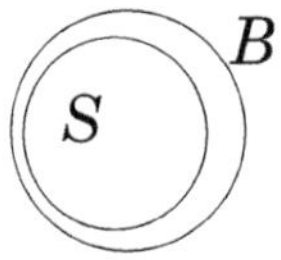

The equation $s = sb$ may now be applied to the definition of distinctive contrast.

Both isolating contrast and distinctive contrast, when applied to a class x, call forth the supplementary class $1 - x$ (or $\bar{x}$). As has been noted above, distinctive contrast calls forth $1 - x$ by implying that an additional statement is possible about $1 - x$. In other words, the message of distinctive contrast, as applied to a class x, is as follows. If something is true about x, then something else must be true about $1 - x$, whatever that something else may be and whatever $1 - x$ may be. There are two possibilities. Either that something else and $1 - x$ may be implied and left unstated. Or they may be explicitly specified. It will be useful to formalize the definition of distinctive contrast in Boolean fashion. In order to do so, a statement involving x will be first formulated and it will then be specified what exactly happens when distinctive contrast is applied to x.

Consider a proposition involving x expressed by the equation

$$x = xy \qquad \text{“(All) } x \text{ is } y\text{.”} \tag{5}$$

It follows naturally from equation (5) that

$$1 - x = 1 - xy. \quad \text{(Subtract both sides from 1.)} \tag{6}$$

Furthermore, in relation to x and y, the universe (1) is made up of four combinations classes: (1) things that are both x and y (xy); (2) things that are y but not x ($\bar{x}y$); (3) things that are x but not y ($x\bar{y}$); and (4) things that are neither x nor y ($\bar{x}\bar{y}$). That may be expressed in the equation

$$1 = xy + \bar{x}y + x\bar{y} + \bar{x}\bar{y}. \tag{7}$$

It follows from this equation that

$$1 - xy = \bar{x}y + x\bar{y} + \bar{x}\bar{y}. \qquad (8)$$

$1 - xy$ also occurs in equation (6). Inserting the equivalent of $1 - xy$ into equation (6), one obtains

$$1 - x = \bar{x}y + x\bar{y} + \bar{x}\bar{y}. \qquad (9)$$

But it is also clear that an equation such as $x = xy$ involves the absolute denial

$$x\bar{y} = 0. \qquad (10)$$

Indeed, from

$$x = xy,$$

it follows that

$$x - xy = 0.$$

Or also,

$$x(1 - y) = 0,$$

which is the same in notation as

$$x\bar{y} = 0.$$

Entering (10) into (9), one obtains

$$1 - x = \bar{x}y + \bar{x}\bar{y}. \qquad (11)$$

Or also

$$1 - x = (1 - x)(y + \bar{y}). \qquad (12)$$

In other words, it follows from

$x = xy$ "All x is (some) y." (1)

that

$\bar{x} = \bar{x}(y + \bar{y})$. "All not-$x$ is (some of) either y or not-y." (12)

This is easily confirmed by a mere look at the area outside X in the following diagram. That area falls either inside Y or outside Y.

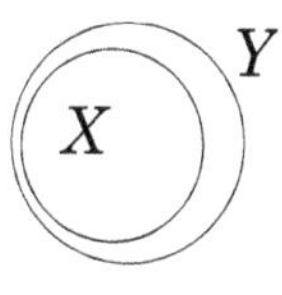

It may be concluded that distinctive contrast does not place constraints on the implied logical statement. Consider a simple statement such as "John is a carpenter," which is of the type $x = xy$. If one applies distinctive contrast to John, as in "*John (on his part)* is a carpenter," then someone else (on his part) may be either a carpenter or not a carpenter, which is of the type $\bar{x} = \bar{x}(y + \bar{y})$.

[*Old version of §A2.4.4, published in 1999*]
When properly intoned for distinctive contrast, "*Jack* won" implies "as opposed to someone else who did something else" (as when implying: "And what about Jeff?"). Like isolating contrast, distinctive contrast *makes explicit reference to the supplementary class 1 – x* ("not Jack"). What is the difference with isolating contrast?

With distinctive contrast, $1 - x$ is *part of another proposition.* "*Jack* won" invokes another proposition that is a logical inference from the proposition concerning x. This other proposition is implied in "as opposed to someone else *who did something else*," an expression used above to paraphrase the effect of distinctive contrast.

If distinctive contrast explicitly invokes the closest logical inference involving $1 - x$, what is this logical inference? It seems the equation closest to $x = \ldots$, while also containing $1 - x$, would simply be $1 - x = \ldots$.

Slight differences occur in the inferred proposition depending on the original proposition about x. For instance, either (1) others won besides Jack or (2) Jack was the only winner. The corresponding equations are

(1) $x = vy$;

(2) $x = y$.

The symbol v is the indefinite class. According to the laws of thought, the equations inferred from (1) and (2) are

(1') $v(1 - x) = 1 - y$;

(2') $1 - x = 1 - y$.

Equation (1) means "Some people (v) other than Jack ($1 - x$) did something else than win ($1 - y$)." This follows logically if it is not known whether Jack was the only winner or others also won. Equation (2) means "Everyone else but Jack ($1 - x$) did something else than win ($1 - y$)." This follows logically if Jack was the only winner. The difference between (1) and (2) is not signaled by "*Jack* won." In this example, it must be inferred from the context.

Whereas isolating contrast does not exceed *one* equation, distinctive contrast involves *two* equations. "*Jack* won," intoned for distinctive contrast, may be represented as follows:

PAIR OF EQUATIONS TOGETHER	$x = vy$
EXHIBITING DISTINCTIVE CONTRAST	$v(1 - x) = 1 - y$

The first equation corresponds to "Jack won." The second equation corresponds to the intonation characteristic of distinctive contrast affecting "Jack" in the sense of "as opposed to someone else who did something else."

Incidentally, the double equation involved in distinctive contrast may account for the fact that Greek μέν … δέ …, which is a marker of distinctive contrast, often involves not one statement but a *pair* of statements.

A2.4.5 The difference between isolating contrast and distinctive contrast

Both isolating contrast and distinctive contrast explicitly mark the presence of $1 - x$. But it is obviously not possible to make a single statement about x and $1 - x$ at the same time. According to the fundamental law described in §A2.2.1, namely $x(1 - x) = 0$, there is no class that is both Jack (x) and not Jack ($1 - x$).

However, $1 - x$ ("not Jack") *can* enter a proposition containing x ("Jack") if it is part of $1 - (1 - x)$ ("not not Jack"). This is exactly what happens in isolating contrast. It is in a sense the *closest possible logical path* to allowing $1 - x$ to enter the very same proposition. The result is $x[1 - (1 - x)]$, the class of those objects that are both Jack and not not Jack, or, in ordinary speech, "*Jack*, not someone else."

Distinctive contrast does what isolating contrast cannot do. It invokes $1 - x$ itself directly. But distinctive contrast exhibits other limitations. As noted, $1 - x$ itself cannot be produced in the same proposition. Distinctive contrast therefore chooses the *closest possible path* in a different direction. It invokes $1 - x$ by producing the most immediate logical inference involving $1 - x$, as follows:

$$1 - x = 1 - y.$$

Everyone else but Jack ($1 - x$) did something else than win ($1 - y$).

$$v(1 - x) = 1 - y$$

Some people (v) other than Jack ($1 - x$) did something else than win ($1 - y$).

A2.4.6 The complementary character of isolating contrast and distinctive contrast

In all brevity, when applied to x, isolating contrast additionally calls forth $1 - (1 - x)$, distinctive contrast additionally calls forth $1 - x = \ldots$.

isolating contrast applied to x states	$1 - (1 - x)$ in addition to x
distinctive contrast applied to x states	$1 - x = \ldots$ in addition to $x =$

The double reference to x invokes an impression of forcefulness. This is perhaps the origin of the name "emphasis."

I have elsewhere (Depuydt 1991a), rather intuitively and tentatively, suggested that isolating contrast is marked and distinctive contrast unmarked. If one compares $1 - (1 - x)$ in isolating contrast and $1 - x$ in distinct contrast, $1 - \ldots$ (the second "not") could be considered the mark. But this leaves the equation element (=) in distinctive contrast out of consideration.

Both types of contrast exhibit *directness* in one respect and *indirectness* in another respect. What is more, both types make up for their indirectness in one respect by their directness in the other respect. Isolating contrast has $1 - x$ *directly* in the same proposition, but in the indirect form $1 - (1 - x)$. Distinctive contrast has the *direct* form $1 - x$, but only *indirectly* in another proposition.

Both types of contrast seek fast and immediate access to explicit reference to the supplementary class. Apparently, in speaking about any class x, human beings sometimes become aware more vividly of $1 - x$, which is always lurking just under the surface anyway. Expression is then given to $1 - x$ by means of isolating or distinctive contrast.

Together, the two types of emphasis wonderfully complement one another in a perfect system. Logically speaking, they are both immediate expansions of an original statement about x by referring to $1 - x$. They are both one step away, so to speak, but each in a different direction. Isolating contrast is *intra-propositional.* Distinctive contrast is *extra-propositional.*

A2.4.7 Contrast and thought

These definitions of isolating contrast and distinctive contrast are, I believe, as simple as can possibly be. They are comprehensive. They rely on a basic property of reasoned speech, the logical closeness of $1 - x$ in every statement about x. Boole demonstrated the closeness of $1 - x$ by showing that 1

– x is present in every logical inference derived from propositions about x. The phenomenon of contrast demonstrates the closeness of $1 - x$ in a more direct and explicit way than by logical inference. In contrast, the presence of $1 - x$ is signaled directly and explicitly by empirical markers such as cleft sentences and special types of intonation.

In short, I believe the conclusion anticipated at the outset is not exaggerated, namely that isolating contrast and distinctive contrast lay bare the fiber of the most fundamental laws of human thought.

The fact remains that invoking $1 - x$ *does not bring extra information.* No new classes of objects are introduced. The meaning of contrast is therefore a type of meaning that is very different from the meaning 'house' of the word *pr* or the meaning 'past tense' in a verb form.

It is not certain whether Boole ever contemplated contrast in connection with his work on logic. Nor can we now know whether he would have placed contrast under the "remaining elements of speech" mentioned in the following extract from his *Investigation* (Boole 1854: 38).

> As to any remaining elements of speech, it will, upon examination, be found that they are used either to give a more definite significance to the terms of discourse, and thus enter into the interpretation of the literal symbols already considered, or to express some emotion or state of feeling accompanying the utterance of a proposition, and thus do not belong to the province of the understanding, with which alone our present concern lies.

At all events, Boole's definition of his "remaining elements" does not suit isolating and distinctive contrast. They ought not to be treated together with vague expressions of emotion such as *ooh* and *ah*. Their meaning is discrete and well delineated.*

* The only explicit reference to isolating contrast that I have been able to find so far in the work of prominent logicians is in Schröder's *Vorlesungen*. I believe, with Whitehead (1897: x), that Schröder is one of three scholars, alongside Boole and Venn, who have done the most to establish solid foundations for modern scientific logic. But it is clear that Schröder treats isolating contrast as little more than an emotional frill, a far cry from the claim made in the present article that isolating contrast lays bare the basic fiber of human thought. The relevant citation is as follows (Schröder 1890: 143):

> I am not thereby speaking of cases in which emphasis goes as far as changing the very meaning of a sentence, in the way that the meaning of the well-known expression "I did not know you yet from *that* side" is changed if a bad actor declaims it with the following emphasis: "I did not know you from that *side*." I am only speaking here of effects that can mark a sentence independently of its logical contents.

> Ich will dabei nicht reden von Fällen, wo die Betonung geradezu den Sinn des Satzes selbst verändert, wie der bekannte Ausdruck: "von *der* Seite kannt' ich dich noch nicht" dies erfuhr, als ein schlechter Schauspieler mit der Betonung: "von der *Seite* kannt' ich dich noch nicht" denselben deklamirte. Ich will nur reden von den Wirkungen des Satzes, die unbeschadet seines logischen Gehaltes nebenher gehen können.

Earlier, MacFarlane (1879: 39 §112 and 46 §128) had discussed the self-evident equation

$$x = 1 - (1 - x),$$

explicitly describing (§112) the expression $1 - (1 - x)$ as "not not x" and noting that "it is customary with grammarians to point out, that an adjective, with an even number of *nots* before it, is equivalent to the simple adjective." But MacFarlane does not make the connection with language that I would make, in the sense that, if x is "sheep," then x means "sheep" and $1 - (1 - x)$ means "*sheep!* (and not not-sheep, or something else but sheep)." MacFarlane adduces the equation mainly to observe that some confuse it with the Principle of Contradiction, which corresponds to a different equation, namely

$$x(1 - x) = 0,$$

which means something else: "There is no such thing (= 0) that is x and not-x at the same time."

APPENDIX 3

HOW THE MIND DRAWS INFERENCES THAT LEAD IT TO ACT AN ENGINEERING APPLICATION

Calculemus
Leibniz

The aim of this appendix is to design an electronic circuit consisting of electromagnetic coils and switches that accurately represents the mental acts involved in producing conditions and premises and the balanced sentence, treated at length in the chapters of this book. The interplay between condition and premise has much to do with how the mind draws inferences that lead it to act. The validity of the structures described in this book's chapters depends on their being reproducible in a functional physical mechanism. The following engineering application is meant to support the proposition that, if an operational apparatus can be constructed that adequately replicates how we draw inferences that lead us to act, a similar mechanism of the biochemical kind must exist in the brain, even if the precise nature of that mechanism remains at present undetermined. One cannot predict how closely a working electronic circuit might approximate what happens in the brain. But as long as an artificial model works, its structure ought to be at least similar to what happens in reality. Showing that an artificial mechanism *can* exist does much to suggest that an analogous one actually does exist.

A3.1 THE THREE LEVELS OF THOUGHT

Much of this book revolves around what happens when someone is in the process of forming resolve when making a statement such as "If it is in fact the case that it is raining, then I am staying inside." It is easy to sense that such a thought is meant to lead to action, even if action may not always materialize. How does the mind progress from observing that it is raining (or even just assuming that it is true that is raining) to adopting the resolve to stay inside? How do we acquire the determination to act based on what we observe? Chapter 6 describes the logical structure of this mental progression. The task at hand is to construct an equivalent electronic circuit (for some earlier bibliography on the construction of such circuits, see Craven 1955: 129–30).

"It rains" (*r*) and "I stay inside" (*i*) are two statements on the first level of thought, that of things (§2.2). The first level need not further concern us here. It suffices to assume, with Boole, that the mind is somehow capable of turning statements on the first level of thought into the components of a statement on the second level of thought, that of events. Ultimately, the first level would also need to be represented in an artificial electronic circuit. But the present focus is on the second and third levels of thought and how they interact.

An example of a statement on the second level of thought, that of condition and contingent statement, is "When it rains (*r*), I stay inside (*i*)." Its symbolic representation is

$$r = r \times i \text{ or } ri.$$

The class of events exhibiting the attribute *r* is the same as the class of events exhibiting both attributes *r* and *i*. This means that there are no *r* that are not also *i*. The corresponding diagram is as follows.

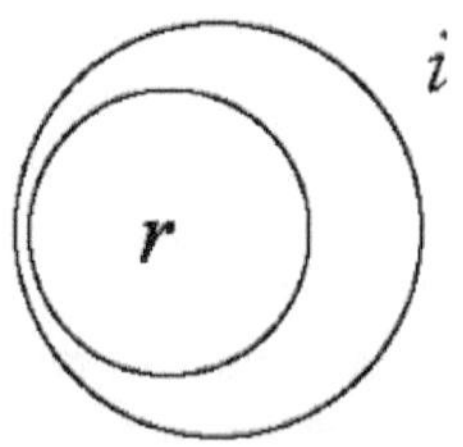

This diagram is not strictly speaking a Venn diagram. The circles represent classes, in this case classes of events on the second level of thought. In Venn diagrams, the circles rather represent "compartments into which classes may be put" (Venn 1894: 119). In Venn's method, empty compartments do not vanish, as in the diagram above, but are "scratched out." The Venn diagram equivalent to the diagram above is therefore as follows.

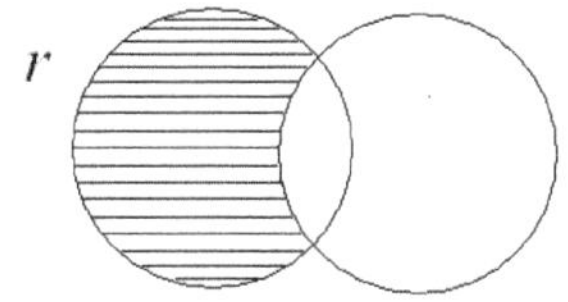

The part of r that is not i is empty and is therefore scratched out. The benefit of scratching out compartments only becomes apparent when three or even more classes are in play. It is therefore germane that the iconic representation of the Venn diagram holds three circles, such as the one depicted in a stain glass window in the hall of Gonville and Caius College of the University of Cambridge, Venn's college for sixty-six years (Edwards 2004: 7). Conditional sentences involve just two classes of events. With only two events, placing one circle inside the other fulfills at least as much the "main requirement" of a diagram, namely that of "being an aid to the eye" (Venn 1894: 116). That is why the diagram representing one circle inside another is preferred here and has been throughout this book.

An example of a statement on the third level of thought, that of premise and consequence or inference, is exemplified by "If it is raining, then I am staying inside." The third level is applied to the second level by assuming one of the two components of the second level—in the present case "it rains" (r) and "I stay inside" (i)—to be either true (1) or false (0). Consider, for example, the assumption that it is true that it is raining, or $r = 1$. It does not matter whether or not it is actually raining. What matters is that the mind assumes or imagines it to be true that it is. The information $r = 1$ can be entered into the equation given above, $r = ri$, as follows:

$1 = 1 \times i$. Or: $i = 1$, which represents the inference: "It is true that I stay inside."

The drawing of an inference has thus been accomplished: "If it is the case that it is raining ($r = 1$), then I am staying inside ($i = 1$).

The third level presupposes the existence of the second level, just as the second level presupposes the existence of the first level. The second level describes which combinations of events exist and which do not. Thus, in $r = ri$, one of the four possible combinations between two events r and i and their complementary classes $\bar{r}$ (not-r) and $\bar{i}$ (not-i) does not exist:

$r \times \bar{i}$ or $r\bar{i} = 0$.

It does not happen (= 0) that it rains and I do not stay inside ($r\bar{i}$). The three other combinations do exist:

$$1 = ri + \bar{r}i + \bar{r}\bar{i}.$$

> "The totality of events (1) consists of (=) instances in which it rains and I am inside as a class (ri), instances in which it does not rain and I am inside as a class ($\bar{r}i$), and instances in which it does not rain and I do not stay inside as a class ($\bar{r}\bar{i}$)."

The second level of thought does not concern what actually happens. It is therefore without verbal tense, past, present, or future. Verbal tense is added to the third level, which does refer to particular events and is therefore situated in time, as in "If (it is the case that) it has rained," "If (it is the case that) it is now raining," and "If (it is the case that) it will rain." Nor does the second level concern what should happen. Unlike the third level, it does not include a call to action.

A statement on the third level is composed of two statements on the second level. Consequently, a simply test that easily sets apart a statement made on the second level such as "When it rains, I stay inside" from a statement made on the third level such as "If it is the case that it is raining, I am staying inside" is that two true statements on the second level of thought can be derived from the latter but not from the former, namely "It (is true that it) is raining" and "(It is true that) I am staying inside."

An electronic circuit duplicating how we draw inferences in order to act should reflect the two layers of the second and third levels of thought. In such a circuit, activations of electric currents represent acts of the mind on both levels. One activation of a current operates on the second level, another operates on the third level.

A3.2 BASIC POINTS AND CONNECTIONS OF AN ELECTRONIC CIRCUIT REPRESENTING INFERENCES

In constructing an electronic circuit, the second level of thought needs to be constructed first, in order to represent what happens when someone states, "When it rains, I stay inside." The second level concerns classes of events, such as "all the instances in which it rains" (r). Evidently, the mind is somehow able to call forth classes of events. What is more, the mind is unable to think logically about any class of events without also taking into consideration its complementary class. When the mind conceives of instances in which it rains (r), the class of all instances in which it does not rain ($\bar{r}$) is also present. A class apparently cannot gain definition in the mind without setting itself off against all that it is not. It is somehow not difficult to imagine that something is as much defined by what it is not as by what it is. The contemplation of two classes *r* and *i* therefore involves four classes r, $\bar{r}$, i, and $\bar{i}$. They can be arranged as follows on a circuit board.

r	i
$\bar{i}$	$\bar{r}$

Eight connections are possible between these four points. Each connection represents a conditional sentence. Thus, the connection that is made from *r* in the direction of *i* represents the sentence "When it rains, I stay inside" ($r = ri$). "When I stay inside, it rains" ($i = ir$) represents the connection in the opposite direction, made from *i* in the direction of *r*. The eight connections are as follows: from r to i, from r to $\bar{i}$, from $\bar{r}$ to i, from $\bar{r}$ to $\bar{i}$, and their four inversions.

These eight possible connections can be represented on the circuit board by installing four cords allowing two connections in either direction from each of the four points. The pattern is as follows.

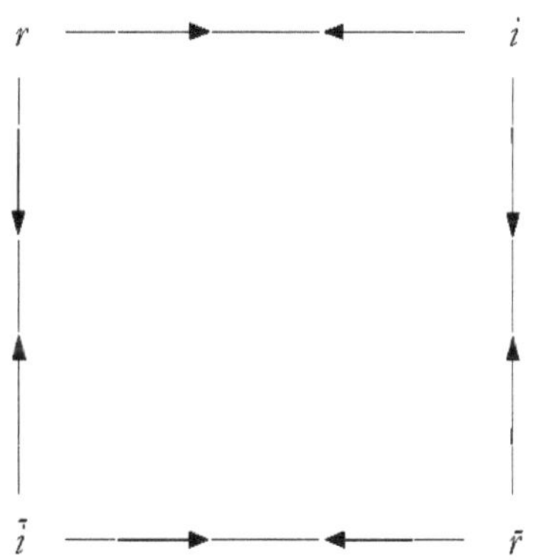

The eight arrows mark the eight possible directions in which connections can be made. The diagonal positioning of r and $\bar{r}$ and of i and $\bar{i}$ prevents the cords from crossing. Alternative arrangements are obviously possible.

Six other conceivable connections between two of the six points r, $\bar{r}$, i, and $\bar{i}$ are not represented on the circuit board: rr, ii, $\bar{r}\bar{r}$, $\bar{i}\bar{i}$, $r\bar{r}$, and $i\bar{i}$. The circuit is designed as it is because, in Boolean algebra, the six combinations rr, ii, $\bar{r}\bar{r}$, $\bar{i}\bar{i}$, $r\bar{r}$, and $i\bar{i}$ do not represent additional classes of events. In four of the six combinations, namely rr, ii, $\bar{r}\bar{r}$, and $\bar{i}\bar{i}$, the combination of a class with itself is again that class. Thus, $rr = r$ "the instances in which it both rains and rains (rr) are the same as the instances in which it rains (r)." In two of the six combinations, $r\bar{r}$ and $i\bar{i}$, the fundamental axiom of thought applies that something cannot be something and not be that same something at the same time. It cannot rain and not rain all else being the same. That is: $r\bar{r} = 0$.

A3.3 OPERATION OF THE CIRCUIT ON THE SECOND LEVEL OF THOUGHT: THE MENTAL ACT INVOLVED IN STATING "WHEN IT RAINS, I STAY INSIDE"

A3.3.1 Fastening one connection and its purport in Boolean algebra

The mental act involved in formulating a conditional sentence, whether the sentence is explicitly stated or not, can be represented on the circuit board by *fastening a connection at one of four points in one of two directions.*

For example, the mental act involved in "When it rains, I stay inside" ($r = r \times i$ or ri) corresponds to fastening the connection at r in the direction of i. Fastening a connection has strictly definable consequences for the seven other possible connections between the four points. Two of the connections are closed as an inevitable result. For example, the fastening of the connection at r in the direction of i closes the connection from r to $\bar{i}$ as well as the connection in the opposite direction on the same cord from $\bar{i}$ to r.

In Boolean algebra, the fastening of the connection at r in the direction of i and the corresponding mental act of formulating "When it rains, I stay inside" can be represented as

$$r = r \times i \text{ or } r\,i. \qquad (1)$$

This equation may be paraphrased as follows: Connecting at r in the direction of i involves both r and i (ri) at the same time. From (1), it follows that

$$r - ri = 0, \qquad (2)$$

or also that

$$r\,(1 - i) = 0. \qquad (3)$$

Since $1 - i$, or the universe (1) or everything one can possibly think about minus (–) i can also be represented as $\bar{i}$, equation (3) can also be written as

$$r\bar{i} = 0. \qquad (4)$$

Equation (4) conveys in Boolean algebra that the connection from r in the direction of $\bar{i}$ is closed. And because equation (4) is equivalent to

$$\bar{i}r = 0, \qquad (5)$$

the connection in the opposite direction from $\bar{i}$ towards r is also closed. The equation $r\bar{i} = 0$ therefore conveys that two connections are closed on the circuit board. They are the connection from point r to point $\bar{i}$ and the inverted connection from point $\bar{i}$ to point r.

The fastening of the connection from r in the direction of i is represented by ➔ in the following drawing of the circuit board. The symbol "➔" corresponds to the mental act involved in formulating "When it rains, I stay inside."

"When it rains, I stay inside" ($r = ri$)

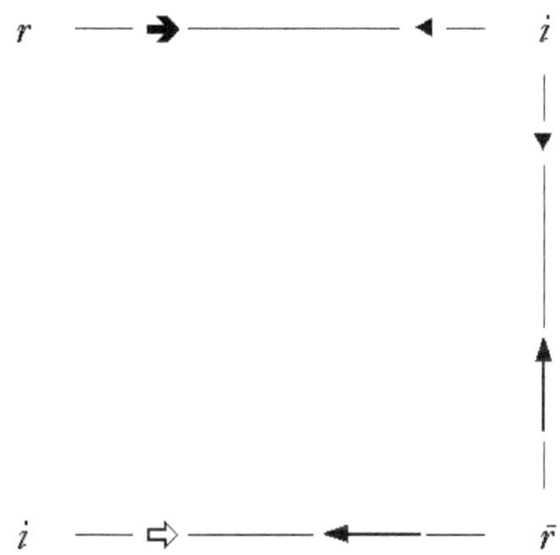

As a consequence of ➔, the connections between r and $\bar{i}$ are closed. The cord between r and $\bar{i}$ is therefore not represented in the drawing above.

The other five connections remain open. Their behavior is subject to strict rules. One of the five connections is fastened secondarily in one of two directions. In the present case, the connection from $\bar{i}$ is fixed in the direction of $\bar{r}$. The fastening of this connection is represented above by ⇨. It is expressed in Boolean algebra as

$$\bar{i} = \bar{i}\bar{r}. \qquad (6)$$

The fact that both connections are fastened at the same time in effect means that equation (1) and equation (6) have the same logical purport.

In Boolean algebra, equation (6) is in fact equivalent to equation

$r = ri$, (1)

which denotes the initial fastening of a connection. It has been noted above that equation (1) leads to

$$\bar{i}r = 0. \qquad (5)$$

Equation (6) can be deduced from equation (5) as follows. Because $r = 1 - \bar{r}$, equation (5) can be rewritten as

$$\bar{i}\,(1 - \bar{r}) = 0,$$

or also as

$$\bar{i} - \bar{i}\bar{r} = 0,$$

which is the same as the required result,

$$\bar{i} = \bar{i}\bar{r}. \qquad (6)$$

The four remaining connections exhibit the following pattern. They form two pairs of two connections that run in two directions from the same point. From i, connections can run in the direction of either r or $\bar{r}$. These two paired connections are denoted by the symbol "▸" in the diagram above. From $\bar{r}$, connections can run in the direction of either i or $\bar{i}$. These two paired connections are denoted by the symbol "→" in the diagram above. In electronic circuits, such paired connections are in effect parallel connections, represented by "+" or "OR" in Boolean algebra. A typical diagram of a parallel connection is as follows.

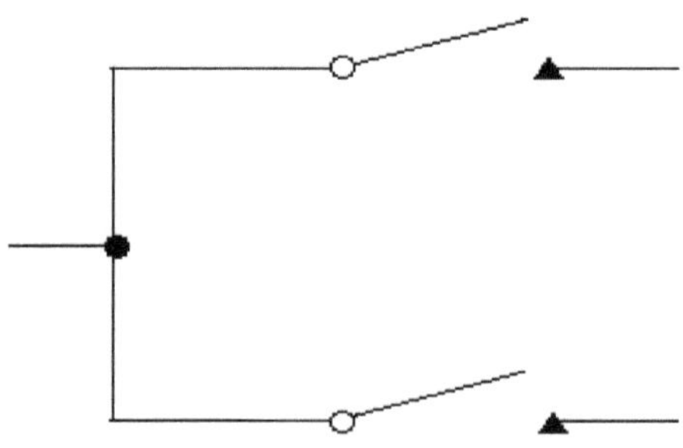

The counterpart is a connection in series, represented by "×" or "AND." A typical diagram of a series connection is as follows:

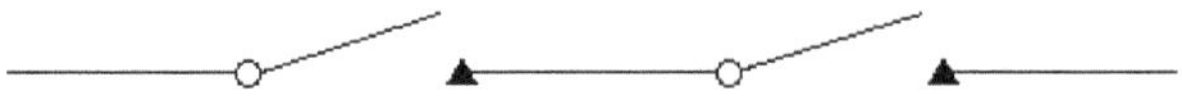

The two pairs of connections can be represented by the following equations:

$i = i(r + \bar{r})$ and

$\bar{r} = \bar{r}(i + \bar{i})$

The connection from i is in the direction of either r or $\bar{r}$. The connection from $\bar{r}$ is in the direction of either i or $\bar{i}$.

In sum, fastening *one* of eight connections triggers the closing of *two* connections. *Five* other connections remain open. *One* is secondarily fastened and *four* form *two* pairs of *two* running in *two* directions from *one and the same* point. The six open connections run in both directions on three of the four cords.

What are the equivalents of this pattern in linguistic expression?

A3.3.2 The linguistic expression equivalent to fastening a connection

Each fastening of one of the eight connections in the circuit board above represents a mental act conveyed in language by *a conditional sentence*. For example, fastening the connection from r to i is equivalent to the mental act expressed in language by the conditional sentence "When it rains, I stay inside."

The consequences of fastening one connection on the circuit board for the other connections have been described in A3.3.1. Two connections are closed and five remain open. Fastening the connection from r to i closes the two opposed connections from r towards $\bar{i}$ and from $\bar{i}$ towards r. These two connections are expressed in language by the conditional sentences "When it rains, I do not stay inside" ($r = r\bar{i}$) and "When I do not stay inside, it rains" ($\bar{i} = \bar{i}r$). The fact that their corresponding connections have been closed in effect means that they can be said to *contradict* the conditional sentence "When it rains, I stay inside" ($r = r\,i$).

The five remaining connections remain open. Accordingly, none of the corresponding linguistic expressions can be said to contradict "When it rains, I stay inside." One of the five connections is also fastened, secondarily, namely from $\bar{i}$ to $\bar{r}$, corresponding to the conditional sentence "When I do not stay inside, it does not rain" ($\bar{i} = \bar{i}\bar{r}$). Accordingly, "When I do not stay inside, it does not rain" has the exact same logical purport as "When I stay inside, it rains."

The four other connections come in two pairs. Two connections run from point i to either r or $\bar{r}$. The corresponding linguistic expression is the composite conditional sentence "When I stay inside (i), it rains (r) or it does not rain ($\bar{r}$)." Two connections also run from point $\bar{r}$ to either i or $\bar{i}$. The corresponding linguistic expression is the composite conditional sentence "When it does not rain ($\bar{r}$), I stay inside (i) or I do not stay inside ($\bar{i}$)."

A3.3.3 Mechanical device for fastening connections

The need is for four copies of a device allowing three actions at each of the four points r, $\bar{r}$, i, and $\bar{i}$ on the circuit board: (1) absence of a connection in either direction; (2) a connection in one direction precluding a connection in the other direction; (3) a connection in the latter direction precluding a connection in the former direction. A switch of the following design, here installed at r, fulfils these requirements.

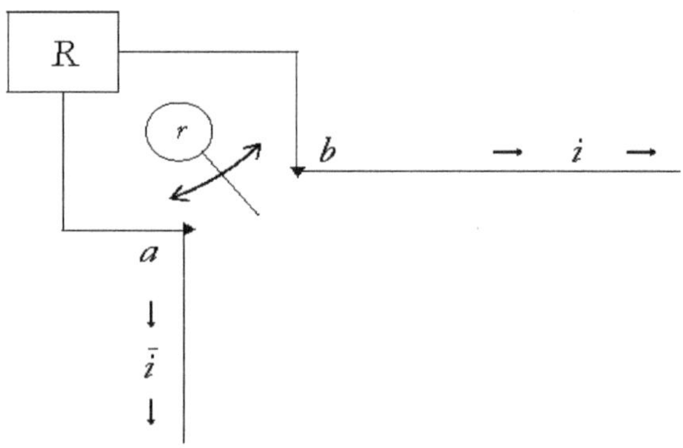

The switch has three positions. In the first position, its arm touches contact point b and a connection towards i is established. In the second position, its arm touches a and a connection towards $\bar{i}$ is established. In a third position, no contact is made with either a or b. It is crucial to the design of the switch that contact cannot be established with *a* and *b* at the same time. In accordance with the fundamental axiom of thought, one cannot state, "(When it rains,) I stay both inside and not inside." Accordingly, in Boolean algebra: $i\bar{i} = 0$.

In an electronic circuit, a mental act is represented by the activation of an electric current. In one possible design, that act of the mind may be represented by an electromagnetic coil linked to two points of contact at each of the four points. The activation of an electric current in a coil induces a magnetic field in either of the two points of contact but not in both at the same time. The magnetic field attracts the arm of the switch and contact is established in one of two directions.

For example, at point *r*, coil R may be activated to induce a magnetic field in *b* so that the arm of the switch at *r* is attracted to *b* and the connection is fastened in the direction of *i*. The equivalent linguistic expression is the conditional sentence "When it rains, I stay inside." The equivalent expression in Boolean algebra is $r = ri$. The position of the switch is now as follows.

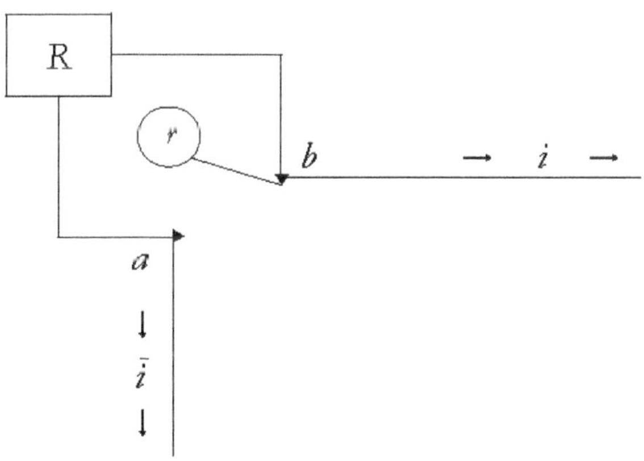

A3.4 OPERATION OF THE CIRCUIT ON THE THIRD LEVEL OF THOUGHT: THE MENTAL ACT INVOLVED IN STATING "IF IT IS THE CASE THAT IT IS RAINING, I AM STAYING INSIDE"

The second level of thought includes no inferences from premises and no calls to action. The second level of thought is static. It is a map of relations between events that is given before mental acts on the third level of thought are activated. Thus it is easy to imagine the mind having made its peace, as it were, with the relation between instances of "It rains" and instances of "I stay inside" expressed in the precept "When it rains, I stay inside." Indeed, there is no compelling reason to be outside when it rains. The precept involves the mental act of canceling the combination class of instances in which it rains and I am inside. In an electronic circuit, a mental act corresponds to the activation of an electric current. The activation of a current on the second level of thought should result in a static relation between events. In the circuit described above, an electromagnetic coil is activated in order to fasten or fix exactly one connection of the eight possible connections that are involved in a relation between two events. Once fastened, the connection is static. The link of the fastening holds. The mind has settled into acceptance of the proposition "When it rains, I stay inside."

It is against the background of the second level of thought that the third level of thought is activated. The third level of thought presupposes the second level of thought just as the second level of thought presupposes the first level of thought. The interaction between the second level and the third level explains how the mind draws inferences that lead it to act. It seems reasonable to suspect beforehand that the mind does not draw inferences all at once and out of the blue. When the mind is induced to form a resolve to act, it would seem that such resolve is formed in the context of a prior state of the mind that forms the basis for whatever inferences may have to be drawn as a certain situation arises.

On the second level of thought, mental acts are induced typically by a notion of how one views relations between events generally. On the third level of thought, mental acts are induced typically by a certain situation that arises in actual fact. One linguistic expression of a thought on the third level is "If it is the case that it is raining, then I am staying inside." A certain situation has arisen. The mind is led to believe, either by direct observation or otherwise, that it is actually raining. There is little difference to the workings of the mind between accepting something to be true by observing it and assuming it to be true without observation in fact.

"If it is the case that it is raining" is a premise. This premise leads by a certain mental act to the inference "then I am staying inside." This inference constitutes a call to action. In the circuit board proposed above, the mental act of drawing an inference from a premise should be represented by an electric current of its own. On the second level of thought, an electric current generated by coil *R* has already locked the switch at *r* in a position of contact with *b* in the direction of *i*. This mental act is part of what might be called the way in which one looks at the world. As a result, the circuit board is in the following static configuration.

"When it rains, I stay inside" ($r = ri$)

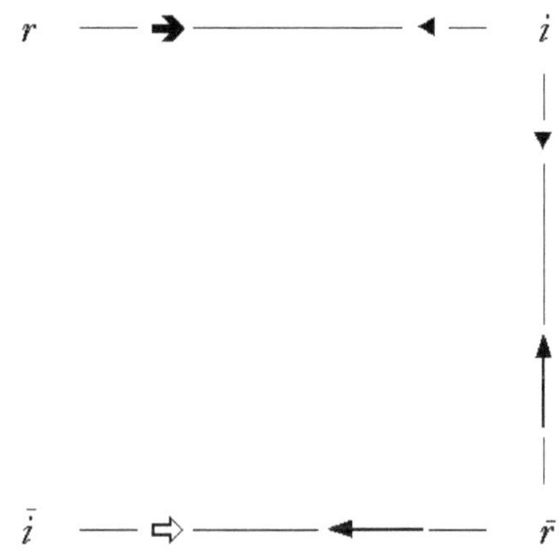

A mental act on the third level of thought comes about in a distinctly different manner. It is typically caused by a perception that reaches the mind through the bodily senses. The senses trigger the mind to be activated on the third level of thought.

On the circuit board, the mental act expressed by "If it is the case that it is raining" activates an electric current at point *r* and this current runs its course by the following design.

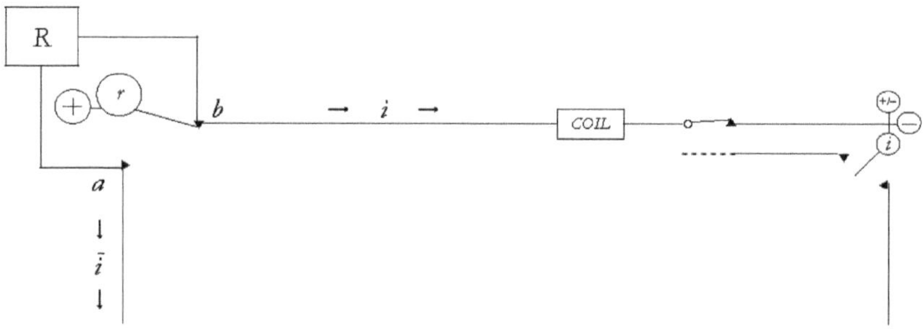

Because of the prior fastening of the connection at *r* in the direction of *i*, a current activated at *r* will need to run in the direction of *i*. The inference drawn from "It is actually raining" is therefore "I am staying inside." Different designs seem possible. In the above design, an additional coil closes a switch towards *i*.

It is the interplay between the second and third levels of thought that gives a certain level of stability and sense of resolve to thought. Without the second level of thought, acting on premises on the third level of thought would be left without any firm foundation. When a specific instance of "it rains" comes about, verbal tense (past, present, future) is added because specific instances happen at a specific time.

Conversely, when "if (it is the case that) I am staying inside" is the premise, an electric current is generated at *i* in the electronic circuit. With "When it rains, I stay inside" being given on the second level of thought, this current can run either towards *r* or towards $\bar{r}$. Indeed, if it is the case that I am inside, it may be the case that it is raining or that it is not raining. The switch at *i* is not fixed by the coil that controls it. An electronic current generated at *i* is not committed to either direction.

To serve as an equivalent of the logical purport of the balanced sentence in earlier Egyptian, the electronic circuit needs to be adjusted. Consider the classic example *mrr.f jrr.f* "he likes (something) (*l*), he does (it) (*d*) (and vice versa)." The equivalent equation in Boolean algebra is $l = d$. The circuit board would look as follows.

l —➔———⬅— d

$\bar{d}$ —⇨———⇦— $\bar{l}$

In conclusion, we are still far removed from fulfilling Leibniz's dream as described by Feys (1955: 98).

> Leibniz dreamed of a time where not only matters of mathematics proper, but also all scientific matters whatever and even matters of philosophy, of ethics, of politics, could be settled by the use of a calculus. Instead of losing themselves in seemingly infinite vain disputations, wise men would quietly take their place at a table, deciding without more ado: 'Calculemus' ['Let us calculate']. *

But that does not diminish the fact that all thought in as far as it is reasonable must ultimately be definable in mathematical equations.

* Leibniz writes as follows in his *De scientia universali seu calculo philosophico* (p. 84 in Erdmann's edition [Leibniz 1959]):

> Quo facto, quando orientur controversiae, non magis disputatione opus erit inter duos philosophos, quam inter duos Computistas. Sufficiet enim, calamos in manus sumere, sedereque ad abacos, et sibi mutuo (accito si placet amico) dicere: c a l c u l e m u s.

APPENDIX 4

ERRATA IN PREVIOUS PUBLICATIONS

A4.1 WORKS BY LEO DEPUYDT

References in the Errata are to (page,*line*) numbers.

1993a

(viii–ix [heading]) for "Preface" read "Contents"; (xii,*24*) for "one another" read "another event"; (xiii,*31*) after "*yaktubu*" add ", which"; (xv,*15*) for "operate" read "operates"; (12 n. 12) for " " " read " " "; (15,*2*) read *mʿḥʿt*, (70) the second example is badly chosen (see Shisha-Halevy 1995: 307 n. 36); (70,*16*) delete "consists"; (93,*3,6,9,12*) for "n. 105" read "n. 106"; (117 n. 1,*5*) for "sparkled" read "sparked"; (117 n. 1,*5*) for "see n. 9" read "see n. 10"; (123,*27*–124,*2*) for "Likewise ..." read "Likewise, the single form *nta⁻f-sôtm* in Coptic covers functions expressed by both substantival *sḏm.n=f* and adjectival *sḏm.n=f* in Middle Egyptian" (Satzinger 1994b: 397); (125,*10*) for "*consecutive*" read "*successive*"; (152,*last*) for "section C" read "section D"; (159,*3*) for "set" read "sat"; (167,*8*) delete "expressing contingency"; (202,*1*) for "the unity of the" read "there is unity of"; (203,*20*) add comma after "syllogism"; (220,*8*) for "Example 7" read "Example 6"; (221,*4*) for "Example 8" read "Example 7"; (226, n. 28,*1*) for "or" read "nor"; (235,*1*) remove "*m–xt.T*" from margin; (269,*9*) after "Neugebauer" add "Otto"; (269,*32*) for "79 n. 80" read "79 n. 90." Additional bibliography: Satzinger 1994a; Shisha-Halevy 1995.

1993b

(24,*6–11*) the entries under "*Form A*" and "*Form B*" need to switch columns from the line "second geminate verbs" onward; (25a,*13*) for "If" read "As soon as" or "The moment that"; (25b,*4*) to the substantival uses (a)–(f) of *sḏmw.f*, add "(g) After genitive *n*: *n jb n h3w snf.f* (pEbers 91,15–16) 'because of the desire that its blood might come down'"; (25b n. 45,*7*) for "*Jr.f* 'I-do-it'" read "*Jr.j* 'I-do-it'"; (27,*4–6*) switch "If he will desire that you live, then you will live" and "If he will desire that you die, you will die"; (30a,*21*) for "267–85" read "52–70"; (30a) to the bibliography of research on future *sḏm.f* forms up to 1993, add Polotsky 1990: 770–71 §§7–8.

1995c

(19,*11*) For "A.H. Gardiner's" read "A.H. Gardiner"; (26 n. 9,*4*) for "*CE*" read "*CdÉ*"; (33,*7*) for "*Nmty-nḥt*" read "*Nmty-nḫt*."

1996a

See Depuydt 1999d: 647–48.

1996b

(20,*10*) for "*jrwtj n rnpwt.k*" read "*jrwtj n rnpwt.k jm*." [See now also Schenkel 2006c: The construction passive participle plus suffix pronoun (see Depuydt 1996b: 18) may well not exist.]

1999d

(xxxvi,*8*) for "Shisha-Halevy" read "Ariel Shisha-Halevy"; (xlvii,*24*) for "makes it" read "makes its"; (xlviii,*29*) for "that what" read "that which"; (100,*3*) for "heard said" read "heard or said"; (102,*1*) for "also expressed" read "also be expressed"; (111,*9*) for "" read ""; (117,*2*) for 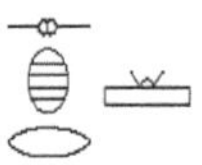 (*sḫr* "plan") read 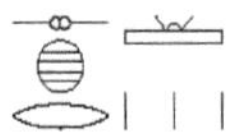 (*sḫrw* "plans"); (263,*13–14*) for "His name is not *Mmj* / *Mmjjt*" read "His/her name …"; (574,*27*; 575,*6*) as regards the writing of the royal name Ramses that Champollion saw in a cartouche from Abu Simbel on September 14, 1822, for "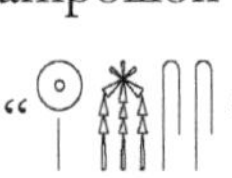" read "" (see Schenkel 2002b, and now also Schenkel 2004 and 2006b); (584,*11*) for "*šesher*" read "*sheser*"; (588,*9* end) delete one of the two sets of plural strokes; (588,*11*) for "*k3w ḏd3(w)*" (as translation of "fat oxen" at 111,*6–7*) read *jwꜥ ḏd3*; (589,*1*) for "*ḏd3wt k3w*" (as translation of "who have fat oxen" at 116,7) read "*ḏd3wt jw3*"; (599,*9*) for "your smell is like Re" read "your smell is sweet like Re."

In addition, the following were communicated by Paul Chapman on July 27, 2004: (451,*11–12*) for "… all three of them future tense" better read "… all three verb forms exhibiting *tj* being future tense"; (505,*9*) for "follows" read "precedes"; (531,*10*) for "The entity performs the choosing and does not undergo it" read "The entity undergoes the choosing and does not perform it"; (559,*5*) for "INDIRECT" read "DIRECT"; (588,*11*) for "*k3w ḏd3*(*w*)" ("fat bulls") read *jwꜥ ḏd3* ("fat oxen"); (589,*1*) for "*ḏd3wt k3w*" ("who have fat bulls") read "*ḏd3wt jw3*" ("who have fat oxen"); (620,*8* [ex. 7]) for "5(c)" read "5(b)"; (626,*7* [ex. 12]) for "*m3.n.ṯ*" read "*m3n.ṯ*"; (629,*2* [ex. 12]) for "*jr.k3.f* 'then he (masc. sing.) will do'" read "*jr.k3.k* 'then you (masc. sing.) will do'."

2002a

(first footnote [*]): for "320" read "520."

2002/3

(15,*6*) for "his" read "this"; (17,*7*) for "non-singular" read "plural."

2003a

(40 n. 19,*4*) for "461–520, at 589" read "416–514, at 489"; (43,*9*) for "Orientalist" read "Orientalists."

Six additional remarks are in order. First (24), as regards the "cornerstone" (*Eckstein*) of the passive participle theory, a reference to Polotsky 1984: 120 n. 10 might have been appropriate. Second (45), the form *wnt.k jr.k* could also easily be the conversion of *jw.k jr.k* 'you do', with *wnt* as the particle-like equivalent of *ntt*, and not necessarily of *wn.k jr.k* 'you used to do'. *Wnt.k jr.k* would then be equivalent to *ntt.k jr.k*. 'You do what you generally do' would then be a natural rendering of *i.jr.k wnt.k jr.k*. Third, as Matthias Wappes has observed to me, Sethe (1899: 331 §762) interprets *jrt* in *jrt jtwj* 'what my fathers have done' as a past passive participle and not as a past relative verb form as I had assumed. However, Sethe does postulate the existence of the so-called "perfective relative" verb form—that is, the one that is past without exhibiting affix *n*—as the examples in his §764 show. The relevant paragraph is therefore §764 rather than §762. Fourth, it might have been emphasized more that the very non-existence of *passive* adjectival (relative) verb forms is an argument that the suffix conjugation derives from passive participles. Fifth, for completeness' sake, Vycichl 1959 should have been taken into consideration. Sixth, Polotsky (1996: 16–17 §2.1.3) makes three observations: (1) the function of the dative following passive participles in Syriac was not yet clear to Theodore Nöldeke; (2) Maspero (1871: 5) already understood the relation in question between passive and possessive for ancient Egyptian; (3) the linking of passive and possessive in European languages (as in derivation of French *il m'a embarassé* from late Latin *ille me habet basiatum*) gained currency owing to Benveniste (1952), who otherwise denied that certain possessive constructions in Old Persian and in Armenian are passive in character.

2006a

(171,*5*) for "eine Art" read "einer Art." [For **ⲚⲔⲞⲦⲔ** "lie down" versus **ⲱⲂϣ** "sleep," see also **ⲠⲈⲚⲔⲞⲦⲔ ⲘⲠⲱⲂϣ** "lying down while (literally: of) sleeping" (John 11:13).]

2006b

(27,*5–6*) For "two versions (W and M of Pyramid Texts §§ 1650a and 1651c" read "Pyramid texts §492b"; (27,*10–11*) for "When N is reviled, Atum is reviled" read "When N will be reviled, Atum will be reviled".

A4.2 NOTES ON WINAND 1989

I take the occasion to make the following seven suggestions regarding this interesting study.

(1) For "**ϯ** [*ti*], **Ⲕ**- [*k*-], **ϥ**-[*f*-], **Ⲥ**- [*s*-]" (159), read "**ϯ**- [*ti*-], **Ⲕ**- [*k*-], **ⲦⲈ**- [*te*-], **ϥ**- [*f*-], **Ⲥ**- [*s*-]."

(2) For "les articles indéfinis **ⲞⲨⲀ** [*wa*], **ⲞⲨⲞⲚ** [*won*], **ϩⲞⲈⲒⲚⲈ** [*hoyne*]" (159), read "les articles indéfinis **ⲞⲨ**- [*u*-], **ϩⲈⲚ**- [*hen*-]" ("the indefinite articles **ⲞⲨ**- *u*-, **ϩⲈⲚ**- *hen*-"), because **ⲞⲨⲀ** *wa*, **ⲞⲨⲞⲚ** *ouon*, and **ϩⲞⲈⲒⲚⲈ** *hoyne* are normally classified as "indefinite pronouns."

(3) For "Une heure vint" ('An hour came') as the translation of **ⲞⲨⲚ ⲞⲨⲞⲨⲚⲞⲨ ⲚⲎⲨ** *wn u-unu nêw* (159), read "Une heure vient (viendra)" ('An hour is coming').

(4) Statements about the frequency of occurrence of **ⲞⲨⲚ** *wn* (159, 171) apply only to *sentence converters* (see Polotsky 1960: 410–11 [1971: 256–57] §35), not to the entire present system (cf. ibid. 401 [247] §§19–21).

(5) For "*ỉr.t.ỉ* … 'mon œuil'," "*ḏr.t.f* … 'la main'," and "*ỉr.t=f* 'son œuil'" (163–64), read "mes yeux" ('my eyes'), "les mains" ('his hands'), and "ses yeux" ('his eyes') (see Depuydt 2002a: 813–14, 2002/3: 17; *jrt.j*, *ḏrt.f*, and *jrt.f* are not singular, not plural, and not dual, but rather something in between singular and dual)

(6) For "*ẖr jr wʿ rmt ḏrḏr* (*m*) *ỉy* 'et quand un étranger vint'" ('and when a stranger came') (164), read "*ẖr jr wʿ rmt ḏrḏr* (*ḥr*) *ỉy* …."

(7) For **ⲞⲨⲱⲚ** *wôn* (166, 167), read **ⲞⲨⲞⲚ** *won.*

Furthermore, I am not sure that I agree with the statement that "the suffix pronoun in itself does not have defining value"* (163). Rather, I believe that expressions consisting of a noun and an attached suffix pronoun are definite owing precisely to the suffix pronoun. Nor am I certain that the interrogative pronoun "is considered indefinite in Coptic"* (171). True, the interrogative **ⲟⲩⲏⲣ** *wêr* 'how many?' would be. But **ⲟⲩ** *u* 'what?' (Egyptian *jḫ* 'what?'), to which the statement in question makes reference, is not.

Finally, both Winand (1989: 168–70) and Simpson (1996: 159) bring concepts such as subject and predicate, theme and rheme, and topic and comment to bear on the problem of existential sentences. I do not think that these concepts are relevant to the matter. In fact, I believe that they are ghost concepts.

* Le pronom suffixe n'a pas en lui-même de valeur définissante.
* Est considérée en copte comme indéfinie.

DEFINITIONS OF SOME KEY CONCEPTS

For the sake of conceptual clarity, an attempt has been made throughout this work to assign just one term to each phenomenon. What follows is a list of the principal terms along with a description of the single phenomenon that each term denotes. The present list includes some general terms such as "causality" and "knowledge."

adjectival: making reference to a property and mostly in addition to an entity, like the adjective *nfr*, which sometimes means 'good' but mostly 'good one.' Synonymous to "relative" when modifying "clause" and "verb form," as in "adjectival verb form" and "adjectival clause"; makes up "nominal" together with "substantival." Counterpart: *substantival*.

"algebra of logic": Name used in past work on logic for what is here called attribute mathematics. Counterpart: *numerical algebra*.

apodosis: Main clause of a conditional sentence; contains a contingent statement or statement depending on a condition. Counterpart: *protasis*.

Aristotelian logic: Logical system of Aristotle's writings on logic and any developments derivative of that logical system; valid to the extent that it goes but incomplete and therefore ultimately inadequate; fully supplanted in the mid nineteenth century by attribute mathematics, so-called "mathematical logic." Counterpart: *attribute mathematics*.

attribute: Feature of observable reality whose presence or absence unites things into a class of things on the level of things or unites events into a class of events on the level of events. For example, on the level of things, the presence of the attribute "car" unites all the things that exhibit that attribute into the class of cars and its absence unites all the things that lack that attribute into the class of non-cars. Likewise, on the level of events, the presence of the attribute "it rains" unites all the events that exhibit that attribute into the class of events when it rains and its absence unites all the events that lack that attribute into the class of events when it does not rain.

attribute mathematics: Branch of mathematics formulating all the laws concerning attributes. Counterpart: *quantity mathematics*.

axiom: Immediately apparent observation of reality that cannot be proven but whose strength derives from its overwhelming obviousness and from which deductive inference can be drawn. See *deductive inference.*

balanced sentence: String of words consisting of two halves that are presented as equivalent to one another, as in "You're OK, I'm OK."

balanced sentence broadly defined: String of words consisting of two halves that are presented as equal to one another. Counterpart: *balanced sentence narrowly defined.*

balanced sentence narrowly defined: Balanced sentence whose two halves contain a substantival verb form; attested in Old and Middle Egyptian. Counterpart: *balanced sentence broadly defined.*

causality: A much discussed concept of uncertain definition; viewed here provisionally as nothing more than a metaphor and a product of the imagination, inspired by transitive actions visibly performed by an actor upon a direct object. For example, in "The boy kicked the ball," the boy's action visibly appears to affect the ball. Accordingly, in "When it rains, I stay inside," one might imagine metaphorically that rain causes me to stay inside in the same way that kicking visibly affects a ball. But the logically equivalent statement, namely "When I do not stay inside, it does not rain" shows that this effect is imagined. The fact that it does not rain does not cause me not to be inside.

certainty: Feeling or mental state pertaining to a proclaimed presence of a match between what is inside the mind and what is outside the mind (that is, a truth) or absence of a match (that is, a falsehood). For example, one can be certain that something does or does not match, that is, is or is not true. Counterpart: *doubt.*

class: Group whose member or members share one or more presences or absences of attributes—attributes of things on the level of things or attributes of events on the level of events; here an exact synonym of *set.*

cleft sentence: Common name for sentence constructions in which an element is singled out for contrastive emphasis in a way that metaphorically evokes the image of a cleft in the sentence, as in the English construction "It is … that …" in "*It is* with great pleasure *that* I welcome you here today."

combination class: Group of one or more things on the level of the things or of one or more events on the level of the events sharing two or more presences or absences of attributes.

conditio sine qua non: Condition denoting events that encompass the events represented in the main clause, like "Only when the sun shines" in "Only when the sun shines do I lie on the beach"; as opposed to a regular condition, which denotes events that *are* encompassed by the events represented in the main clause, like "When the sun shines" in "When the sun shines, I lie on the beach." The difference is best represented by diagrams in which circle *S* represents the occasions when the sun shines and circle *B* represents the occasions when I lie on the beach. In the following two diagrams, *S* is either a regular condition or a *conditio sine qua non.*

S as regular condition *S* as *conditio sine qua non*

When *S* is a regular condition, as in "When the sun shines, I lie on the beach," it is possible for me to lie at the beach when the sun does *not* shine: circle *B* is larger than circle *S*. When *S* is a *conditio sine qua non*, as in "Only when the sun shines do I lie on the beach," it is *not* possible for me to lie at the beach when the sun does *not* shine: circle *S* is larger than circle *B*. Counterpart: *(regular) condition.*

condition: Event joined to another event called a contingent event in such a way that exactly one of the four combinations of the two events and their supplements or contradictories is presented as non-existent, namely the combination of the presence of the condition and the absence of the contingent event. For example, the event "It rains" is a condition in "When it rains, then I stay inside" because it is joined to another event, namely "I stay inside," in such a way that the combination of the presence of "It rains" and the absence of "I stay inside" is said to be non-existent. Only three combinations exist: either it rains and I stay inside; or it does not rain and I stay inside; or it does not rain and I do not stay inside. Counterpart: *contingent event.*

conditional clause: Subordinate clause denoting a condition. Counterpart: *contingent sentence.*

conditional sentence: Sentence consisting of a conditional clause and its main clause.

consequence: What follows inevitably from a premise on the tertiary level of thought, in addition to the level of things and the level of events; here an exact synonym of result. Counterpart: *premise.*

contingency: Dependency on conditions, quintessential property of contingent events. See *contingent event.*

contiguity: Relation of distinctive closeness between two events that are neither fully simultaneous nor discretely successive but rather follow one another quickly and often even overlap. In Egyptian and elsewhere, two contiguous events can be signaled by the joint application of (1) contrastive emphasis and (2) presenting the events as simultaneous or even in reverse order (most detailed explanation of the mechanism in Depuydt 1999d: 693–729), as in *jj.n(.j)* … *r ẖrt-nṯr š3d.n(.j) š*… 'When I came to the necropolis, I dug a pool' or 'No sooner did I come to the necropolis than I dug a pool' (lit., 'It is after having dug a pool that I came to the necropolis' [that is, 'so soon after coming did I dig that it seemed as if I dug before coming']) (tomb of Qereri at El-Hawawish [Kanawati 1986: Figure 20c]; see pp. 202–4 above). Compare, in the dimension of space as distinct from time, "If I'd hold you any closer, I'd be in back of you" (Groucho Marx, *A Day at the Races*), that is, "so close am I to you that it almost seems as if it is *in back of you* [emphasized as it is unexpected] that I am."

contingent event: Event joined to another event called a condition in such a way that exactly one of the four combinations of the two events and their supplements or contradictories is presented as non-existent, namely the combination of the presence of the condition and the absence of the contingent event. For an example, see the entry *condition* above. Counterpart: *condition.*

contingent sentence: Sentence denoting an event that depends on a condition; often functions as main clause to which a conditional clause is subordinated. Counterpart: *conditional clause.*

contingent tense: Verb form whose function it is to denote contingent events, such as Old and Middle Egyptian verb forms exhibiting the morphemes *ẖr*, *k3*, and *jn*.

contradictory: Here an exact synonym of supplement, q.v.

contrastive emphasis: Linguistic device by which a thing or an event is presented as the supplement of its supplement, here also called isolating contrast. For example, the supplement of "sheep" is "everything but sheep" or "not-sheep." The class of sheep can be presented as the supplement of its supplement by pronouncing "sheep" with a higher pitch of the voice, as in "*sheep!*," that is, "not-not-sheep" or "sheep and not anything else." In Boolean notation, "not-sheep" is $1 - s$, that is, the universe (1) or anything one could think about minus (–) sheep (s). "Not-not-sheep" then must be $1 - (1 - s)$. Obviously, the reference is again to sheep: $1 - (1 - s) = 1 - 1 + s = 0 + s = s$. The supplement of "It rains" are the occasions when it does not rain. The supplement of the supplement is presented in "It *does* rain," that is, it is not the case that it does not rain.

deductive inference: Conclusion that follows with inevitability from an axiom or from another deductive inference. Counterpart: *inductive inference.*

deductive thought: Type of thought of which every carefully articulated step is derived from a preceding step according to a rigorous and unerring internal consistency, without consideration of empirical fact. The initial steps of chains of deductive steps are axioms, q.v. Counterpart: *inductive thought.*

definite: Allowing of only one solution; said of a deductive inference, as in mathematics. For example, addition is definite: adding 5 to 15 has one solution only, namely 20. But taking a square root is indefinite. The square root of 4 is either +2 or –2. Counterpart: *indefinite.*

dependency on conditions: Quintessential property of contingent events. See *contingent event.*

event: Member of a class of events on the secondary level of thought, namely the level of events. Counterpart: *thing.*

falsehood: Lack of match between what is inside the mind and what is outside the mind. Counterpart: *truth.*

final: Said of an explanation that is traced back to the fundamental laws of thought, which define the limits of human thought and beyond which thought is not possible.

indefinite: Allowing of more than one solution; said of a deductive inference, as in mathematics. For example, taking a square root is indefinite. The square root of 4 is either +2 or –2. But addition is definite: adding 5 to 15 has one solution only, namely 20. Counterpart: *definite.*

inductive inference: Probable conclusion following from repeated observation, with the senses, of patterns of regular behavior. Counterpart: *deductive inference.*

inductive thought: Type of thought in which outside reality is observed to note regularities in behavior, with repeated observation leading to ever greater but never absolute certainty (in contrast with the absolute certainty of the steps of deductive thought). Counterpart: *deductive thought.*

isolating contrast: Here an exact synonym of contrastive emphasis, q.v.

knowledge: Provisionally defined here as any outcome of efforts to assimilate what is inside the mind to what is outside the mind.

level: The principal articulation of thought and reasoning, the three levels being the primary level of things, the secondary level of events, and the tertiary level of premise and consequence.

level of events: Secondary level of thought and reasoning, presupposes the existence of the primary level of the things. An example of a statement on the secondary level is "When it rains, I stay inside."

level of premise and consequence: Tertiary level of thought and reasoning, presupposes the existence of the primary level of things and the secondary level of events. An example of a statement on the tertiary level is "If it is the case that it is raining, then I am staying inside."

level of things: Primary level of thought and reasoning. An example of a statement on the primary level is "I stay inside."

logic: Traditional name of what is here called attribute mathematics, q.v.

"mathematical logic": Name used in past work on logic for what is here called attribute mathematics, q.v.

mathematics: Here understood as a name for any kind of deductive thought, q.v. Counterpart: *inductive thought.*

necessary condition: Here an exact synonym of *conditio sine qua non*, q.v.

nexus: Term encountered occasionally in linguistic writings, apparently used to denote the link between the two parts of a logical statement represented in Boolean algebra by the equation mark (=).

nominal: Term encompassing all that is substantival and all that is adjectival.

occasion: Here an exact synonym of event, q.v.

postulate: Here an exact synonym of axiom, q.v.

premise: Assumption that a condition is either true (1) or false (0), leading to an inevitable and necessary result or a consequence. Counterpart: *consequence* or *result.*

primary proposition: Statement on the primary level of thought and reasoning, that of things. An example is "I stay inside." Counterparts: *secondary proposition*, *tertiary proposition.*

protasis: Subordinate clause of a conditional sentence; contains a condition. Counterpart: *apodosis.*

"pure mathematics": Name used in the past for the type of thought that encompasses both what is called mathematics and what is called logic.

quantity mathematics: Branch of mathematics that studies anything that is capable of increase or diminution. Counterpart: *attribute mathematics.*

reasoning: Any effort of the mind that produces inductive or deductive thought.

regular condition: Here an exact synonym of condition, q.v.

relative: Here an exact synonym of adjectival when modifying "clause" or verb form," q.v.

result: Here an exact synonym of consequence, q.v.

reversibility: Basic property of an equation and its two halves, in the sense that the first half relates in the same way to the second half as the second half relates to the first half.

right and wrong: Terms that evaluate certainty and doubt, which are themselves terms that describe a state of mind regarding that which is true or false.

rule of absorption: Rule of logic according to which the following is valid: $a + ab = a$.

Scholastic logic: Type of logic practiced in the Middle Ages and based on Aristotelian logic.

secondary proposition: Statement on the secondary level of thought and reasoning, that of events. An example is "When it rains, I stay inside." Counterparts: *primary proposition*, *tertiary proposition*.

set: Here an exact synonym of class, q.v.

substantival: Anything referring to an entity.

supplement: Class of things on the level of things or of events on the level of events whose members all share the absence of an attribute; here an exact synonym of supplementary class and contradictory.

supplementary class: Here an exact synonym of supplement, q.v.

"symbolic logic": Name used in past work on logic for what is here called attribute mathematics.

tertiary proposition: Statement on the tertiary level of thought and reasoning, that of relations between events. An example is "If it is the case that it is raining, then I am staying inside." Counterparts: *primary proposition*, *secondary proposition*.

thing: Member of a class of things on the primary level of thought, namely the level of things. Counterpart: *event*.

thought: Any activity of the mind.

truth: Match between what is inside the mind and what is outside the mind. Counterpart: *falsehood*.

"universal algebra": Term used in the past, particularly by Alfred North Whitehead, for the branch of thought that encompasses both logic and mathematics, also sometimes called pure mathematics.

universe: Here defined, in Boole's sense, as anything one could possibly think about.

Wechselsatz: German equivalent of "balanced sentence."

BIBLIOGRAPHY

Aegyptische Inschriften aus den Königlichen Museen zu Berlin, vol. 1: *Inschriften von der Ältesten Zeit bis zum Ende der Hyksoszeit*. Leipzig: Hinrichs, 1913.

Allen, James P. 1979a. "Synthetic and Analytic Tenses in the Pyramid Texts." In *L'Égyptologie en 1979, axes prioritaires de recherche*, 1:19–27. Colloques internationaux du Centre national de la recherche scientifique 595. Paris: CNRS.

———. 1979b. "Is the 'Emphatic' Sentence an Adverbial-Predicate Construction?" *Göttinger Miszellen* 32: 7–15.

———. 1984. *The Inflection of the Verb in the Pyramid Texts*. Bibliotheca Aegyptia 2.1–2. Malibu: Undena.

———. 2002. *The Heqanakht Papyri*. Egyptian Expedition 27. New York: Metropolitan Museum of Art.

Anthes, Rudolf. 1930. "Eine Polizeistreife des Mittleren Reiches in die westliche Oase." *Zeitschrift für ägyptische Sprache und Altertumskunde* 65: 108–14.

Barns, John W. B. 1956. *Five Ramesseum Papyri*. Oxford: Griffith Institute.

Bennett, Deborah J. 2004. *Logic Made Easy*. New York: Norton.

Benveniste, Émile. 1952. "La construction passive d parfait transitif." *Bulletin de la Société linguistique* 48: 52–62.

Bilabel, Friedrich, and Adolf Grohmann, with a contribution by Georg Graf. 1934. *Griechische, koptische und arabische Texte zur Religion und religiösen Literatur in Ägyptens Spätzeit*. Veröffentlichungen aus den badischen Papyrus-Sammlungen 5. Heidelberg: Universitätsbibliothek.

Blackman, Aylward M. 1932. *Middle-Egyptian Stories*. Bibliotheca Aegyptiaca 2. Brussels: Egyptologische Stichting Koningin Elisabeth.

Boole, George. 1854. *An Investigation of the Laws of Thought, On Which Are Founded the Mathematical Theories of Logic and Probabilities*. London: Walton and Maberly. Reprint, New York: Dover, 1958.

Brockelmann, Carl. 1982. *Arabische Grammatik*. 21st ed., edited by Manfred Fleischhammer. Leipzig: VEB.

Brovarski, Edward. 1989. "The Inscribed Material of the First Intermediate Period from Naga ed-Dêr." 2 vols. Ph.D. diss., University of Chicago.

Budge, E. A. Wallis. 1898. *The Book of the Dead: The Chapters of Coming Forth by Day.* 3 vols. London: Kegan Paul, Trench, Trübner.

Cassonnet, Patricia. 2000. *Études de néo-égyptien: Les Temps seconds ỉ-sḏm.f et ỉ-ỉrỉ.f sḏm entre syntaxe et sémantique.* Paris: Cybèle.

Černý, Jaroslav. 1939. *Late Ramesside Letters.* Bibliotheca Aegyptiaca 9. Brussels: Egyptologische Stichting Koningin Elisabeth.

Church, Alonzo. 1936. "A Bibliography of Symbolic Logic." *Journal of Symbolic Logic* 1, no. 4: 121–218.

Clère, J.J. 1949. "Une nouvelle forme verbale relative en égyptien." In *Actes du XXIe Congrès Internationale des Orientalistes.* Pp. 64–66. Paris Imprimerie nationale.

Craven, T.L. 1955. "An Engineering Application of Boolean Algebra." In *Celebration of the Centenary of* The Laws of Thought *by George Boole.* Proceedings of the Royal Irish Academy 57.A.6 (pp. 63–130), 120–30. Dublin: Hodges, Figgis & Co., Ltd.

de Buck, Adriaan. 1935–61. *The Egyptian Coffin Texts.* 7 vols. Oriental Institute Publications 34, 49, 64, 67, 73, 81, 87. Chicago: Oriental Institute.

———. 1963. *Egyptian Readingbook: Exercises and Middle Egyptian Texts.* 2nd ed. Leiden: Nederlands Instituut voor het Nabije Oosten.

de Cenival, Jean Louis. 1972. "Sur la forme *sḏm.f* à redoublement ou *mrr.f.*" *Revue d'Égyptologie* 24: 40–45.

De Morgan, Augustus. 1872. *A Budget of Paradoxes.* London: Longmans, Green.

Depuydt, Leo. 1985a. "Semantik und Religion: Der Fall des 'Hintersichschauers'." In *Fourth International Conference of Egyptology: Abstracts of Papers*, ed. by Sylvia Schoske, 51–52. Munich: Lipp GmbH. This is the abstract for Depuydt 1988c and 1992b.

———. 1985b. "'Voir' et 'regarder' en copte: étude synchronique et diachronique." *Revue d'Égyptologie* 36: 35–42.

———. 1986. "Specificity or Emphasis in Egyptian and Coptic Nominal Sentences?" Review of *Studies in the Nominal Sentence in Egyptian and Coptic*, by John Callender. *Chronique d'Égypte* 61: 358–67.

———. 1988a. "New Horizons in Coptic and Egyptian Linguistics." Review of *Coptic Grammatical Categories: Structural Studies in the Syntax of Shenoutean Sahidic*, by Ariel Shisha-Halevy. *Chronique d'Égypte* 63: 391–406.

———. 1988b. "The End of *ḫr=f sḏm=f* in the Heqanakhte Letters." *Revue d'Égyptologie* 39: 204–8.

———. 1988c. "Die 'Verben des Sehens': Semantische Grundzüge am Beispiel des Ägyptischen." *Orientalia* 57: 1–13.

———. 1989a. "The Contingent Tenses of Egyptian." *Orientalia* 58: 1–27.

———. 1989b. Review of *The Coptic* share *Pattern and Its Ancient Egyptian Ancestors*, by Michael Green. *Chronique d'Égypte* 65: 186–92.

———. 1991a. "On Distinctive and Isolating Emphasis in Egyptian and in General." *Lingua Aegyptia: Journal of Egyptian Language Studies* 1: 33–56.

———. 1991b. "Late Egyptian *inn*, 'if,' and the Conditional Clause in Egyptian." *Journal of Egyptian Archaeology* 78: 69–78.

———. 1992a. Review of *"Emphasis" and Sentential Meaning in Middle Egyptian*, by Friedrich Junge. *Journal of the American Oriental Society* 112: 330–33.

———. 1992b. "Der Fall des 'Hintersichschauers'." *Göttinger Miszellen* 126: 33–38.

———. 1993a. *Conjunction, Contiguity, Contingency: On Relationships between Events in the Egyptian and Coptic Verbal Systems.* New York: Oxford University Press.

———. 1993b. "A History of Research on the Prospective *Sḏm.f* Forms in Middle Egyptian." *Journal of the American Research Center in Egypt* 30: 11–31.

———. 1993c. "Zur Bedeutung der Partikeln  *ỉsk* und  *ỉs*." *Göttinger Miszellen* 136: 11–25.

———. 1993d. "'Wisdom Made a Weapon': On Manichaeism in Egypt." *Chronique d'Égypte* 68: 301–15.

———. 1995a. "Sentence Pattern and Verb Form: Egyptian Grammar since Polotsky." *Le Muséon* 108: 39–48.

———. 1995b. "Condition and Premise in Middle Egyptian." *Revue d'Égyptologie* 46: 81–88.

———. 1995c. "On the Empirical Distinctness of Certain Adverbial Clauses in Old and Middle Egyptian." *Chronique d'Égypte* 80: 18–33.

———. 1996a. *Materials for Egyptian Grammar: Catalogue of Coordinates and Satellites of the Middle Egyptian Verb.* Leuven: Peeters.

———. 1996b. "Twixt Relative Verb Form and Passive Participle in Egyptian." *Zeitschrift der Deutschen Morgenländischen Gesellschaft* 146: 1–24.

———. 1996c. "On an Egyptianism in Akkadian." *Orientalia Lovaniensia Periodica* 27: 23–27.

———. 1997. "On Contiguity in Middle and Late Egyptian." *Zeitschrift für ägyptische Sprache und Altertumskunde* 124: 23–37.

———. 1999a. "Condition and Premise in Egyptian and Elsewhere and the Laws of Thought in Expanded Boolean Algebra." *Zeitschrift für ägyptische Sprache und Altertumskunde* 126: 97–111. Revised reprint, appendix 1 above.

———. 1999b. "Contrast in Egyptian and in General and the Laws of Thought in Boolean Algebra." *Göttinger Beiträge zur Sprachwissenschaft* 2: 37–60. Revised reprint, appendix 2 above.

———. 1999c. "Analyzing the Use of Idioms Past (with Special Focus on Sovereign Nubia)." *Studien zur altägyptischen Kultur* 27: 33–63.

———. 1999d. *Fundamentals of Egyptian Grammar*, vol. 1: *Elements.* Norton, Mass.: Frog.

———. 1999e. Review of *Texts from the Amarna Period in Egypt*, by William J. Murnane, edited by Edmund S. Meltzer. *Journal of Near Eastern Studies* 58: 311–13.

———. 2000. "Late Egyptian Sentences with Two Subordinate Clauses but No Main Clause." *Lingua Aegyptia: Journal of Egyptian Language Studies* 7: 125–140.

———. 2001a. "The Meaning of the Coptic Particle ⲣⲱ." *Journal of Coptic Studies* 3: 113–28.

———. 2001b. Review of *Étude grammaticale du papyrus Bremner-Rhind*, by Jacqueline Lustman. *Bibliotheca Orientalis* 58: 80–83.

———. 2002a. "A New Coptic Grammar." Review of *A Coptic Grammar*, 1st ed., by Bentley Layton. *Journal of the American Oriental Society* 122: 807–15.

———. 2002b. Review of *Mittelägyptisch für Anfänger*, 6th ed., by Erhart Graefe. *Journal of the American Oriental Society* 122: 602–4.

———. 2002/3. "Demotic Script and Demotic Language (III): *R-ḥr.f* 'On Them' in *Mythus* 18,7." *Enchoria: Zeitschrift für Demotistik und Koptologie* 28: 7–18.

———. 2003a. "The Origin of the Egyptian Suffix Conjugation." *Lingua Aegyptia: Journal of Egyptian Language Studies* 11: 23–47.

———. 2003b. Review of *Ägyptische Algorithmen: Eine Untersuchung zu den mittelägyptischen mathematischen Aufgabentexten*, by Annette Imhausen. *Journal of the American Oriental Society* 123: 877–80.

———. 2006a. "Von 'schlafen' zu 'liegen': Markierungsverlust als ein Hauptgesetz der ägyptischen Sprachentwicklung." In *Coptica – Gnostica – Manichaica: Mélanges offerts à Wolf-Peter Funk*, ed. by Louis Painchaud and Paul-Hubert Poirier, 123–33. Bibliothèque copte de Nag Hammadi, Section « Études » 7. Louvain: Peeters.

———. 2006b. "Our Knowledge of Egyptian since Polotsky." In: *"After Polotsky": New Research Trends in Egyptian and Coptic Liguistics. Akten des Kolloquiums im Gedenken an den 100. Geburstag von Hans Jakob Polotsky (13.9.1905–10.8.1991) in Bad Honnef bei Bonn 09.–11. September 2005*, ed. by Verena M. Lepper, in cooperation with Peter Nagel & Wolfgang Schenkel, 17–29. *Lingua Aegyptia: Journal of Egyptian Language Studies* 14. Göttingen: Seminar für Ägyptologie und Koptologie.

———. 2007. "Questions and Related Phenomena in Coptic and in General: Final Definitions Based on Boole's Laws of Thought." In *The World of Early Egyptian Christianity: Language, Literature, and Social Context. Essays in Honor of David W. Johnson*, ed. by James E. Goehring and Janet Timbie, 72–94. Washington, D.C.: Catholic University of America Press.

———. 2008. "The Function of the Particle 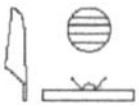 in Old and Middle Egyptian." In *Egypt and Beyond: Essays Presented to Leonard H. Lesko*, ed. by Stephen E. Thompson and Peter Der Manuelian, 91–108. Providence, Rhode Island: Department of Egyptology and Ancient Western Asian Studies.

———. in press. "Contrastive *jw.jr*, Conditional *jw.jr*, Temporal *jw.jr*: On Separating a Grammatical Siamese Triplet." In *Acts of the Eighth International Conference of Demotic Studies, Würzburg, August 2002*.

———. also in press. "From 'My Body' to 'Myself' to 'As For Me' to 'Me Too' in 3000 Years: A Peculiar Triple Shift in Meaning in Egyptian and Its Explanation."

Doret, Eric. 1980. "A Note on the Egyptian Construction Noun + *sḏm.f*." *Journal of Near Eastern Studies* 39: 37–45.

Eco, Umberto. 1980. *Il nome della rosa*. Milano: Bompiani.

Edel, Elmar. 1955/64. *Altägyptische Grammatik*. Analecta Orientalia 34, 39. Rome. Pontifical Biblical Institute.

———. 1984. "Neue Belege für die aktivische *Sḏmw.f*–Form." In Junge 1984: 25–38.

———. 1995. "Ein Graffito ungewöhnlichen Inhalts mit einer aktivischen *Sḏmw.f* Form. In *Gedenkschrift für Winfried Barta*, ed. by Dieter Kessler and Regine Schulz, 125–32, plate 1. Münchener ägyptologische Untersuchungen 4. Frankfurt: Peter Lang.

Edwards, Anthony William Fairbank. 2004. *Cogwheels of the Mind: The Story of Venn Diagrams*. Foreword by Ian Stewart. Baltimore: Johns Hopkins University Press.

Edwards, I. E. S. 1982. "The Bankes Papyri I and II." *Journal of Egyptian Archaeology* 68: 126–33.

El-Aguizy, Ola. 1997. "The particle *kꜣ* and Other Related Problematic Passages in Papyrus Westcar." *Bulletin de l'Institut français d'Archéologie orientale* 97: 157–63.

Elanskaya, Alla I. 1979. La construction dite "construction adverbiale" et le rôle de ⲟⲩⲛ̄/ⲙ̄ⲛ en copte." In *Ž F. Šampol'on i dešifrovka egipetskich ieroglifov*, 29–33. Moscow: I. S. Kacnel'son. n.v.

Englund, Gertie, and Paul John Frandsen, eds. 1986. *Crossroad: Chaos or the Beginning of a New Paradigm. Papers from the Conference on Egyptian Grammar, Helsingør 28–30 May 1986.* CNI Publications 1. Copenhagen: The Carsten Niebuhr Institute of Ancient Near Eastern Studies.

Engsheden, Åke. 2003. *La reconstitution du verbe en égyptien de tradition 400–30 avant J.-C.* Uppsala Studies in Egyptology 3. Uppsala: Uppsala University, Department of Archaeology and Ancient History.

Erichsen, Wolja. 1933. *Papyrus Harris I: Hieroglyphische Transkription.* Bibliotheca Aegyptiaca 5. Brussels: Egyptologische Stichting Koningin Elisabeth.

Erman, Adolf. 1881. "Altägyptische Studien." *Zeitschrift für ägyptische Sprache und Altertumskunde* 19: 41–66.

———. 1884. "Spuren eines alten Subjunktivs im Koptischen." *Zeitschrift für ägyptische Sprache und Altertumskunde* 22: 28–37.

———. 1896. *Gespräch eines Lebensmüden mit seiner Seele aus dem Papyrus 3024 der Königlichen Museen.* Abhandlungen der Königlichen Akademie der Wissenschaften zu Berlin aus dem Jahre 1896, Philosophisch-historische Klasse II. Berlin: Georg Reimer.

———. 1904. *Aegyptische Chrestomathie zum Gebrauch auf Universitäten und zum Selbstunterricht.* Berlin: Reuther & Richard.

———. 1929. *Mein Werden und Mein Wirken: Erinnerungen eines alten Berliner Gelehrten.* Leipzig: Quelle & Meyer.

———. 1933. *Neuaegyptische Grammatik.* 2d ed. Leipzig: Engelmann.

Farina, Jules. 1927. *Grammaire de l'Ancien Égyptien (Hiéroglyphes).* Translated from the Italian by René Neuville. Paris: Payot.

Feys, Robert. 1955. "Boole as a Logician." In *Celebration of the Centenary of* The Laws of Thought *by George Boole.* Proceedings of the Royal Irish Academy 57.A.6 (pp. 63–130), 97–106. Dublin: Hodges, Figgis & Co., Ltd.

Firchow, Otto. 1953. *Untersuchungen zur ägyptischen Stilistik*, vol. 2: *Grundzüge der Stilistik in den altägyptischen Pyramidentexten.* Deutsche Akademie der Wissenschaften zu Berlin, Institut für Orientforschung, Veröffentlichung 21. Berlin: Akademie-Verlag.

Frandsen, Paul John. 1974. *An Outline of the Late Egyptian Verbal System.* Copenhagen: Akademisk Forlag.

———. 1975. *At-Former.* Copenhagen: Egyptological Institute.

Gardiner, Alan H. 1909. *The Admonitions of an Ancient Egyptian Sage from a Hieratic Papyrus in Leiden.* Leipzig: Hinrichs. Reprint, Hildesheim: Olms, 1969.

———. 1923. "The Eloquent Peasant." *Journal of Egyptian Archaeology* 9: 5–25.

———. 1932. *Late-Egyptian Stories.* Bibliotheca Aegyptiaca 1. Brussels: Egyptologische Stichting Koningin Elisabeth.

———. 1934. "Two Employments of the Independent Pronouns." *Journal of Egyptian Archaeology* 20: 13–17.

———. 1937. *Late-Egyptian Miscellanies.* Bibliotheca Aegyptiaca 7. Brussels: Egyptologische Stichting Koningin Elisabeth.

———. 1947. Review of *Études de syntaxe copte*, by H. J. Polotsky. *Journal of Egyptian Archaeology* 33: 95–101.

———. 1957 (1927). *Egyptian Grammar: Being an Introduction to the Study of Hieroglyphs.* 3rd ed. Oxford: Griffith Institute. 1st ed., 1927.

Gardiner, Alan H., and F. Vogelsang. 1908. *Die Klagen des Bauern.* Literarische Texte des Mittleren Reiches 1 (Hieratische Papyrus aus den Königlichen Museen zu Berlin 4). Leipzig: Hinrichs.

Gasser, James, ed. 2000. *A Boole Anthology: Recent and Classical Studies in the Logic of George Boole.* Synthese Library 291. Dordrecht: Kluwer.

Gilula, Mordechai. 1987. "The 'Emphatic' Form of the Verb *iw* 'to come'." In *Form und Mass: Beiträge zur Literatur, Sprache und Kunst des alten Ägypten (Festschrift Gerhard Fecht)*, ed. by Jürgen Osing and Günter Dreyer, 137–41. Wiesbaden: Harrassowitz.

Goedicke, Hans. 1970. *The Report about the Dispute of a Man with his Ba: Papyrus Berlin 3024.* Baltimore and London: The Johns Hopkins Press.

Goldstine, Herman H. 1972. *The Computer from Pascal to von Neumann.* Princeton, N.J.: Princeton University Press.

Grandet, Pierre. 1994–99. *Le Papyrus Harris I (BM 9999).* 3 vols. Bibliothèque d'Étude 109.1–2, 129. Cairo: Institut français d'Archéologie orientale.

Grattan-Guinness, Ivor. 2000. *The Search for Mathematical Roots, 1870–1940: Logics, Set Theories and the Foundations of Mathematics from Cantor through Russell to Gödel.* Princeton and Oxford: Princeton University Press.

Grattan-Guinness, Ivor, and Gérard Bornet. 1997. *George Boole: Selected Manuscripts on Logic and Its Philosophy.* Science Networks Historical Studies 20. Basel: Birkhäuser.

Griffith, Francis Llewelyn. 1889. *The Inscriptions of Siût and Dêr Rîfeh.* London: Trübner.

Gunn, Battiscombe. 1924. *Studies in Egyptian Syntax.* Paris: Geuthner.

Hailperin, Theodore. 1976. *Boole's Logic and Probability: A Critical Exposition from the Standpoint of Contemporary Algebra.* Studies in Logic and the Foundations of Mathematics 85. Amsterdam: North-Holland.

———. 1981. "Boole's Algebra Isn't Boolean Algebra." *Mathematics Magazine* 54: 172–84. Reprint, Gasser 2000: 61–77.

Hussein, Ramadan B. 2004. "The Texts of the Coffin of *Ppy-im3*: Translation and Annotations." M.A. thesis, Brown University.

Jansen-Winkeln, Karl. 1997. "Intensivformen und 'verbale Pluralität' im Ägyptischen." *Lingua Aegyptia: Journal of Egyptian Language Studies* 5: 123–36.

Jéquier, Gustave. 1911. *Le Papyrus Prisse et ses variantes.* Paris: Geuthner.

———. 1928. *La Pyramide d'Oudjebten.* Service des Antiquités de l'Égypte: Fouilles à Saqqara. Cairo: Institut français d'Archéologie orientale.

———. 1933. *Les Pyramides des reines Neit et Apouit.* Service des Antiquités de l'Égypte: Fouilles à Saqqara. Cairo: Institut français d'Archéologie orientale.

Johnson, Janet H. 1976. *The Demotic Verbal System.* Studies in Ancient Oriental Civilization 38. Chicago: Oriental Institute.

———. 1987. "The Use of the Articles and the Generic in Demotic." *Aspects of Demotic Lexicography: Acts of the Second International Conference for Demotic Studies, Leiden, 19–21 September 1984*, ed. by S. P. Vleeming, 41–55. Studia Demotica 1. Leuven: Peeters.

———. 1991. *Thus Wrote 'Onchsheshonqy: An Introductory Grammar of Demotic*, 2nd ed. Studies in Ancient Oriental Civilization 45. Chicago: Oriental Institute. 1st ed., 1986.

Johnson, W. E. 1892. "The Logical Calculus." *Mind* 1: 3–30.

Junge, Friedrich. 1972. "Zur Funktion des *sḏm.ḫr=f*." *Journal of Egyptian Archaeology* 58: 133–39.

———. 1978. *Syntax der mittelägyptischen Literatursprache: Grundlagen einer Strukturtheorie.* Deutsches Archäologisches Institut, Abteilung Kairo. Mainz/Rhein: von Zabern.

———, ed. 1984. *Studien zu Sprache und Religion Ägyptens,* vol. 1: *Sprache (Festschrift Wolfhart Westendorf).* Göttingen: Junge.

———. 1986. "A Study on Sentential Meaning and the Notion of 'Emphasis' in Middle Egyptian." In Englund and Frandsen 1986: 91–117.

———. 1989. *"Emphasis" and Sentential Meaning in Middle Egyptian.* Göttinger Orientforschungen 4.20. Wiesbaden: Harrassowitz.

Kanawati, Naguib. 1986. *The Rock Tombs of El-Hawawish.* The Cemetery of Akhmim 6. Sydney: Macquarie Ancient History Association.

Kant, Immanuel. 1787. *Kritik der reinen Vernunft.* Wiesbaden: Insel Verlag, 1956.

Kitchen, Kenneth A. 1975–90. *Ramesside Inscriptions: Historical and Biographical.* 8 vols. Oxford: Blackwell.

———. 1996. *Ramesside Inscriptions: Translated & Annotated*, vol. 2: *Ramses II, Royal Inscriptions.* Oxford.: Blackwell.

Klenk, Virginia. 1983. *Understanding Symbolic Logic.* Englewood Cliffs, N.J.: Prentice-Hall.

Kloth, Nicole. 2002. *Die (auto-)biographischen Inschriften des ägyptischen Alten Reiches: Untersuchungen zu Phraseologie und Entwicklung.* Studien zur altägyptischen Kultur, Beihefte 8. Hamburg: Buske.

Kruchten, Jean-Marie. 1981. "Comment on écrit l'histoire égyptienne." *Annuaire de l'Institut de Philologie et d'Histoire Orientales et Slaves* 25: 51–64.

———. 1994. "*Ir wnn śḏm.f* (*śḏm.n.f*) et *ir śḏm.n.f*: une approche structuraliste." *Journal of Egyptian Archaeology* 80: 97–108.

———. 1997. "About *iw* and *wn(n)* in Late Egyptian." *Zeitschrift für ägyptische Sprache und Altertumskunde* 124: 57–70.

———. 1999. "From Middle Egyptian to Late Egyptian." *Lingua Aegyptia: Journal of Egyptian Language Studies* 6: 1–97.

Laita, Luis M. 1977. "The Influence of Boole's Search for a Universal Method in Analysis on the Creation of His Logic." *Annals of Science* 34: 163–76. Reprint, Gasser 2000: 45–59.

Lange, Hans O., and Heinrich Schäfer. 1908. *Grab- und Denksteine des Mittleren Reiches im Museum von Kairo, No. 20001–20780,* vol. 2: *Text zu No. 20400–20780.* Catalogue général des antiquités égyptiennes du Musée du Caire. Berlin: Reichsdruckerei.

Lapp, Günther. 1997. *The Papyrus of Nu (BM EA 10477).* With a contribution by Thomas Schneider. Catalogue of Books of the Dead in the British Museum 1. London: British Museum.

Layton, Bentley. 2004. *A Coptic Grammar.* Porta Linguarum Orientalium 20. Wiesbaden: Harrassowitz. 1st ed., 2000.

Leclant, Jean. 1968. "Fouilles et travaux en Egypte et au Sudan." *Orientalia* 37: 94–140.

Leibniz, Gottfried Wilhelm. 1959. *Opera philosophica quae extant latina gallica germanica omnia.* Aalen: Scientia. Reprint of Johann Eduard Erdmann's edition of 1840, with additional texts and an introduction supplied by Renate Vollbrecht.

Loprieno, Antonio. 1986. "Egyptian Grammar and Textual Features." In Englund and Frandsen 1986: 255–87.

———. 1995. *Ancient Egyptian: A Linguistic Introduction.* Cambridge: Cambridge University Press.

Luft, Ulrich. 1992. *Das Archiv von Illahun: Briefe 1.* Hieratische Papyri aus den Staatlichen Museen zu Berlin – Preussischer Kulturbesitz 1. Berlin: Akademie-Verlag.

Macfarlane, Alexander. 1879. *Principles of the Algebra of Logic, with Examples.* Edinburgh: David Douglas.

MacHale, Desmond. 1985. *George Boole: His Life and Work*. Dublin: Boole Press.

Malaise, Michel. 1985. "La conjugaison suffixale dans les propositions conditionelles introduites par *ir* en ancien et moyen égyptien." *Chronique d'Égypte* 60: 152–67.

Maspero, Gaston. 1871. *Des formes de la conjugaison en égyptien antique, en démotique et en copte*. Bibliothèque de l'École des Hautes Études, Sciences philologiques et historiques 6. Paris: A. Franck.

Menu, Bernadette. 1973. "Le prêt en droit égyptien (Nouvel Empire et Basse Époque)." *Cahier de recherches de l'Institut d'Égypte et de Papyrologie de Lille* 1: 61–141.

Nagel, Peter. 1980. "Die Determination des Subjektpronomens im Präsens I und das Problem der Satztypen des Koptischen." *Hallesche Beiträge zur Orientforschung* 2: 77–93. n.v.

Nasar, Sylvia. 1998. *A Beautiful Mind*. New York: Simon & Schuster.

Navailles, R., and François Neveu. 1988. "Une ténébreuse affaire: P. Bankes I." *Göttinger Miszellen* 103: 51–60.

Naville, Edouard. 1901. *The Temple of Deir El Bahari*, part 4 (plates LXXXVII–CVIII). London: Egypt Exploration Fund.

———. 1908. *The Funeral Papyrus of Iouiya*. Theodore M. Davis' Excavations: Bibân el Molûk. London: Constable.

Niccacci, Alviero. 1980. "Su una formula dei 'Testi dei Sarcofagi'." *Liber Annuus: Studium Biblicum Franciscanum* 30: 197–224, with plates 5–6.

———. 1982. Review of *Syntax der mittelägyptischen Literatursprache*, by Friedrich Junge 1978. *Liber Annuus: Studium Biblicum Franciscanum* 32: 529–37.

Osing, Jürgen. 1987. "New Light on the Vocalisation of Egyptian Verb Forms." In *Lingua Sapientissima: A Seminar in Honour of H. J. Polotsky*, ed. by J. D. Ray, 65–71. Cambridge: Faculty of Oriental Studies.

———. 1998. *Hieratische Papyri aus Tebtunis I*. The Carlsberg Papyri 2. 2 vols. CNI Publications 17. Copenhagen: Carsten Niebuhr Institute.

Parker, Stephen T. 1996. "Nominal and Prospective Uses of the *Sḏm.f* in the Coffin Texts." Ph.D. diss., University of Chicago.

Parkinson, Richard B. 1991. *The Tale of the Eloquent Peasant*. Oxford: Griffith Institute.

———. 1996. "On a Proposed Survival of the 'Lebensmüder' into the Christian Era." *Chronique d'Égypte* 71: 389–92.

Peirce, Charles S., ed. 1883. *Studies in Logic by Members of the Johns Hopkins University*. Boston: Little, Brown. Reprint, with introduction by Max H. Fisch and preface by Achim Eschbach. Foundations of Semiotics 1. Amsterdam and Philadelphia: John Benjamins, 1983.

Peust, Carsten. 1997. Review of *Das Verbalsystem der königlichen und privaten Inschriften, XVIII. Dynastie bis einschießlich Amenophis III*, by Thomas Ritter. *Zeitschrift der Deutschen Morgenländischen Gesellschaft* 147: 488–94.

Piccione, Peter. 1980. "On the Use of the *Sḏm.n.f* in the Historical Texts of Ramesses III at Medinet Habu." *Serapis* 6: 103–16.

Polotsky, Hans Jakob. 1935. "Zu einigen Heidelberger koptischen Zaubertexten." *Orientalia* 4: 416–25. Reprint, 1971: 275–84.

———. 1939. Review of *Die Texte aus den Gräbern der Herakleopolitenzeit von Siut*, by Hellmut Brunner. *Orientalistische Literaturzeitung* 42: 155–60. Reprint, 1971: 28–30.

———. 1940. "Une règle concernant l'emploi des formes verbales dans la phrase interrogative en néo-égyptien." *Annales du Service des Antiquités de l'Égypte* 40: 241–45. Reprint, 1971: 33–37.

———. 1944. *Études de syntaxe copte.* Cairo: Société d'Archéologie copte. Reprint, 1971: 102–207.

———. 1957. "The 'Emphatic' *Sḏm.n.f* Form." *Revue d'Égyptologie* 11: 109–17. Reprint, 1971: 43–51.

———. 1960. "The Coptic Conjugation System." *Orientalia* 29: 392–422. Reprint, 1971: 238–68.

———. 1964. "Ägyptische Verbalformen und ihre Vokalisation." Review of *De oplossing van een gewichtig probleem: de vocalisatie van de egyptische werkwoordsvormen*, by Jozef Vergote. *Orientalia* 33: 267–85. Reprint, 1971: 52–70.

———. 1965. *Egyptian Tenses.* The Israel Academy of Sciences and Humanities, Proceedings, Vol. II, No. 5. Jerusalem: The Central Press. Reprint, Polotsky 1971: 71–95.

———. 1969. "Zur altägyptischen Grammatik." Review of *Altägyptische Grammatik*, by Elmar Edel. *Orientalia* 38: 465–81. Reprint Polotsky 2007: 15–31.

———. 1971. *Collected Papers.* Jerusalem: Magnes.

———. 1976. "Les transpositions du verbe en égyptien classique." *Israel Oriental Studies* 6: 1–50. Reprint Polotsky 2007: 55–104.

———. 1984. "'Perfekt' und 'Imperfekt': von Sethe zu Gardiner." In *Studies in Egyptology Presented to Miriam Lichtheim*, ed. by Sarah Israelit-Groll, 768–72. Jerusalem: Magnes. Reprint Polotsky 2007: 168–72.

———. 1996. "Notes on a Neo-Syriac Grammar." Edited posthumously by Olga Kapeliuk. *Israel Oriental Studies* 16: 11–48.

———. 2007. *Scripta Posteriora on Egyptian and Coptic*, ed. by Verena M. Lepper and Leo Depuydt. Lingua Aegyptia: Studia monographica 7. Göttingen: Seminar für Ägyptologie und Koptologie. Errata in editorial parts: (page v,*13*) for "languages" read "langages"; (2,*32*) for "the study Egyptian" read "the study of Egyptian"; (2,*34*) for "see p. 53 above" read "see p. 53 below"; (205,*14*) for "To do what for you…?" read "So that you might do what for him…?"; (205,*15*) for "so as to do what for you…?" read "in return for your doing what for him…?

Porter, Bertha, and Rosalind L. B. Moss, assisted by Ethel W. Burney. 1972. *Topographical Bibliography of Ancient Egyptian Hieroglyphic Texts, Reliefs, and Paintings*, vol. 2: *Theban Temples.* 2nd ed. Oxford: Oxford University Press.

Quack, Joachim Friedrich. 1994. *Die Lehren des Ani: Ein neuägyptischer Weisheitstext in seinem kulturellem Umfeld.* Orbis Biblicus et Orientalis 141. Freiburg, Switzerland: Universitätsverlag; Göttingen: Vandenhoeck & Ruprecht.

———. 2001. Ein neuer Versuch zum Moskauer literarischen Brief. *Zeitschrift für ägyptische Sprache und Altertumskunde* 128: 167–81.

———. 2003. Review of *Fundamentals of Egyptian Grammar*, vol. 1: *Elements*, by Leo Depuydt. *Lingua Aegyptia: Journal of Egyptian Language Studies* 11: 259–65.

Quecke, Hans. 1979. Review of *The Demotic Verbal System*, by Janet H. Johnson. *Orientalia* 48: 435–47.

———. 1983. Review of *Hallesche Beiträge zur Orientwissenschaft* 1, including Nagel 1980. *Orientalia* 52: 296–302.

———. 1988. Review of *Thus Wrote 'Onchsheshonqy*, by Janet H. Johnson. *Orientalia* 57: 220–23.

Ritter, Thomas. 1995. *Das Verbalsystem der königlichen und privaten Inschriften, XVIII. Dynastie bis einschließlich Amenophis III.* Göttinger Orientforschungen 4.30. Wiesbaden: Harrassowitz.

Russell, Bertrand. 1919. *Mysticism and Logic and Other Essays.* Third Impression. London: Longmans, Green, and Co.

———. 1955. Letter to Felix E. Hackett. In *Celebration of the Centenary of* The Laws of Thought *by George Boole.* Proceedings of the Royal Irish Academy 57.A.6 (pp. 63–130), 64. Dublin: Hodges, Figgis & Co., Ltd.

Sander-Hansen, C. E. 1941. *Über die Bildung der Modi im Altägyptischen.* Kongelige Danske Videnskabernes Selskab, Historisk-filologiske Skrifter 1.3. Copenhagen: Munksgaard.

Sandman, Maj. 1938. *Texts from the Time of Akhenaten.* Bibliotheca Aegyptiaca 8. Brussels: Egyptologische Stichting Koningin Elisabeth.

Satzinger, Helmut. 1993. "Emphase" oder die Rhematisierung eines adverbialen Komplements im Ägyptischen (Rezensionsartikel). *Wiener Zeitschrift für die Kunde des Morgenlandes* 83: 189–206.

———. 1994a. "'Die Protasis *ir sḏm.f* ...'—Some Afterthoughts." *Lingua Aegyptia: Journal of Egyptian Language Studies* 4: 271–74.

———. 1994b. Review of Depuydt 1993a. In *Anthropological Linguistics* 36: 395–97.

Sauneron, Serge. 1989. *Un traité égyptien d'ophiologie: Papyrus du Brooklyn Museum No 47.218.48 et .85.* Bibliothèque générale 11. Cairo: Institut français d'Archéologie orientale.

Schenkel, Wolfgang. 1981. "*Sḏm=f* und *sḏm.w=f* als Prospektivformen." In *Studies Presented to Hans Jakob Polotsky*, ed. by Dwight W. Young, 506–27. East Gloucester, Mass.: Pirtle & Polson.

———. 1985. "Zur Verbalflexion der Pyramidentexte." Review of *The Inflection of the Verb in the Pyramid Texts*, by James P. Allen. *Bibliotheca Orientalis* 42: 481–94.

———. 1987. "Königsmutter *Nfr.w*: Phantom oder Realität." *Göttinger Miszellen* 96: 93–94.

———. 1988. "Aktuelle Perspektiven der ägyptischen Grammatik." Review of *Crossroad: Chaos or the Beginning of a New Paradigm*, edited by Gertie Englund and Paul John Frandsen. *Bibliotheca Orientalis* 45: 269–89.

———. 1990. *Einführung in die altägyptische Sprachwissenschaft.* Darmstadt: Wissenschaftliche Buchgesellschaft.

———. 1992. Review of *Future at Issue. Tense, Mood and Aspect in Middle Egyptian*, by Pascal Vernus. *Orientalistische Literaturzeitung* 87: 377–80.

———. 1994. *Tübinger Einführung in die klassisch-ägyptische Sprache und Schrift.* Tübingen: Schenkel.

———. 1997. *Tübinger Einführung in die klassisch-ägyptische Sprache und Schrift.* Tübingen: Schenkel.

———. 1998. "Standardtheorie und invertierte Standardtheorie." *Zeitschrift für ägyptische Sprache und Altertumskunde* 125: 140–60.

———. 2000. "Die Endungen des Prospektivs und des Subjunktivs (*śčm=f*, *śčm.w=f*, *śčm.y=f*) nach Befunden der Sargtexte. Mit einem Anhang zum prospektiven Partizip *śčm.t*(*i*)=*f*(*i*)." *Lingua Aegyptia: Journal of Egyptian Language Studies* 7: 27–112.

———. 2002a. "Zur Formenbildung des prädikativen *śčm=f* der Verben II.gem., vornehmlich nach dem Zeugnis der Sargtexte." *Göttinger Miszellen* 189: 89–98.

———. 2002b. "Ramses: Die Erfindung einer Graphie in der Nacherzählung der Entzifferungsgeschichte der Hieroglyphen." *Göttinger Miszellen* 191: 85–88.

———. 2004. "Ptolemaios und die Sprache der Hieroglyphen: Noch einmal zur Nacherzählung der Entizifferungsgeschichte der Hieroglyphen." *Göttinger Miszellen* 198: 105–12.

———. 2004/5. "Das *sčm(.w)=f*-Passiv, Perfekt vs. Futur, nach dem Zeugnis der Sargtexte." *Zeitschrift für ägyptische Sprache und Altertumskunde* 131: 173–88 and 132: 40–54

———. 2005. *Tübinger Einführung in die klassisch-ägyptische Sprache und Schrift.* Tübingen: Schenkel.

———. 2006a. "Prädikatives und abstrakt-relativisches *sčm.n=f*: Beobachtungen an den Verben II. gem. und ult. *n* im Korpus der Sargtexte." In *Egyptian, Semitic and General Grammar: Workshop in Memory of H. J. Polotsky (8–12 July 2001)*, ed. by Gideon Goldenberg and Ariel Shisha-Halevy, 40–60. Jerusalem: Israel Academy of Sciences and Humanities.

———. 2006b. "Ramses, Thutmosis und Henry Salt." *Göttinger Miszellen* 208: 89–94.

———. 2006c. "Ein vermeintlicher Fall der agenslosen Relativkonstruktion." *Göttinger Miszellen* 209: 75–79.

Schröder, Ernst. 1880. Review of *Begriffsschrift*, by Gottlob Frege. *Zeitschrift für Mathematik und Physik, historisch-literarische Abteilung* 25: 81–95.

———. 1890. *Vorlesungen über die Algebra der Logik (Exakte Logik)*, vol. 1. Leipzig: Teubner.

Sethe, Kurt. 1899. *Das aegyptische Verbum im Altaegyptischen, Neuaegyptischen, und Koptischen*, vol. 2. Leipzig: Hinrichs.

———. 1907. "Mißverstandene Inschriften." *Zeitschrift für ägyptische Sprache und Altertumskunde* 44: 35–41.

———. 1908–22. *Die altägyptischen Pyramidentexte.* 4 vols. Leipzig: Hinrichs. Reprint, Hildesheim: Olms, 1960.

———. 1927. *Erläuterungen zu den Aegyptischen Lesestücken.* Leipzig: Hinrichs.

———. 1928. *Aegyptische Lesestücke zum Gebrauch im akademischen Unterricht.* 2nd ed. Leipzig: Hinrichs.

———. 1932–33. *Urkunden des Alten Reichs.* 4 vols. Urkunden des ägyptischen Altertums 1. Leipzig: Hinrichs.

Sethe, Kurt; and Wolfgang Helck. 1906–58. *Urkunden der 18. Dynastie.* 22 vols. Urkunden des ägyptischen Altertums 4. Leipzig: Hinrichs.

Shannon, Claude E. 1938. "A Symbolic Analysis of Relay and Switching Circuits." *Transactions of the American Institute of Engineers* 57: 713–23. Reprint, Shannon 1993: 471–95.

———. 1948. "The Mathematical Theory of Communication." *Bell System Technical Journal*, July/October. Reprint, Shannon and Weaver 1963 and in Shannon 1993: 5–83.

———. 1987. Claude E. Shannon interviewed by Anthony Liversidge (entitled "Father of the Electronic Information Age" at p. 4). *Omni* 9, no. 11: 60–62, 64–66, 110.

———. 1993. *Collected Papers.* Edited by N. J. A. Sloane and Aaron D. Wyner. New York: IEEE.

Shannon, Claude E., and Warren Weaver. 1963. *The Mathematical Theory of Communication.* Urbana: University of Illinois Press.

Shisha-Halevy, Ariel. 1984. "Existential Statements in the Sahidic New Testament: Work Notes." *Göttinger Miszellen* 77: 67–79.

———. 1986. "(*i*)*rf* in the Coffin Texts: A Functional Tableau." *Journal of the American Oriental Society* 106: 641–58.

———. 1989. "Work-Notes on Demotic Syntax, I." *Orientalia* 58: 28–60.

———. 1995. "Some Reflections on the Egyptian Conjunctive." In *Divitiae Aegypti: Koptologische und verwandte Studien zu Ehren von Martin Krause*, ed. by Cäcilia Fluck et al., 300–14. Wiesbaden: Reichert.

Silverman, David. 1985. "Verbal Nominal Clauses in Middle Egyptian." In *Mélanges Gamal Eddin Mokhtar*, ed. by Paule-Posener Kriéger, 2: 269–85. Bibliothèque d'Étude 97.1–2. Cairo: Institut français d'Archéologie orientale.

Silverstein, Michael, ed. 1971. *Whitney on Language: Selected Writings by William Dwight Whitney.* With an introductory essay by Roman Jakobson. Cambridge: MIT Press.

Simpson, R. S. 1996. *Demotic Grammar in the Ptolemaic Sacerdotal Decrees.* Griffith Institute Monographs. Oxford: Griffith Institute.

Smith, G. C. 1982. *The Boole–De Morgan Correspondence.* Oxford Logic Guides. Oxford: Clarendon.

Spiegelberg, Wilhelm. 1899. "Der sogenannte "Arsu" des Papyrus Harris: Eine Vermutung." *Orientalistische Literaturzeitung* 8: 263–65.

———. 1917. "Briefe der 21. Dynastie aus el-Hibeh." *Zeitschrift für ägyptische Sprache und Altertumskunde* 53: 1–30.

Styazhkin, Nikolai I. 1969. *History of Mathematical Logic from Leibniz to Peano.* Cambridge: MIT Press. Russian original, 1964.

Thacker, Thomas W. 1954. *The Relationship of the Semitic and Egyptian Verbal Systems.* Oxford: Clarendon.

Valbelle, Dominique. 1977. "Remarques sur les textes néo-égyptiens non littéraires (§6–10)." *Bulletin de l'Institut français d'Archéologie orientale* 77: 129–36.

Vandier, Jacques. 1936. "Une tombe inédite de la VIe dynastie à Akhmîm." *Annales du Service des Antiquités de l'Égypte* 36: 33–44.

van Evra, James W. 1977. "A Reassessment of George Boole's Theory of Logic." *Notre Dame Journal of Formal Logic* 18: 363–77. Reprint, Gasser 2000: 87–99.

Venn, John. 1894. *Symbolic Logic.* 2nd ed. London: Macmillan. 1st ed., 1881.

Vergote, Jozef. 1957. "Une nouvelle interprétation de deux passages du Naufragé." *Mitteilungen des Deutschen Archäologischen Instituts, Abteilung Kairo* 15: 275–87.

———. 1957–58. "Naar een hernieuwing van de Egyptische grammatica." *Jaarbericht Ex Oriente Lux* 15: 34–46.

———. 1965. "Les formes verbales égyptiennes et leur vocalisation." *Orientalia* 34: 345–71.

———. 1970. "Egyptian." In *Current Trends in Linguistics*, ed. by Thomas A. Sebeok, vol. 6: *Linguistics in South West Asia and North Africa*, ed. by Charles E. Ferguson et al., 531–57. The Hague: Mouton.

Vernus, Pascal. 1981. Formes "emphatiques" en fonction ["that are functionally speaking ..."] non "emphatiques" dans la protase d'un système corrélatif. *Göttinger Miszellen* 43: 73–88.

———. 1985. "Études de philologie et de linguistique (IV): XIII. Non-existence et définition du sujet: *bn* prédicatif en Néo-égyptien." *Revue d'Égyptologie* 36: 153–68.

———. 1990. *Future at Issue. Tense, Mood and Aspect in Middle Egyptian: Studies in Syntax and Semantics.* Yale Egyptological Studies 4. New Haven: Yale Egyptological Seminar.

Vycichl, Werner. 1959. "Ein passives Partizip *qatīl* im Ägyptischen und Semitischen. Der Ursprung der periphrastischen Konjugation *sǧm.n.f* als Parallele zu aramäisch *šemīʿ leh* 'er hat gehört'." *Zeitschrift der Deutschen Morgenländischen Gesellschaft* 109: 253–58.

Wente, Edward F. 1977. Review of *An Outline of the Late Egyptian Verbal System*, by Paul John Frandsen 1974. *Journal of Near Eastern Studies* 36: 310–12.

Westendorf, Wolfhart. 1953. *Der Gebrauch des Passivs in der klassischen Literatur der Ägypter.* Deutsche Akademie der Wissenschaften zu Berlin, Institut für Orientforschung, Veröffentlichung 18. Berlin: Akademie-Verlag.

———. 1962. *Grammatik der medizinischen Texte.* Grundriss der Medizin der Alten Ägypter 8. Berlin: Akademie-Verlag.

Whitehead, Alfred North. 1897. *A Treatise on Universal Algebra with Applications.* New York: Hafner Publishing Company, 1960. I have used the reprint of the original edition published by Cambridge University Press in 1897.

Whitney, William Dwight. 1867. *Language and the Study of Language: Twelve Lectures on the Principles of Linguistic Science.* Reprint, New York: AMS Press.

Willeitner, Joachim. 2003. *Die ägyptischen Oasen: Städte, Tempel, Gräber in der Libyschen Wüste.* Zaberns Bildbände zur Archäologie. Mainz am Rhein: Philipp von Zabern.

Wilmet, Michel. 1957–58. *Concordance du Nouveau Testament sahidique*, vol. 2: *Les mots autochtones.* 2 vols. Corpus Scriptorum Christianorum Orientalium 173, 183 = Subsidia 11, 13. Leuven: Secrétariat du CSCO.

Winand, Jean. 1989. "L'expression du sujet nominal au présent I en néoégyptien." *Chronique d'Égypte* 64: 159–71.

———. 1992. *Études de néo-égyptien*, 1: *La morphologie verbale.* Ægyptiaca Leodiensia 2. Liège: Centre Informatique de Philosophie et Lettres.

Wreszinski, Walter. 1923–35. *Atlas zur altägyptischen Kulturgeschichte.* Leipzig. Reprint, Geneva: Slatkine, 1988.

Yamamoto, Tsunetomo. 1716. *Hagakure: The Book of the Samurai.* Trans. William Scott Wilson. Tokyo: Kodansha International; New York: distributed in the United States by Harper and Row, 1979.

Youssef, A. A.-H., Ch. Leblanc; and M. Maher. 1977. *Le Ramesseum*, vol. 4: *Les batailles de Tounip et de Dapour.* Cairo: Centre d'Étude et de documentation sur l'ancienne Égypte, Collection scientifique.

INDEXES

1 INDEX OF PASSAGES CITED

The pages to which reference is made are those on which the citation itself is found. The actual discussion of the text to which the citation refers may extend to the previous or following page or pages. The following abbreviations found in this index have not been resolved on pages xxi–xxii above:

BD = Book of the Dead
BM = British Museum
r. = recto
v. = verso

1.1 Old and Middle Egyptian (ca. 2500 B.C.E.–ca. 1500 B.C.E.)

1.2 Late Egyptian (ca. 1500 B.C.E.–1000 B.C.E.)

1.3 Coptic (from ca. 300 B.C.E.)

1.3.2 Other

2 INDEX OF AUTHORS CITED

Please note the following:

(1) An author's name may occur more than once on a page to which reference is made. Scanning the whole page, including the footnotes, to establish whether there is more than one mention is therefore advisable.

(2) If an author is mentioned on more than two consecutive pages, the page range is signified by connecting the first and last pages by an En-dash (–). But the references are not necessarily connected in terms of contents.

(3) Authors are referenced in this index only when an opinion of theirs on a certain matter is explicitly cited or described or at least clearly implied.

3 INDEX OF SUBJECTS TREATED

Egyptian and Coptic words are grouped at the end. An indexed term may be mentioned more than once on the page to which reference is made. Scanning the whole page, including the footnotes, to establish whether more than one mention occurs is hence advisable.

EGYPTIAN AND COPTIC WORDS

Egyptian

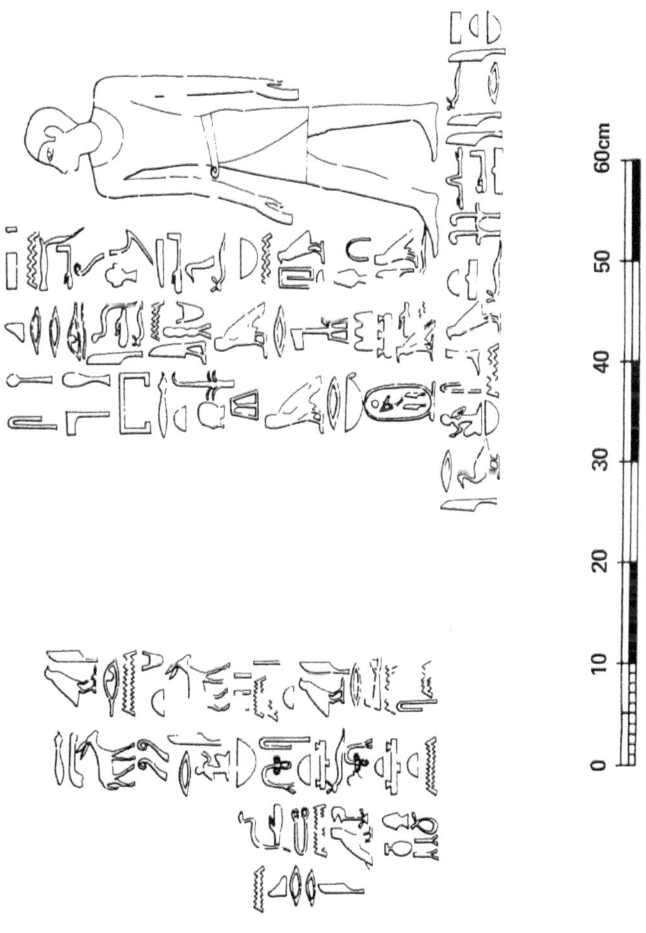

A striking instance of contiguity (see pp. 202–4) found in two short inscriptions north and south of the entrance to Qereri's tomb at El-Hawawish (Figure 20b and 20c in Kanawati 1986)

www.ingramcontent.com/pod-product-compliance
Ingram Content Group UK Ltd.
Pitfield, Milton Keynes, MK11 3LW, UK
UKHW041641031026
15587UKWH00001B/431